FEM *and* Micromechatronics *with* ATILA Software

FEM *and* Micromechatronics *with* ATILA Software

KENJI UCHINO

The Pennsylvania State University, University Park, USA

CRC Press is an imprint of the
Taylor & Francis Group, an **informa** business

CRC Press
Taylor & Francis Group
6000 Broken Sound Parkway NW, Suite 300
Boca Raton, FL 33487-2742

Printed in the United States of America on acid-free paper
10 9 8 7 6 5 4 3 2 1

International Standard Book Number-13: 978-1-4200-5878-9 (Softcover)

Library of Congress Cataloging-in-Publication Data

Uchino, Kenji, 1950-
FEM and micromechatronics with ATILA software / Kenji Uchino.
p. cm.
Includes bibliographical references and index.
ISBN 978-1-4200-5878-9 (alk. paper)

1. Mechatronics. 2. Microelectromechanical systems. I. Title.
TJ163.12.U23 2008 621--dc22

Visit the Taylor & Francis Web site at
http://www.taylorandfrancis.com

and the CRC Press Web site at
http://www.crcpress.com

About the Author

Kenji Uchino, a pioneer in piezoelectric actuators, is the director of the International Center for Actuators and Transducers (ICAT) and professor of electrical engineering at Pennsylvania State University. He is currently teaching graduate courses at Penn State entitled "Ferroelectric Devices," "Micromechatronics," and "Application of FEM," using this textbook and his previous texts, *Ferroelectric Devices* and *Micromechatronics,* published by Marcel Dekker in 2000 and 2003, respectively. He is also the senior vice president and CTO of Micromechatronics, Inc., State College, Pennsylvania, which is the spin-off company from ICAT, and an official distributor of ATILA FEM Computer Software (www.mmech.com).

After earning his Ph.D. from the Tokyo Institute of Technology, Uchino worked as a research associate in the physical electronics department at that university (1976). He joined the faculty of Sophia University (Tokyo) as an associate professor of physics in 1985. He then moved to Penn State in 1991. Professor Uchino was a member of the Space Shuttle Utilizing Committee of NASDA (Japan) from 1986 to 1988 and was the vice president of NF Electronic Instruments (USA) from 1992 to 1994. He has been a consultant for more than eighty U.S., Japanese, and European industries that develop and manufacture piezoelectric actuators. Since 1986 he has been the chairman of the Smart Actuator/Sensor Study Committee, which is partially sponsored by the Japanese government (MITI). He is a fellow of the American Ceramic Society and a member of IEEE, the Materials Research Society, and the Applied Physics Society of Japan. He is also the executive associate editor of the *Journal of Materials Technology* (*Matrice Technology*), and was associate editor of both the *Journal of Intelligent Materials and Structure Systems* and the *Japanese Journal of Applied Physics.*

Professor Uchino's research interests are in the area of solid-state physics, focused primarily on the study of dielectric, ferroelectric, and piezoelectric materials. More specifically, his work in this area has been directed toward the characterization of new materials, device design, and fabrication processes, as well as the development of solid-state actuators to be employed as precision positioners, ultrasonic motors, and high power transducers. He has authored 500 papers, 50 books, and 20 patents related to smart actuators.

Professor Uchino is the recipient of several academic awards including the Adaptive Structures Prize from American Society of Mechanical Engineers, the Smart Product Implementation Award 2007 from SPIE, Outstanding Research Award presented by the Pennsylvania State University Engineering Society, and the Nissan Research Scholarship, Japan. In addition to his academic accomplishments, he is an honorary member of KERAMOS (a national professional ceramic engineering fraternity) and has received the Best Movie Memorial Award for outstanding directing and producing of scientific documentary films at the Japan Scientific Movie Festival (1989) in recognition for several educational videotapes he produced on piezoelectric actuators.

Preface

Remarkable developments have taken place in the field of mechatronics in recent years. As reflected in the "Janglish" (Japanese English) word *mechatronics,* the concept of integrating electronic and mechanical devices in a single structure is one that is already well accepted in Japan. Robots were employed to manufacture many products in Japanese factories, and automated systems for sample displays and sales are utilized routinely in the supermarkets of Japan. Smart morphing technologies have been introduced in aerospace and mechanical areas in large-scale engineering. On the contrary, in small-scale engineering, the rapid advances in semiconductor chip technology have led to the need for micro displacement positioning devices. Further, current micro- and nano-technologies (MEMS) and bio/medical engineering demand precise manipulation mechanisms of the samples in a submicron meter level. Actuators that function through the piezoelectric, magnetostriction, and shape memory effects are expected to be important components in this age of micromechatronic technology, and their designing principle should quickly be established.

This book is the latest in a series of texts concerned with smart actuators written by Kenji Uchino. The first, entitled *Essentials for Development and Applications of Piezoelectric Actuators,* was published in 1984 through Japan Industrial Technology Center. The second, *Piezoelectric/Electrostrictive Actuators*, published in Japanese by Morikita Publishing Company (Tokyo), became one of the best sellers of that company in 1986, and was subsequently translated into Korean. The problem-based text, *Piezoelectric Actuators—Problem Solving*, was also published through Morikita; it included an instructive 60-min video tape. The English translation of that text was completed and published in 1996 through Kluwer Academic Publishers. A material engineer-oriented text, *Ferroelectric Devices,* was published through Marcel Dekker in 2000, and its Japanese translation by Morikita in 2005. The comprehensive text, written in collaboration with Jayne Giniewicz, offered updated chapters on current and future applications of actuator devices in micromechatronic systems. Now, this FEM text is the ninth in Uchino's series, which is being offered to advanced engineers who would like to design smart actuators and transducers practically with FEM computer software.

This textbook introduces the subject of using a finite element method (FEM) with these smart materials and structures, in particular, using the ATILA software code with GiD interface software. We will review briefly the theoretical background of solid-state actuator materials, device designs including ultrasonic motors, and drive/control techniques, first. Then, we learn how to combine the piezoelectric constitutive equations with an FEM formula. Thus, FEM analyzes the quasi-static, resonating, and transient mechanical vibrations of the piezoelectric device under various applied electric field and mechanical boundary conditions. The main focus is on the practical applications such as designing the piezoelectric actuators and transformers, with the ATILA code. A limited educational version, ATILA-Light, is provided as a courtesy of Micromechatronics Inc., which should be installed in the reader's personal computer to be used in learning the simulation process.

During my 30 years of teaching micromechatronics, I found that there was no suitable textbook in this smart (piezoelectric and magnetostrictive) transducer field, except some professional books such as multiauthor paper collections. Despite the mature stage of the piezoelectric industries, there is no simple and useful computer simulation tool for designing these devices. Hence, I decided to author a textbook, *FEM and Micromechatronics*, based on my lecture notes, including various developed device examples. This text introduces the background of piezoelectric devices, practical materials, device designs, drive/control techniques, and typical applications in Part I, and in Part II we learn how to use the ATILA FEM software code in seven practical device examples. You are encouraged to upload the ATILA-Light FEM software today, and start to train by yourself in designing the devices.

To introduce the detailed contents: Part I, "Fundamentals," introduces the reader to the overall background of piezoelectric transducers. Chapter 1, "Trend of Micromechatronics and Computer Simulation," is followed by chapter 2, "Review of Piezoelectricity and Magnetostriction." Chapter 3, "Structures of Smart Transducers," and chapter 4, "Drive/Control Techniques of Smart Transducers," cover practical designing, manufacturing, and controlling essentials of the piezoelectric devices. Chapter 5, "Finite Element Analysis," is a review of FEM basics with example applications of FEA, then, chapter 6, "Design Optimization," follows. Finally, chapter 7 looks at the "Future of the FEM in Smart Structures." Part II, "How to Use ATILA," is a step-by-step instruction section, including seven examples: piezoelectric plates, magnetostrictive rod, composite structures (bimorph, multilayer, cymbal), piezo transformers, several ultrasonic motors, and underwater transducers with acoustic lenses, including the Tonpilz sonar.

The attached CD is designed for a reader to use on two monitors, on one of which the CD HTML file is displayed and the other showing the GiD software opened for the reader's practice. For the reader who does not have two monitors, the printed textbook is useful. A flat-open book style will be convenient to the reader while working on the GiD-ATILA software.

This textbook is intended for graduate students and industrial engineers studying or developing smart actuators, sensors, and their related devices for electrical, optical, medical, mechanical, aerospace, and robotic applications. The text is designed primarily for a graduate course equivalent to thirty 75-minute lectures; however, it is also suitable for self-study by individuals wishing to extend their knowledge in this field. The CD supplement that accompanies the text includes the PowerPoint presentation lecture files, and various examples of finite-element modeling demonstrated step by step. I am indebted to Dr. Philippe Bouchilloux of Samsung Electromechanics–Korea for writing the introduction to the finite element method presented in Part I, chapters 5 and 6 of the text. Though it would be impossible to cover the full range of related materials in a single text, I have selected what I feel are the most important and fundamental topics to an adequate understanding of the design and application of actuator devices for micromechatronic systems.

Finally, I express my sincere gratitude to both academic and industry colleagues at Pennsylvania State University; ISEN; Lille Cedex, France; and Micromechatronics, Inc., State, including the following friends:

Late Prof. Jayne R. Giniewicz, formerly of Indiana University of Pennsylvania
Dr. Philippe Bouchilloux, Samsung Electromechanics, Korea
Dr. Burhanettin Koc, Samsung Electromechanics, Korea
Dr. Jean-Claude Debus, ISEN, France
Dr. Alfredo Vazquez Carazo, Micromechatronics, Inc., Pennsylvania
Mr. Bjorn Andersen, Noliac A/O, Denmark
Mr. Ryan Lee, Sunnytec Electronics, Taiwan

Finally, I also express my appreciation for my wife Michiko's daily support throughout this textbook project.

Kenji Uchino
October 1, 2007
State College, Pennsylvania

Contents

List of Symbols

D	**Electric displacement**
E	**Electric field**
P	**Dielectric polarization**
P_s	**Spontaneous polarization**
α	**Ionic polarizability**
γ	**Lorentz factor**
μ	**Dipole moment**
ε_0	**Vacuum permittivity**
ε	**Relative permittivity**
C	**Curie–Weiss constant**
T_0	**Curie–Weiss temperature**
T_C	**Curie temperature (phase transition temperature)**
G_1	**Gibbs elastic energy**
x	**Strain**
xs	**Spontaneous strain**
X	**Stress**
s	**Elastic compliance**
c	**Elastic stiffness**
v	**Sound velocity**
d, g	**Piezoelectric coefficients**
M, Q	**Electrostrictive coefficients**
k	**Electromechanical coupling factor**
η	**Energy transmission coefficient**
Y	**Young's modulus**
$\tan\delta\,(\tan\delta')$	**Extensive (intensive) dielectric loss**
$\tan\phi\,(\tan\phi')$	**Extensive (intensive) elastic loss**
$\tan\theta\,(\tan\theta')$	**Extensive (intensive) piezoelectric loss**

Suggested Teaching Schedule

(Thirty 75-minute lectures per semester)

I. FUNDAMENTALS

I-1.	Trend of Micromechatronics and Computer Simulation	2 lectures
I-2.	Overview of Piezoelectricity and Magnetostriction	3 lectures
I-3.	Structures of Smart Transducers	2 Lectures
I-4.	Drive/Control Techniques of Smart Transducers	2 Lectures
I-5.	Finite Element Analysis	1 Lecture
I-6.	Design Optimization	2 Lectures
I-7.	Future of the FEM in Smart Structures	1 Lecture

II. HOW TO USE ATILA

Lab Computer Simulation

II-1.	Piezoelectric Plates	
	Rectangular plate	3 Lectures
	Circular disk	1 Lecture
II-2.	Magnetostrictive Rod	1 Lecture
II-3.	Composite Structures	
	Bimorph	1 Lecture
	Cymbal	1 Lecture
II-4.	Piezoelectric Transformers	2 Lectures
II-5.	Ultrasonic Motors	
	L-shape motor	2 Lectures
	Metal-tube motor	1 Lecture
II-6.	Underwater Transducers	
	Langevin Transducer	2 Lectures
	Cymbal under water	1 Lecture
II-7.	Acoustic Lens	1 Lecture

Final Project Presentation 1 Lecture

Prerequisite Knowledge

In order to understand ferroelectric devices, some prerequisite knowledge is necessary. Try to solve the following questions without seeing the answers on the next page.

Q1 Describe the definitions of elastic *stiffness* c and *compliance* s, using a stress X - strain x relation.

Q2 Indicate a *shear stress* X_4 on the following square.

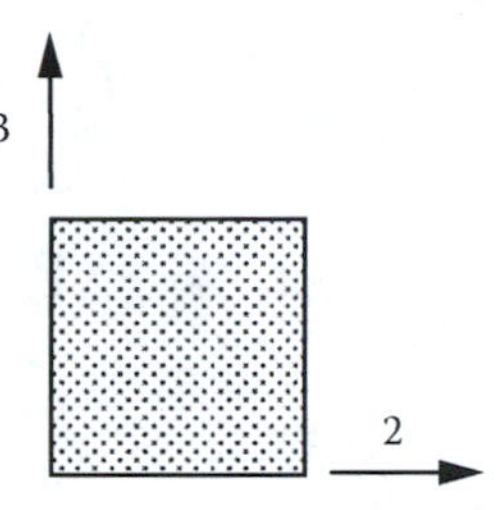

Q3 There is a rod of length L made of a material with sound velocity v. Describe the fundamental extensional *resonance frequency* f.

Q4 Calculate the capacitance C of a capacitor with area S and electrode gap t filled with a material of *relative permittivity* ε.

Q5 Describe the *Laplace transform* of the impulse function $[\delta(t)]$.

Q6 There is a voltage supply with an internal impedance Z_0. Indicate the external impedance Z_1 to obtain the maximum output power.

Q7 Calculate the induced polarization P under an external stress X in a *piezoelectric* with a piezoelectric constant d.

Q8 A connected 4-rod system has three *eigen vibration modes* as illustrated in the figure. Which mode (a, b, or c) corresponds to the highest resonance frequency?

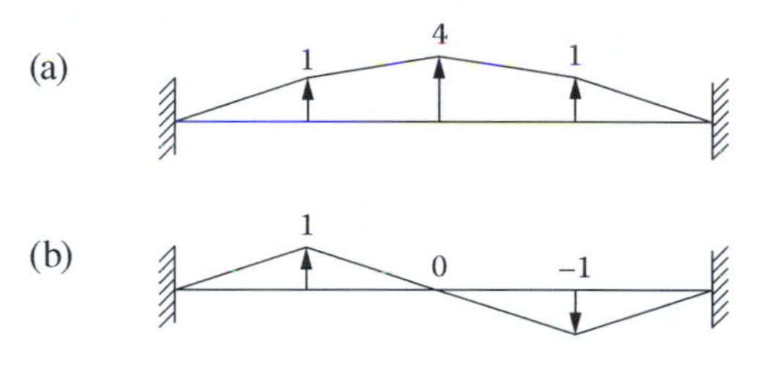

Q9 Obtain y from the following determinant:

$$\begin{vmatrix} 2-3y & -1 & 0 \\ -1 & 2-4y & -1 \\ 0 & -1 & 2-3y \end{vmatrix} = 0$$

Q10 When the damping factor c is not large, and the impulse force f is applied on the following dynamic system:

$$m(d^2x/dt^2) + c\,(dx/dt) + kx = f\,,$$

illustrate roughly the transient response of displacement x as a function of t (you do not need to obtain the equation; just write a rough figure).

ANSWERS

(More than 70% of full score is expected)

Q1 $X = c\,x,\ x = s\,X$

Q2 $x_4 = 2\,x_{23} = 2\,\phi$ (radian)

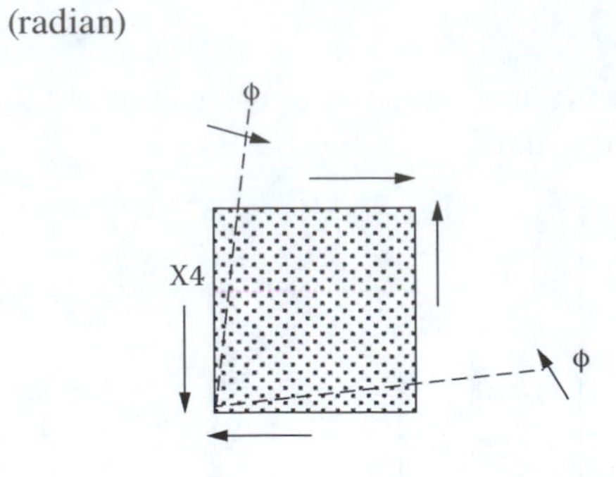

Q3 $f = v\,/\,2L$

Q4 $C = \varepsilon_0\varepsilon\,(S\,/\,t)$ [0.5 point for C = ⊠_(S / t)]

Q5 1

Q6 $Z_1 = Z_0$

On Z_1, current and voltage are given as $V/(Z_0 + Z_1)$ and $[Z_1/(Z_0 + Z_1)]V$, leading to the power:

Power = $V^2 \cdot Z_1/(Z_0 + Z_1)^2 = V^2/(Z_0/\_Z_1^{1/2} + Z_1^{1/2})^2 \le (1/4)\,V^2/\,Z_0$

The maximum is obtained when $Z_0/Z_1^{1/2} = Z_1^{1/2}$. Hence, $Z_1 = Z_0$.

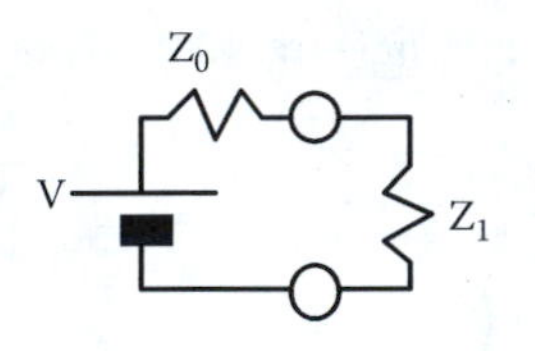

Q7 $P = dX$ (refer to $x = dE$)

Q8 c (Consider the wavelength-like distance.)

Q9 y = 1/6, 2/3, and 1 (These eigen states correspond to a, b, and c in Q8.)

From the determinant, $(2 - 3y)(2 - 4y)(2 - 3y) - (2 - 3y) - (2 - 3y) = 0$

Thus, $(2 - 3y)(1 - 6y)(2 - 2y) = 0$

Q10 Rough sketch for low c (low ξ):

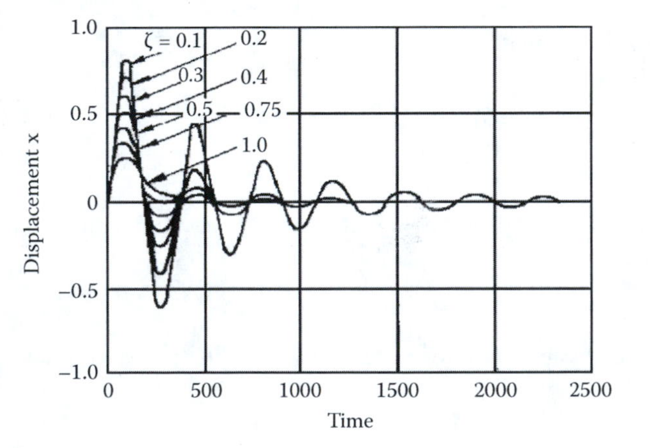

<< DO YOU KNOW ? >>

TITANIC AND LANGEVIN

THE TITANIC.

PAUL LANGEVIN (LEFT) AND ALBERT EINSTEIN.

At 11:45 pm on April 10th, 1912, the wreck of the *Titanic* occurred. If the ultrasonic sonar system had been available then, the tragedy would have been averted. The development of ultrasonic technology was given an impetus by this accident. World War I, which began in 1914, further accelerated the development of this technology; the motivation was to use the technology to locate German U-boats. Dr. Paul Langevin, a professor of the Industrial College of Physics and Chemistry in Paris, who had many friends, including Albert Einstein and Pierre Curie, experimented with ultrasonic signal transmission into the sea, in collaboration with the French navy. Using multiple small single-crystal quartz pellets sandwiched by metal plates (the original Langevin-type transducer), he succeeded in receiving the returned ultrasonic signal from the deep sea in 1917.

Part I

Fundamentals

1 Micromechatronic Trends and Computer Simulation

This chapter summarizes the development trends of micromechatronics and the necessity of computer simulation, in particular, finite element method (FEM).

1.1 THE NEED FOR NEW ACTUATORS

For office equipment such as printers and hard disk drives, market research indicates, tiny motors smaller than 1 cm would be in great demand over the next 10 years. However, using the conventional electromagnetic motor design, it is rather difficult to produce a motor with sufficient energy efficiency. Figure 1.1 shows the smallest electromagnetic motor with reasonable torque level, used in a wristwatch. Although the rotor is as small as 0.3 mm in diameter, and the coil not less than 10 mm in length, it does not rotate. Many turns of coil with a thin conductive wire provide high resistance, leading to Joule heat and low efficiency on the order of only 1%. Unless the conductive wire is replaced by a superconducting-type wire, a significant breakthrough is not expected for electromagnetic motors. Also, the coil sometimes requires a bulky and heavy magnetic shield case. Figure 1.2 illustrates efficiency versus power/size of electromagnetic motors compared to piezoelectric ultrasonic motors. Taking into account the general understanding 30% for 30 W for electromagnetic motors, we can see that the piezoelectric motor/transducer seems to be superior to the electromagnetic types in micro motor areas.

The demand for new actuators has also increased in recent years for *positioner* and *mechanical damper* applications, in addition to the foregoing *miniature motor* segment.

Submicrometer fabrication, common in the production of electronic chip elements, is also becoming important in mechanical engineering. The trends depicted in Figure 1.3 show how machining accuracy has improved over the years.[1] One of the primary reasons for the improved accuracy is the development of more precise position sensors. Sensors utilizing lasers can easily detect nanometer-scale displacements. This could be regarded as another instance of "the chicken or the egg" issue, however, as the fabrication of such precise optical instruments can only be achieved with the submicrometer machining equipment the sensors are designed to monitor.

In an actual machining apparatus comprising translational components (the joints) and rotating components (the gears and motor), error due to backlash will occur. Machine vibration will also lead to unavoidable positional fluctuations. Furthermore, the deformations due to machining stress and thermal expansion cannot be ignored. The need for submicron displacement positioners to improve cutting accuracy is apparent. One example is a prototype for a lathe machine that uses a ceramic multilayer actuator and can achieve cutting accuracy of 0.01 μm.[2]

The concept of *adaptive optics* has been applied in the development of sophisticated new optical systems. Earlier systems were generally designed such that parameters such as position, angle, or the focal lengths of mirror and lens components remained essentially fixed during operation. Newer systems incorporating adaptive optical elements respond to a variety of conditions to essentially adjust the system parameters to maintain optimum operation. The original LIDAR system (a radar system utilizing light waves) was designed to be used on the NASA space shuttle for monitoring the surface of the Earth.[3] A laser beam is projected toward the Earth's surface, and the reflected light is received by a reflection telescope, amplified by a photomultiplier, and the resulting signal is processed to produce an image. The vibration noise and temperature fluctuations of the shuttle make it difficult for a sharp image to be obtained, however, and the use of a responsive positioner was considered to compensate for the detrimental effects.

Active and passive vibration suppression by means of piezoelectric devices is also a promising technology for use in space structures and military and commercial vehicles. Mechanical vibration in a structure traveling through the vacuum of space is not readily damped, and a 10-m-long array of solar panels can be severely damaged simply by the repeated impact of space dust with the structure. Active dampers using shape memory alloys or piezoelectric ceramics are currently under investigation to remedy this type of problem. A variety of *smart skins* designed for military tanks or submarines are illustrated in Figure 1.4. A signal is generated in the sensitive skin with perhaps the impact of a missile on the tank or the complex forces applied to a submarine through turbulent flow, which is fed back to an actuator. The actuator responds by changing its shape to effectively minimize the impact damage on the tank (*structure protection*) or the drag force on the submarine.

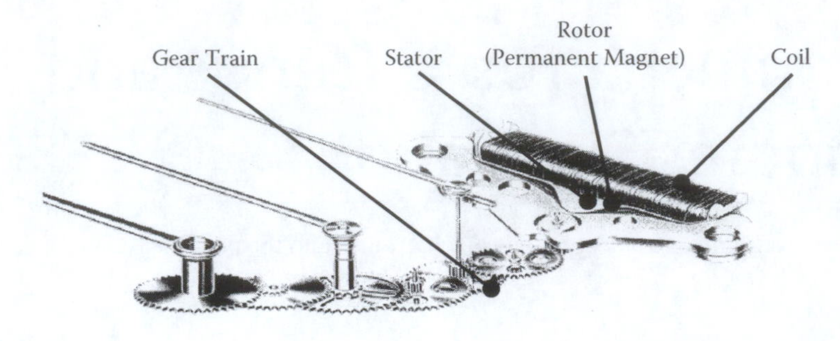

FIGURE 1.1 Smallest electromagnetic motor with reasonable torque level used in a wristwatch. Note that a coil longer than 10 mm is required.

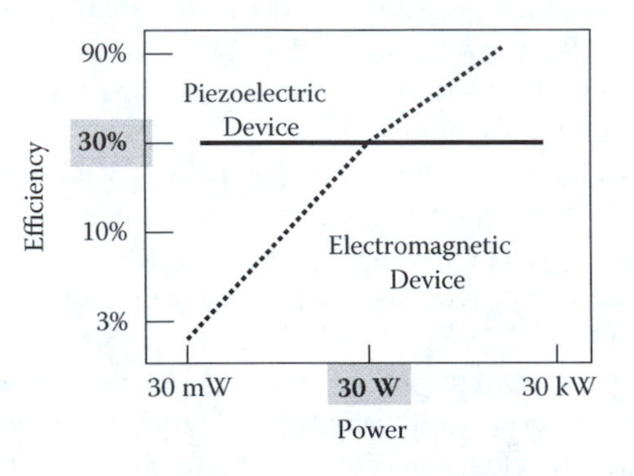

FIGURE 1.2 General efficiency versus power/size relation for the electromagnetic and piezoelectric motors. Note the significant reduction of efficiency for the electromagnetic type.

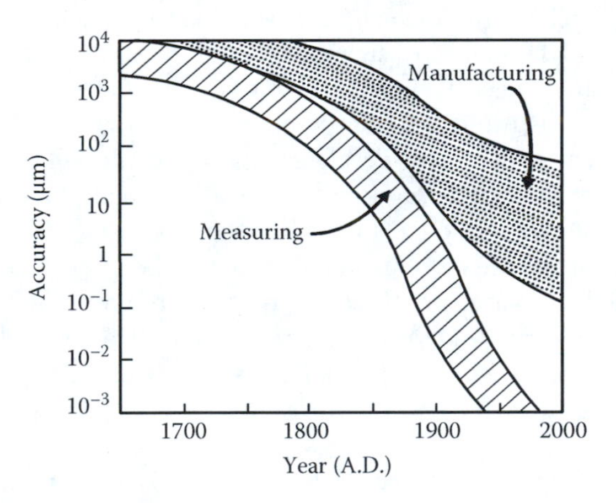

FIGURE 1.3 Obtainable accuracies in manufacturing and measurement over time. Note that current accuracies tend to be on the micrometer and nanometer scales, respectively.

1.2 AN OVERVIEW OF SOLID-STATE ACTUATORS

1.2.1 Smart Actuators

Let us now consider the "smartness" of a material. The various properties relating the input parameters of electric field, magnetic field, stress, heat, and light to the output parameters, charge/current, magnetization, strain, temperature, and light, are listed in Table 1.1. Conducting and elastic materials, which generate current and strain outputs, respectively, with input voltage or stress (diagonal couplings), are sometimes referred to as *trivial materials.* High-temperature superconducting ceramics are

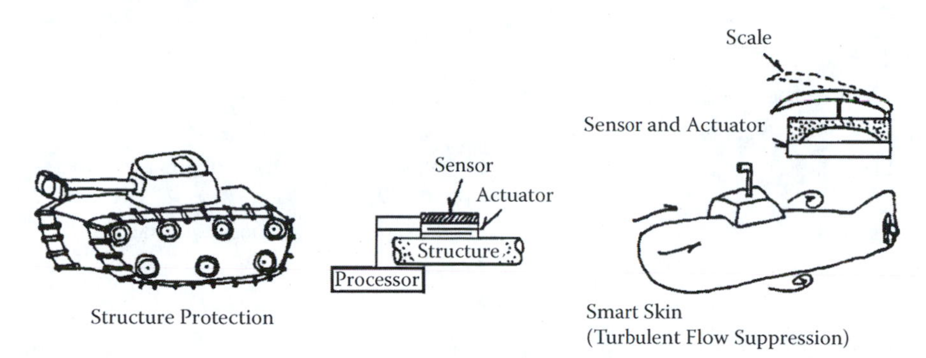

FIGURE 1.4 Smart skin structures for military tanks and submarines.

TABLE 1.1
Various Basic and Cross-Coupled Properties of Materials

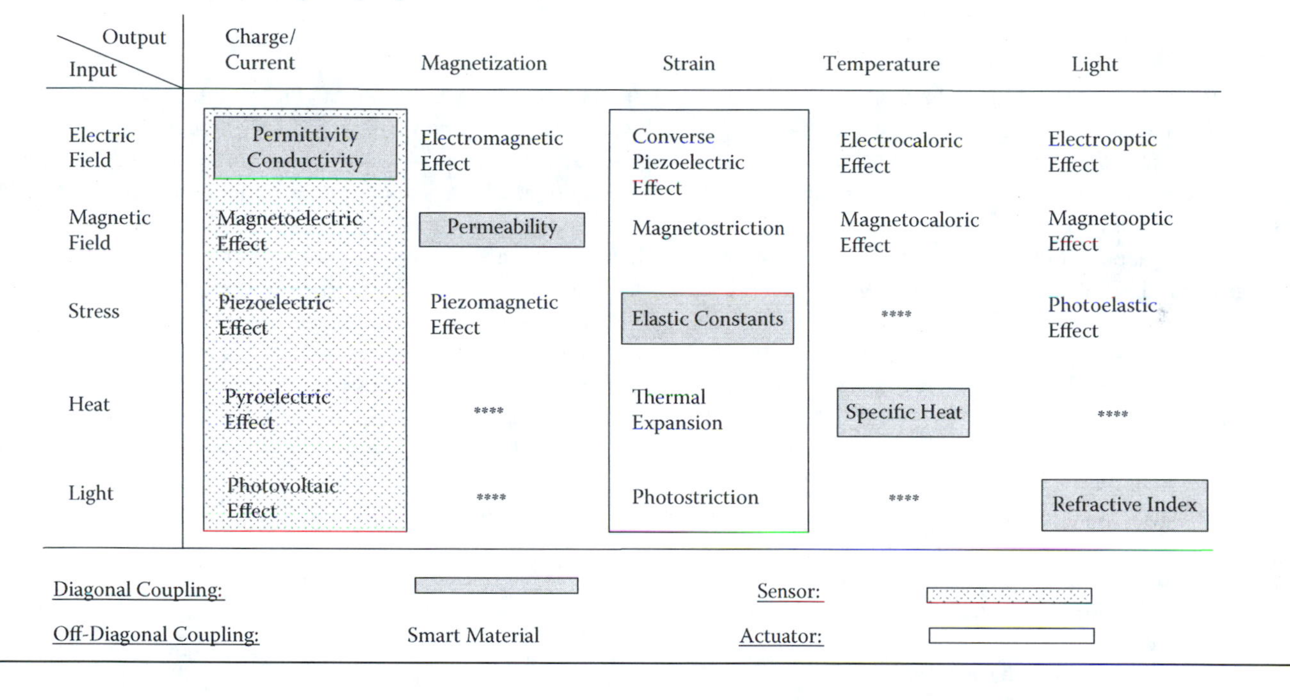

Input \ Output	Charge/ Current	Magnetization	Strain	Temperature	Light
Electric Field	Permittivity Conductivity	Electromagnetic Effect	Converse Piezoelectric Effect	Electrocaloric Effect	Electrooptic Effect
Magnetic Field	Magnetoelectric Effect	Permeability	Magnetostriction	Magnetocaloric Effect	Magnetooptic Effect
Stress	Piezoelectric Effect	Piezomagnetic Effect	Elastic Constants	****	Photoelastic Effect
Heat	Pyroelectric Effect	****	Thermal Expansion	Specific Heat	****
Light	Photovoltaic Effect	****	Photostriction	****	Refractive Index

Diagonal Coupling: [shaded box] Sensor: [dotted box]
Off-Diagonal Coupling: Smart Material Actuator: [open box]

also considered trivial materials in this sense, but the figure of merit (electrical conductivity) exhibited by some new compositions has been exceptionally high in recent years, making them especially newsworthy.

On the other hand, pyroelectric and piezoelectric materials, which generate an electric field with the input of heat and stress, respectively, are called *smart materials*. These off-diagonal couplings have corresponding converse effects, the electrocaloric and converse piezoelectric effects, so that both sensing and actuating functions can be realized in the same material. One example is a tooth brace made of a shape memory (superelastic) alloy. A temperature-dependent phase transition in the material responds to variations in the oral temperature, thereby generating a constant stress on the tooth.

Intelligent materials must possess a drive/control or processing function that is adaptive to changes in environmental conditions in addition to their actuator and sensing functions. Photostrictive actuators belong to this category. Some ferroelectrics generate a high voltage when illuminated (the *photovoltaic effect*). Because the ferroelectric is also piezoelectric, the photovoltage produced will induce a strain in the crystal. Hence, this type of material generates a drive voltage dependent on the intensity of the incident light, which actuates a mechanical response. The self-repairing nature of partially stabilized zirconia can also be considered an intelligent response. Here, the material responds to the stress concentrations produced with the initial development of the microcrack (sensing) by undergoing a local phase to reduce the concentrated stress (control) and stop the propagation of the crack (actuation).

If one could incorporate a somewhat more sophisticated mechanism for making complex decisions into its intelligence, a *wise material* might be created. Such a material might be designed to determine that "this response may cause harm" or "this

TABLE 1.2
New Actuators Classified in Terms of Input Parameter

Input parameter	Actuator type/device
Electric field	Piezoelectric/electrostrictive; Electrostatic (Silicon MEMS); Electrorheological fluid
Magnetic field	Magnetostrictive; Magnetorheological fluid
Stress	Rubbertuator
Heat	Shape memory alloy; Bubble jet
Light	Photostrictive; Laser light manipulator
Chemical	Mechanochemical; Metal-hydrite

action will lead to environmental destruction," and respond accordingly. It would be desirable to incorporate such fail-safe mechanisms in actuator devices and systems. A system so equipped would be able to monitor and detect the symptoms of wear or damage so that it could shut itself down safely before serious damage or an accident occurred.

1.2.2 New Actuators

Actuators that operate by a mechanism different from those found in the conventional AC/DC electromagnetic motors and oil/air pressure actuators are generally classified as *new actuators.* Some recently developed new actuators are classified in Table 1.2 in terms of input parameter. Note that most of the new actuators are made from some type of solid material with properties specifically tailored to optimize the desired actuating function, which is why these actuators are sometimes referred to as just *solid-state actuators.*

The displacement of an actuator element must be controllable by changes in an external parameter such as temperature, magnetic field, or electric field. Actuators activated by changes in temperature generally operate through the *thermal expansion* or dilatation associated with a phase transition, such as the ferroelectric and martensitic transformations. *Shape memory alloys*, such as Nitinol, are of this type. Magnetostrictive materials, such as Terfenol-D, respond to changes in an applied magnetic field. *Piezoelectric* and *electrostrictive* materials are typically used in electric-field-controlled actuators. In addition to these, there are silicon-based micro-electro-mechanical-systems (MEMS), polymer artificial muscles, light-activated actuators (in which the displacements occur through the photostrictive effect or a photoinduced phase transformation), and electro/magnetorheological fluids.

The desired general features for an actuator element include the following:

1. large displacement (sensitivity = displacement/driving power),
2. good positioning reproducibility (low hysteresis),
3. quick response,
4. stable temperature characteristics,
5. low driving energy,
6. large generative force and failure strength,
7. small size and low weight,
8. low degradation/aging in usage, and
9. minimal detrimental environmental effects (mechanical noise, electromagnetic noise, heat generation, etc.).

The current version of ATILA FEM software can treat either piezoelectric or magnetostrictive materials and their coupled transducer systems.

1.3 NECESSITY OF COMPUTER SIMULATION

The application of computer simulation to smart devices and systems began in the late 1970s. Some historical examples are introduced in this section.

1.3.1 Deformable Mirror

Precise wave front control with as small a number of parameters as possible and compact construction is a common and basic requirement for *adaptive optical systems*. Continuous surface-deformable mirrors, for example, tend to be more desirable than segmented mirrors in terms of controllability. A deformable mirror can be used to correct the distortion that occurs in a telescope image because of atmospheric conditions, as illustrated in Figure 1.5. Various electrode configurations were proposed to produce a variety of mirror contours.[4]

A simple multimorph deformable mirror was proposed, which could be simply controlled by a microcomputer.[5,6] In the case of the *two-dimensional multimorph deflector design* (x-y plane), a static deflection along the Z axis $f_i(x,y)$ is generated by the voltage distribution on the i-th layer $V_i(x,y)$, which can be obtained by solving the following system of differential equations:

$$A\left[\frac{\partial^4 f_i(x,y)}{\partial x^4}+2\frac{\partial^4 f_i(x,y)}{\partial x^2 \partial y^2}+\frac{\partial^4 f_i(x,y)}{\partial y^4}\right]+B_i\left[d_{31}\frac{\partial^2 V_i(x,y)}{\partial x^2}+d_{32}\frac{\partial^2 V_i(x,y)}{\partial y^2}\right]=0 \quad (1.1)$$

where $[i = 1, 2, \ldots, I]$, A and Bi are constants, d_{31} and d_{32} are piezoelectric strain coefficients, and I is the total number of layers in the multimorph. This equation is derived specifically for piezoelectric layers, but may also be adapted to describe a device comprising electrostrictive layers, provided the nonlinear relationship between the electric field and the induced strain characteristic of electrostrictive materials is properly taken into account.

When the conditions $d_{31} = d_{32}$ (similar to a polycrystalline PZT composition) and $f_i(x,y) = 0$ under $V_i(x,y) = 0$ $(i = 1, 2, ..., I)$ are assumed, then equation (1.1) reduces to

$$\nabla^2 f_i(x,y) + C_i V_i(x,y) = 0 \quad (1.2)$$

where C_i is a constant that is different for each layer and ∇^2 is the Laplacian operator. Thus, we see in general that the contour of the mirror surface can be changed by means of an appropriate electrode configuration and *applied voltage distribution* on each electroactive ceramic layer.

The contour of the mirror surface can be represented by the Zernike aberration polynomials, whereby an arbitrary surface contour modulation $g(x,y)$ can be described as follows:

$$g(x,y) = C_r(x^2+y^2) + C_c^1 x(x^2+y^2) + C_c^2 y(x^2+y^2) + \ldots.. \quad (1.3)$$

Note that the Zernike polynomials are orthogonal and therefore completely independent of each other. The C_r and C_c terms represent *refocusing* and the *coma aberration*, respectively. As far as human vision is concerned, correction for aberrations up to the second order in this series (representing astigmatism) is typically sought to provide an acceptably clear image.

As examples, let us consider the cases where we wish to refocus and to correct for the coma aberration. A uniform large-area electrode can provide a parabolic (or spherical) deformation with the desired focal length. In the case of coma correction, one could employ the electrode pattern shown in Figure 1.6, which consists of only six elements addressed by voltages applied

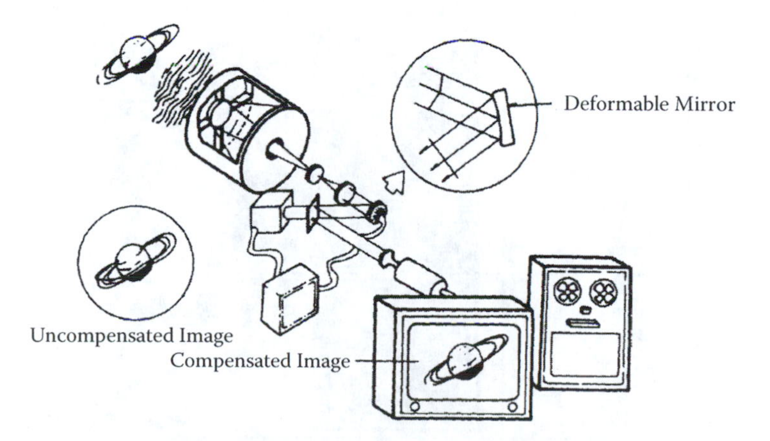

FIGURE 1.5 A telescope image correction system using a monolithic piezoelectric deformable mirror. (From J. W. Hardy, J. E. Lefebre, and C. L. Koliopoulos: *J. Opt. Soc. Am.* **67**, 360 [1977]. With permission.)

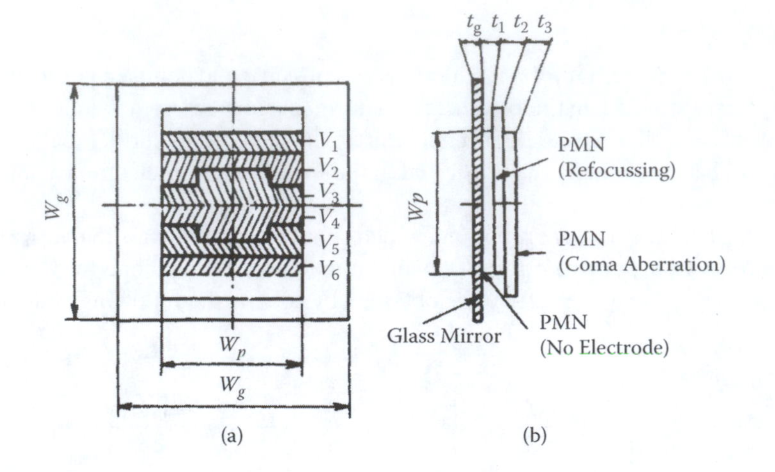

FIGURE 1.6 A two-dimensional multimorph deformable mirror designed for refocusing and coma aberration correction: (a) front view and (b) a cross-sectional view of the structure. (From K. Uchino, Y. Tsuchiya, S. Nomura, T. Sato, H. Ishikawa and O. Ikeda: *Appl. Opt.* **20**, 3077 [1981]. With permission.)

in a fixed ratio. Some experimental results characterizing mirror deformations established for the purpose of refocusing, coma correction, and both refocusing and coma correction are shown in Figure 1.7. These results also demonstrate the effectiveness of achieving a superposition of deformations by means of an appropriate configuration of discrete electrodes.

The younger reader in all likelihood is blissfully unaware of how difficult computer simulation in the early 1980s was. First, we did not have a personal computer in our office but one big university computer. We needed to wait for a couple of days after requesting the calculation. Second, the calculation speed was slow. The aforementioned simulation, such as an 8×8 array mesh with three different voltage levels, took more than 8 h. Third, more than $2,000 was required just for this calculation at that time (which is equivalent to $20,000 nowadays). Also, the analysis was primarily limited to a pseudostatic situation.

Using the ATILA FEM software, you can now design the foregoing deformable mirror in less than 1 h on your personal laptop computer (the author envies the reader very much!).

1.3.2 π-Shaped Ultrasonic Motor

The modal analysis of the ultrasonic motor was demonstrated in the mid-1980s for a π-shaped linear motor, pictured in Figure 1.8.[7] By applying sine and cosine voltages on two multilayer actuators, a π-shaped metal frame vibrates like a trotting horse.

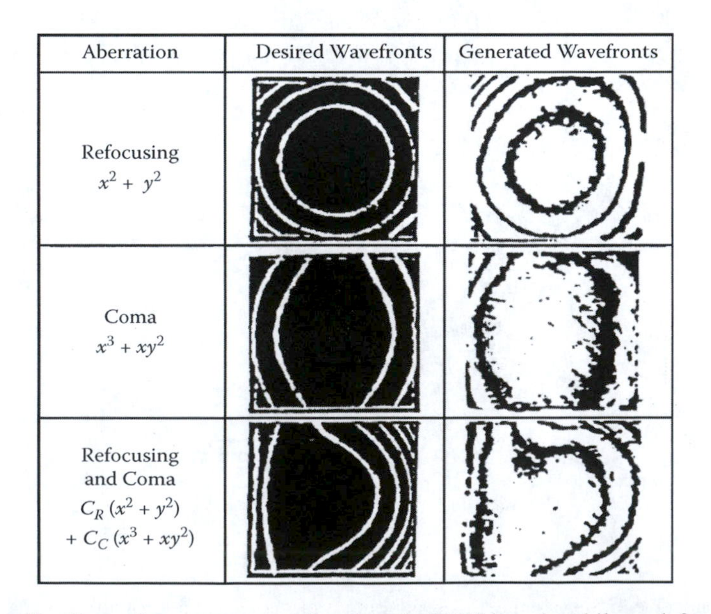

FIGURE 1.7 Interferograms revealing the surface contours produced on the PMN-based multilayer deformable mirror. (From K. Uchino, Y. Tsuchiya, S. Nomura, T. Sato, H. Ishikawa, and O. Ikeda: *Appl. Opt.* **20**, 3077 [1981]. With permission.)

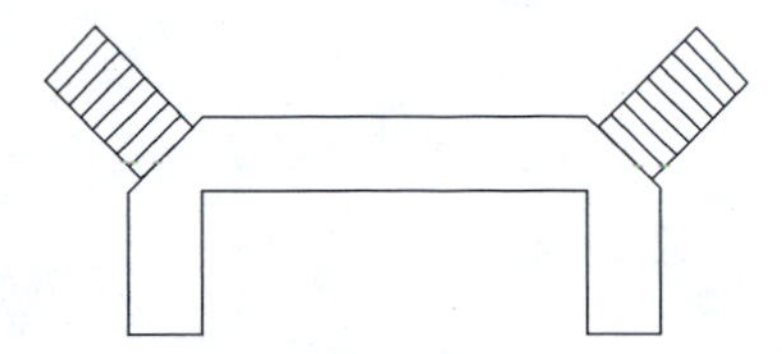

FIGURE 1.8 π-shaped linear motors. (From K. Onishi: Ph.D. thesis, Tokyo Inst. Technology, Japan [1991]. With permission.)

The key to designing this motor is to adjust three resonance frequencies suitably: (a) longitudinal mode of the piezoelectric actuator, (b) bending mode of the metal beam, and (c) bending mode of the metal leg. We will introduce both analytical and FEM simulation approaches.

1.3.2.1 Analytical Approach

The analytical approaches were mostly used in the development in the 1980s. The size of the piezoelectric actuator is determined largely by the resonance frequency at which it will be operated. The π-shaped multilayer motor requires an actuator that will be operated in a thickness longitudinal mode of vibration. The resonance frequency for the desired vibration mode is given by the following equation:

$$f_r = \left(\frac{1}{2l}\right)\sqrt{\frac{1}{\rho s_{33}^D}} \tag{1.4}$$

On the contrary, the two legs and cross bar of the π-shaped structure are referred to as the *cantilevers* and the *free bar*, respectively. The resonance frequency for both is described by a general equation of the form:

$$f = \left(\frac{\alpha_m^2}{2\sqrt{3}\,(2\pi)}\right)\left(\frac{t}{l^2}\right)\sqrt{\frac{Y}{\rho}} \tag{1.5}$$

where l is the length of the element, t is its thickness, Y is Young's modulus, ρ is the mass density, and α_m is a parameter determined by the vibration mode. Values for α_m are given in Figure 1.9 for the first three resonance modes of the cantilever and free bar elements. The second mode resonance frequencies for the cantilevers (a) and the free bar (b) are thus defined as follows:

$$f_{2c} = 1.01\left(\frac{t}{l_c^2}\right)\sqrt{\frac{Y}{\rho}} \tag{1.6a}$$

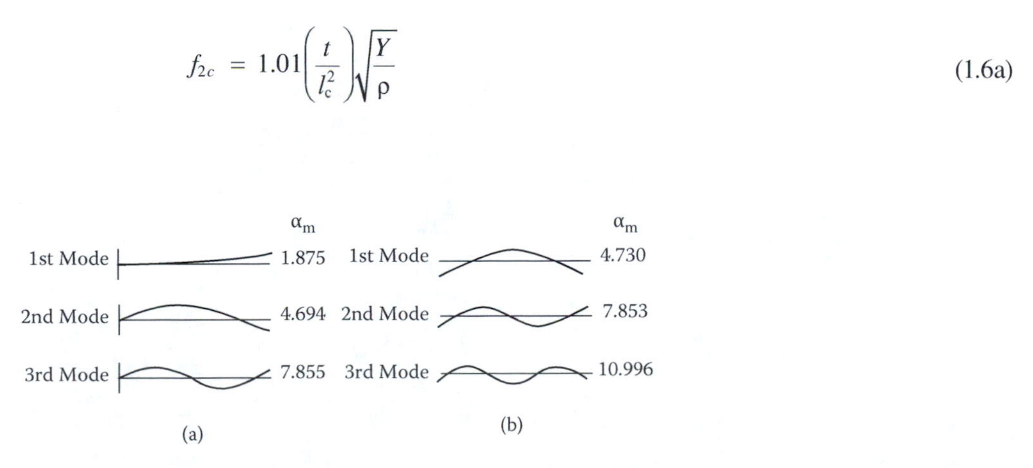

FIGURE 1.9 Values of the parameter α_m corresponding to the first three resonance modes for the elements of a π-shaped linear motor: (a) the cantilevers and (b) the free bar.

$$f_{2b} = 2.83\left(\frac{t}{l_b^2}\right)\sqrt{\frac{Y}{\rho}} \tag{1.6b}$$

1.3.2.2 FEM Approach

Free vibration analysis was performed using an FEM program on the structures. The design of the motor determined by the FEM is presented in Figure 1.10.

Figure 1.11 shows some results of the free vibration FEM on a multilayer π-shaped linear motor. The leg bending is excited mostly at 105 kHz, whereas the transverse beam vibration (or the up-down leg motion) is excited at 108 kHz.

The trace of the leg movement was determined when a sinusoidal drive was applied to one actuator and a cosine drive on the other, as shown in Figure 1.12. Both actuators were driven at 90 kHz with a force of 6 N. The elliptical loci of the legs'

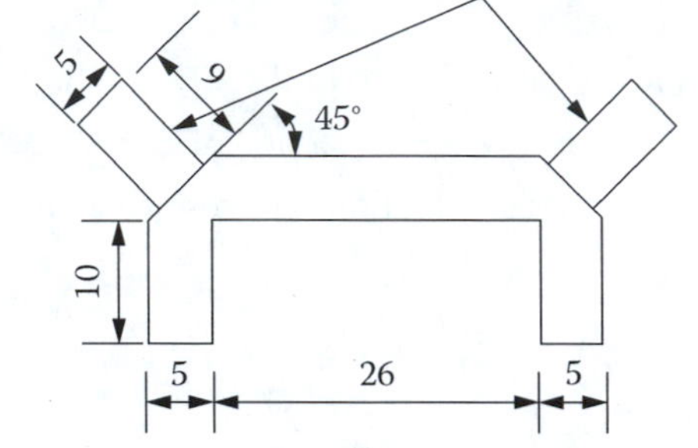

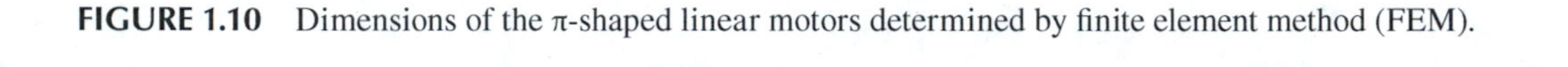

FIGURE 1.10 Dimensions of the π-shaped linear motors determined by finite element method (FEM).

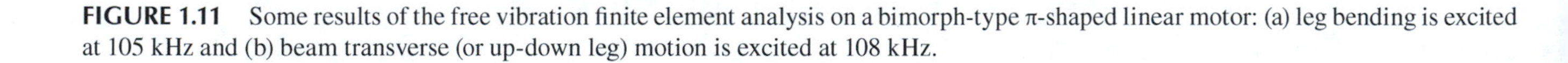

FIGURE 1.11 Some results of the free vibration finite element analysis on a bimorph-type π-shaped linear motor: (a) leg bending is excited at 105 kHz and (b) beam transverse (or up-down leg) motion is excited at 108 kHz.

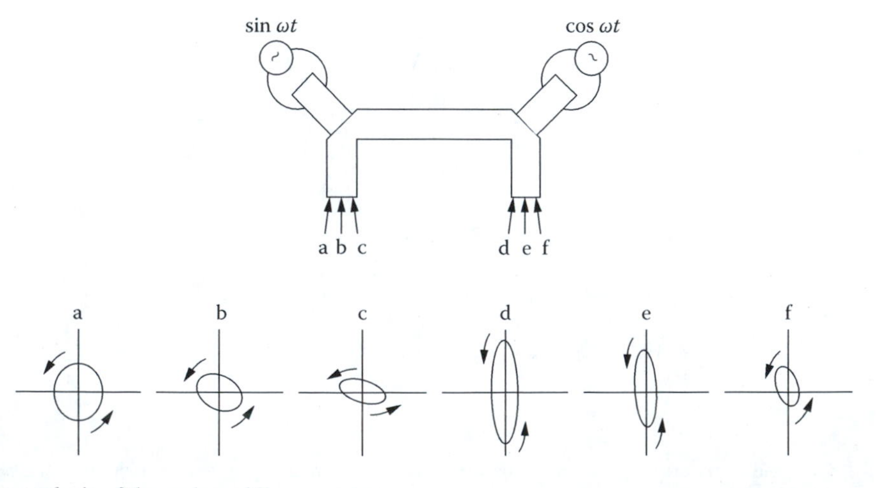

FIGURE 1.12 Vibration analysis of the π-shaped linear motor.

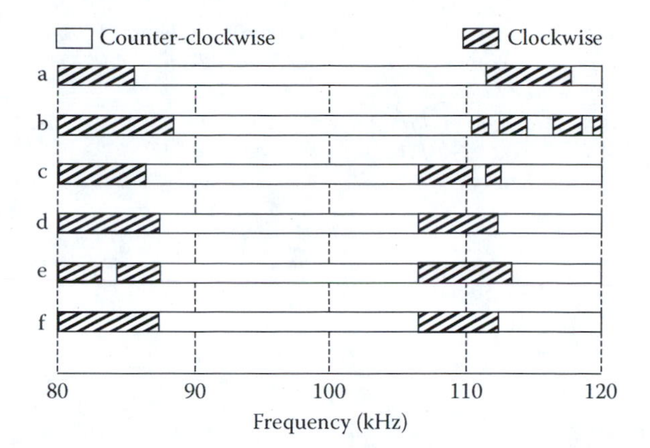

FIGURE 1.13 Rotation direction of the elliptical vibration of the two legs of a π-shaped ultrasonic linear motor as determined by the laser Doppler method. [**a**, **b**, **c**: first leg, **d**, **e**, **f**: second leg].

displacement for various positions on the legs are shown at the bottom of Figure 1.12. The counterclockwise rotation of all points produces an efficient drive force against the rail.

The motor is driven most efficiently over the frequency range that extends from a resonance mode that produces large bending displacements. The drive frequency should also fall between the frequencies of the symmetric and antisymmetric leg motion, ideally such that there is either a 90° or a −90° phase difference between the two legs conducive to the "trotting" motion of the device.

Vibration measurement with a laser vibrometer indicated the counterclockwise elliptical vibration of both legs, which agrees well with FEM predictions (see Figure 1.13). The maximum no-load speed and maximum load were approximately 20 cm/s and 200 gf (20 N), respectively. The propagation direction of the motor was reversed by exchanging the drive signals (sine and cosine) applied to the two multilayer actuators. Tests of the motor's action in both directions resulted in similar speed versus drive voltage curves.

In the mid-1980s, dynamic FEM programs for ultrasonic motor analysis became available. However, as you can see, the analysis was just for 2-D, but took more than a day for this calculation.

1.4 PURPOSE AND CONTENTS OF THIS TEXTBOOK

This textbook was authored as a tutorial tool of the Finite Element Analysis Computer Software ATILA, which is divided into two parts: Part I (Fundamentals) and Part II (How To Use ATILA).

Part I consists of seven chapters; the essential elements of piezoelectricity and magnetostriction are summarized in chapters 1 to 4, the basics of FEM are described in chapters 5 and 6, and chapter 7 is the summary.

Part II consists of seven chapters: seven typical piezoelectric and magnetostrictive device examples are used to teach how to use ATILA in a step-by-step manner, starting with piezoelectric plates, and followed by a magnetostrictive rod, composite structures (bimorph, multilayer, cymbal, etc.), piezoelectric transformers, several ultrasonic motors, underwater transducers (including Tonpilz sonar) and acoustic lenses.

Thanks to Micromechatronics Inc., State College, PA, a finite element analysis software, ATILA-Light (Educational Demo version), is provided as a courtesy tool, which should be uploaded to the reader's personal computer before starting to read this book. ATILA-Light can be found in the companion CD and/or in the Web site, www.MicromechatronicsInc.com. About 60 Mb of hard disk space is required for this program.

REFERENCES*

1. I. Inazaki: *Sensor Technology* **2**, No.2, 65 (1982).
2. K. Uchino: Ultrasonic Techno, No. 8, 19 (1993).
3. *Opt. Spectra*, March, p. 46 (1979).
4. J. W. Hardy, J. E. Lefebre, and C. L. Koliopoulos: *J. Opt. Soc. Am.* **67**, 360 (1977).
5. K. Uchino, Y. Tsuchiya, S. Nomura, T. Sato, H. Ishikawa, and O. Ikeda: *Appl. Opt.* **20**, 3077 (1981).
6. T. Sato, H. Ishikawa, O. Ikeda, S. Nomura, and K. Uchino: *Appl. Opt.* **21**, 3669 (1982).

* References 8 to 11 are recommended for further study.

7. K. Onishi: "π-Shape Ultrasonic Motors." Ph.D. thesis, Tokyo Inst. Technology, Japan (1991).
8. S. Ueha, Y. Tomikawa, M.Kurosawa, and N.Nakamura: *Ultrasonic Motors: Theory and Applications*, Oxford Science Publ., Monographs in EE. Eng. 29 (1993).
9. K. Uchino: *Piezoelectric Actuators and Ultrasonic Motors*, Kluwer Academic Publishers, MA (1996).
10. K. Uchino: *Ferroelectric Devices*, Marcel Dekker, NY (2000).
11. K. Uchino: *Micromechatronics*, Marcel Dekker, NY (2003).

2 Review of Piezoelectricity and Magnetostriction

We will treat the popular smart materials piezoelectric ceramics and magnetostrictive alloys in this chapter. Certain materials produce electric charges on their surfaces as a consequence of the application of mechanical stress. The induced charges are proportional to the mechanical stress. This is called the *direct piezoelectric effect* and was discovered in quartz by Pierre and Jacques Curie in 1880. Materials showing this phenomenon conversely have a geometric strain proportional to an applied electric field. This is the *converse piezoelectric effect*, verified by Gabriel Lippmann in 1881. The root of the word *piezo* means *pressure*; hence, the original meaning of the word piezoelectricity implied "pressure electricity." On the other hand, magnetostriction was discovered by James Joule, also in the 19th century, in magnetic metals, Fe, Ni, and Co. However, its intensive study had to wait until the discovery of "giant" magnetostrictive materials, including rare earth metals, in the 1960s.

Piezoelectricity was first applied in underwater sonar to find U-boats during World War I (1914), and is now extensively utilized in the fabrication of various devices, such as transducers, actuators, surface acoustic wave devices, frequency control, and so on. In this chapter we first overview the piezoelectric and magnetostrictive materials, and then present the basic mathematical treatment (tensor/phenomenology) that is essential to understand the fundamentals of finite element analysis. Then, typical example applications of piezoelectric materials are introduced.[1–4]

2.1 PIEZOELECTRIC MATERIALS: AN OVERVIEW

2.1.1 Piezoelectric Figures of Merit

There are five important figures of merit in piezoelectrics: the *piezoelectric strain constant d*, the *piezoelectric voltage constant g*, the *electromechanical coupling factor k*, the *mechanical quality factor* Q_m, and the *acoustic impedance Z*. These figures of merit are considered in this section.

2.1.1.1 Piezoelectric Strain Constant d

The magnitude of the induced strain x by an external electric field E is represented by this figure of merit (an important figure of merit for actuator applications):

$$x = dE \tag{2.1}$$

2.1.1.2 Piezoelectric Voltage Constant g

The induced electric field E is related to an external stress X through the piezoelectric voltage constant g (an important figure of merit for sensor applications):

$$E = gX \tag{2.2}$$

Taking into account the relation $P = d\,X$, we obtain an important relation between g and d:

$$g = d/\varepsilon_0\,\varepsilon \tag{2.3}$$

where ε_0 is the vacuum permittivity (8.854×10^{-12} [C/Vm]) and ε is the relative permittivity.

2.1.1.3 Electromechanical Coupling Factor k

The terms *electromechanical coupling factor*, *energy transmission coefficient*, and *efficiency* are sometimes confused. All are related to the conversion rate between electrical energy and mechanical energy, but their definitions are different.

(a) The electromechanical coupling factor k

$$k^2 = (\text{Stored mechanical energy/Input electrical energy}) \tag{2.4}$$

or

$$k^2 = \text{(Stored electrical energy/Input mechanical energy)} \tag{2.5}$$

Let us calculate equation (2.4) when an electric field E is applied to a piezoelectric material. Because the input electrical energy is (1/2) $\varepsilon_0 \varepsilon E^2$ per unit volume and the stored mechanical energy per unit volume under zero external stress is given by (1/2) x^2/s = (1/2) $(dE)^2/s$, k^2 can be calculated as follows:

$$\begin{aligned} k^2 &= [(1/2)\,(dE)^2/s]/[(1/2)\,\varepsilon_0 \varepsilon\, E^2] \\ &= d^2/\varepsilon \varepsilon \cdot s \end{aligned} \tag{2.6}$$

(b) The energy transmission coefficient λ_{max}[5]

Not all the stored energy can be actually used, and the actual work done depends on the mechanical load. With zero mechanical load or a complete clamp (no strain), zero output work is done.

$$\lambda_{\max} = \text{(Output mechanical energy/Input electrical energy)}_{\max} \tag{2.7}$$

or

$$\lambda_{\max} = \text{(Output electrical energy/Input mechanical energy)}_{\max} \tag{2.8}$$

Let us consider the case when an electric field E is applied to a piezoelectric under constant external stress X (<0, because a compressive stress is necessary to work to the outside). As shown in Figure 2.1, the output work can be calculated as follows:

$$\int (-X)\, dx = -(dE + s\,X)\,X \tag{2.9}$$

whereas the input electrical energy is given by

$$\int E\, dP = (\varepsilon_0 \varepsilon\, E + dX)\, E \tag{2.10}$$

We need to choose a proper load to maximize the energy transmission coefficient. From the maximum condition of

$$\lambda = -(dE + s\,X)\,X \,/\, (\varepsilon_0 \varepsilon\, E + dX)\, E \tag{2.11}$$

we can obtain

$$\lambda_{\max} = [(1/k) - \sqrt{(1/k^2) - 1}]^2$$

$$= [(1/k) + \sqrt{(1/k^2) - 1}]^{-2} \tag{2.12}$$

Note that

$$k^2/4 < \lambda_{\max} < k^2/2 \tag{2.13}$$

depending on the k value. For a small k, $\lambda_{\max} = k^2/4$, and for a large k, $\lambda_{\max} = k^2/2$.

It is also worth noting that the aforementioned maximum condition does not agree with the condition that provides the maximum output mechanical energy. The maximum output energy can be obtained when the load is half of the maximum generative stress: $-(dE - s\,(dE/2s))\,(-dE/2s) = (dE)^2/4s$. In this case, as the input electrical energy is given by $(\varepsilon_0\varepsilon\, E + d\,(-dE/2s))E$,

$$\lambda = 1/2[(2/k^2) - 1] \tag{2.14}$$

which is close to the value $\lambda_{\max}$, but has a different value from the predicted theoretical value.

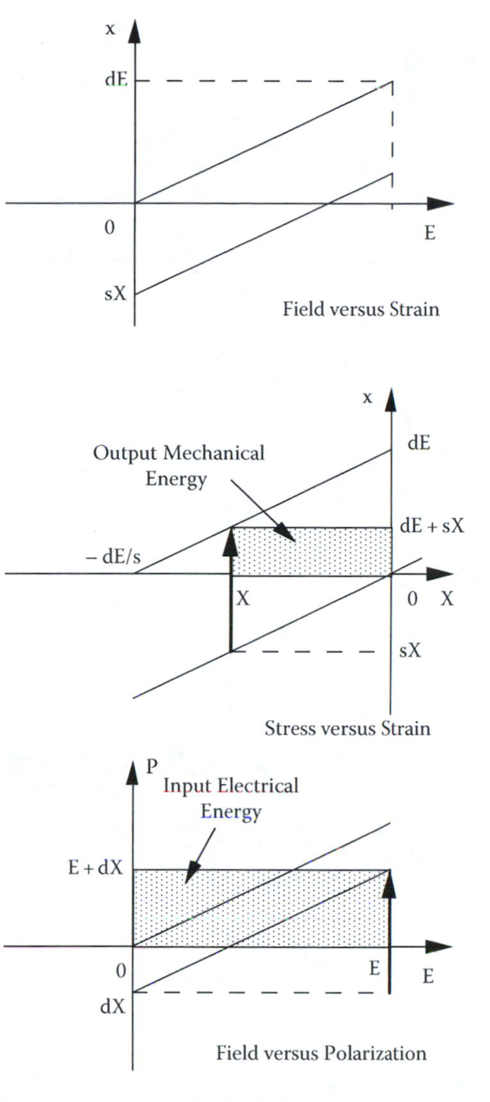

FIGURE 2.1 Calculation of the input electrical and output mechanical energy.

(c) The efficiency η

$$\eta = (\text{Output mechanical energy})/(\text{Consumed electrical energy}) \tag{2.15}$$

or

$$\eta = (\text{Output electrical energy})/(\text{Consumed mechanical energy}) \tag{2.16}$$

In a work cycle (e.g., an electric field cycle), the input electrical energy is transformed partially into mechanical energy and the rest is stored as electrical energy (electrostatic energy, similar to a capacitor) in an actuator. In this way, the ineffective energy can be returned to the power source, leading to near 100% efficiency, if the loss is small. Typical values of dielectric loss in PZT are about 1–3%. This loss generates heat in the piezoelectric components when the cyclic drive frequency is increased.

2.1.1.4 Mechanical Quality Factor Q_m

The mechanical quality factor Q_m is a parameter that characterizes the sharpness of the electromechanical resonance spectrum. When the motional admittance Y_m is plotted around the resonance frequency ω_0, the mechanical quality factor Q_m is defined with respect to the full width [$2\Delta\omega$] at $Y_m/\sqrt{2}$ (3 dB down point) as

$$Q_m = \omega_0/2\Delta\omega \tag{2.17}$$

Also note that Q_m^{-1} is equal to the mechanical loss ($\tan \delta_m$). The Q_m value is very important in evaluating the magnitude of the resonant strain. The vibration amplitude at an off-resonance frequency (*dEL*, *L*: length of the sample) is amplified by a factor proportional to Q_m at the resonance frequency. For a longitudinal vibration rectangular plate through d_{31}, the maximum displacement is given by $(8/\pi^2)Q_m d_{31} EL$.

2.1.1.5 Acoustic Impedance Z

The acoustic impedance Z is a parameter used for evaluating the acoustic energy transfer between two materials. It is defined, in general, by

$$Z^2 = (\text{pressure/volume velocity}) \tag{2.18}$$

In a solid material,

$$Z = \sqrt{\rho c} \tag{2.19}$$

where ρ is the density and c is the elastic stiffness of the material.

In more advanced discussions, there are three kinds of impedances: specific acoustic impedance (pressure/particle speed), acoustic impedance (pressure/volume speed), and radiation impedance (force/speed) (see Reference 6 for the details). For the material data included in ATILA software, see Figure 2.2.

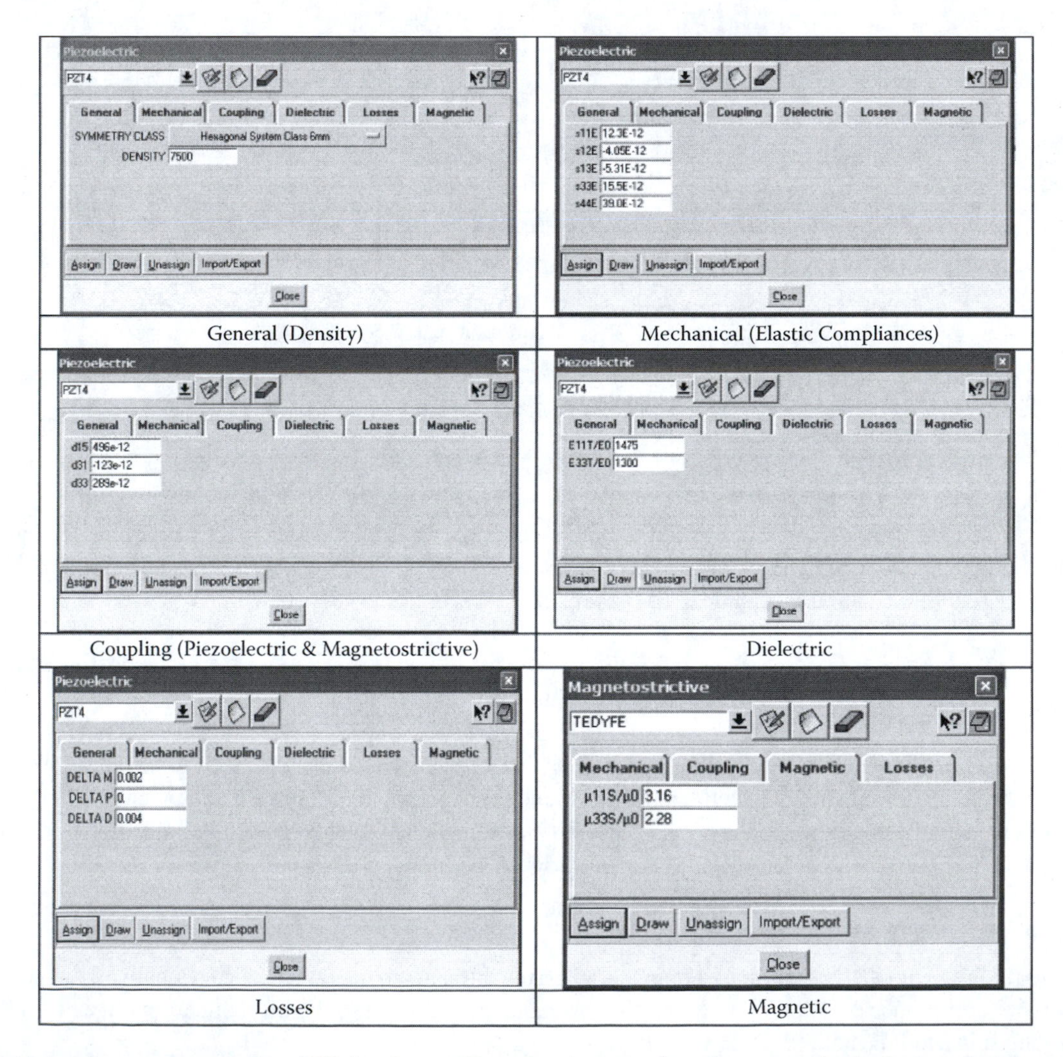

FIGURE 2.2 Material parameters used in ATILA software. Note that there are three losses, mechanical, piezoelectric, and dielectric, in a piezoelectric, which provides better admittance and vibration amplitude at the resonance point.

2.1.2 Piezoelectric Materials

This section summarizes the current status of piezoelectric materials: piezoceramics, piezocomposites, and piezofilms. Table 2.1 shows the material parameters of some of the piezoelectric materials.[7,8]

2.1.2.1 Polycrystalline Materials

Barium titanate ($BaTiO_3$) is one of the most thoroughly studied and most widely used piezoelectric materials. Just below the Curie temperature (120°C), the vector of the spontaneous polarization points in the [001] direction (tetragonal phase); below 5°C it reorients in the [011] direction (orthorhombic phase); and below −90°C, in the [111] direction (rhombohedral phase). The dielectric and piezoelectric properties of ferroelectric ceramic $BaTiO_3$ can be affected by its own stoichiometry, microstructure, and by dopants entering onto the A or B site in solid solution. Modified ceramic $BaTiO_3$ with dopants such as Pb or Ca ions have been developed to stabilize the tetragonal phase over a wider temperature range and are used as commercial piezoelectric materials. The initial application was for Langevin-type piezoelectric vibrators.

Piezoelectric $Pb(Ti,Zr)O_3$ solid-solution [PZT] ceramics have been widely used because of their superior piezoelectric properties. The phase diagram for the PZT system ($PbZr_xTi_{1-x}O_3$) is shown in Figure 2.3. The crystalline symmetry of this solid-solution system is determined by the Zr content. Lead titanate also has a tetragonal ferroelectric phase of perovskite structure. With increasing Zr content, x, the tetragonal distortion decreases, and at $x > 0.52$ the structure changes from the tetragonal 4 mm phase to another ferroelectric phase of rhombohedral 3 m symmetry. The line dividing these two phases is called the *morphotropic phase boundary*. The boundary composition is considered to have both tetragonal and rhombohedral phases coexisting together. Figure 2.4 shows the dependence of several piezoelectric d constants on composition near the morphotropic phase boundary. The d constants have their highest values near the morphotropic phase boundary. This enhancement in piezoelectric effect is attributed to the increased ease of reorientation of the polarization under an applied electric field.

TABLE 2.1
Piezoelectric Properties of Representative Piezoelectric Materials

Parameter	Quartz	$BaTiO_3$	PZT 4	PST5H	$(Pb,Sm)TiO_3$	PVDF-TrFE
d_{33} (pC/N)	2.3	190	289	593	65	33
g_{33} (10^{-3}Vm/N)	57.8	12.6	26.1	19.7	42	380
k_t	0.09	0.38	0.51	0.50	0.50	0.30
k_p		0.33	0.58	0.65	0.03	
$\varepsilon_3^T/\varepsilon_0$	5	1700	1300	3400	175	6
Q_m	$>10^5$		500	65	900	3–10
T_C (°C)		120	328	193	355	

Source: From Y. Ito and K. Uchino: Piezoelectricity, *Wiley Encyclopedia of Electrical and Electronics Engineering*, Vol. **16**, p. 479, John Wiley and Sons, New York (1999); W. A. Smith: *Proc. SPIE—The International Society for Optical Engineering* 1733 (1992). With permission.

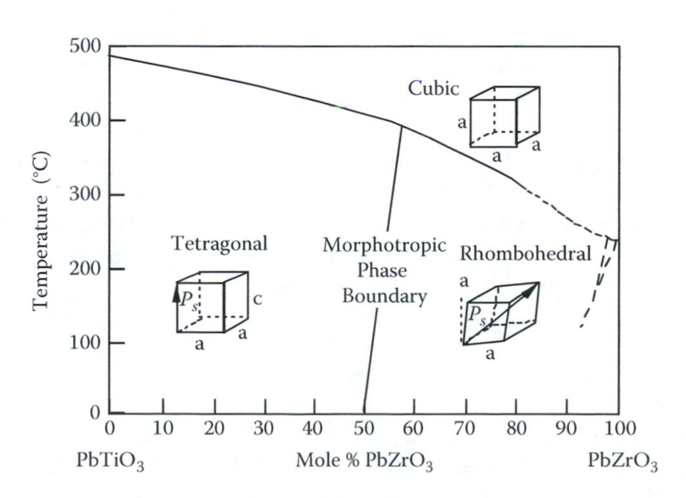

FIGURE 2.3 Phase diagram of lead zirconate titanate (PZT).

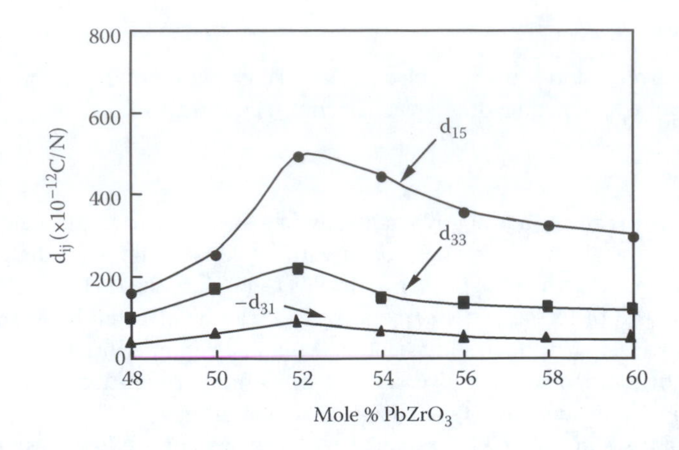

FIGURE 2.4 Dependence of several *d* constants on composition near the morphotropic phase boundary in the PZT system.

Doping the PZT material with donor or acceptor ions changes its properties dramatically. Donor doping with ions such as Nb^{5+} or Ta^{5+} provides soft PZTs, such as PZT-5, because of the facility of domain motion due to the resulting Pb vacancies. On the other hand, acceptor doping with Fe^{3+} or Sc^{3+} leads to hard PZTs, such as PZT-8, because the oxygen vacancies will pin domain wall motion. Subsequently, PZT in ternary solid solution with another perovskite phase has been investigated intensively. Examples of these ternary compositions are PZTs in solid solution with $Pb(Mg_{1/3}Nb_{2/3})O_3$, $Pb(Mn_{1/3}Sb_{2/3})O_3$, $Pb(Co_{1/3}Nb_{2/3})O_3$, $Pb(Mn_{1/3}Nb_{2/3})O_3$, $Pb(Ni_{1/3}Nb_{2/3})O_3$, $Pb(Sb_{1/2}Sn_{1/2})O_3$, $Pb(Co_{1/2}W_{1/2})O_3$, and $Pb(Mg_{1/2}W_{1/2})O_3$, all of which are patented by different companies. The end member of PZT, lead titanate, has a large crystal distortion. $PbTiO_3$ has a tetragonal structure at room temperature with its tetragonality c/a = 1.063. The Curie temperature is 490°C. Densely sintered $PbTiO_3$ ceramics cannot be obtained easily, because they break up into a powder when cooled through the Curie temperature because of the large spontaneous strain. Lead titanate ceramics modified by adding a small amount of additives exhibit a high piezoelectric anisotropy. Either (Pb, Sm)TiO_3[9] or (Pb, Ca)TiO_3[10] exhibits an extremely low planar coupling, that is, a large k_t/k_p ratio. Here, k_t and k_p are thickness-extensional and planar electromechanical coupling factors, respectively. Because these transducers can generate purely longitudinal waves through k_t associated with no transverse waves through k_{31}, clear ultrasonic imaging is expected without "ghosts" caused by the transverse wave. (Pb,Nd)(Ti,Mn,In)O_3 ceramics with a zero temperature coefficient of surface acoustic wave delay have been developed as superior substrate materials for SAW device applications.[11]

2.1.2.2 Relaxor Ferroelectrics

Relaxor ferroelectrics can be prepared either in polycrystalline form or as single crystals. They differ from the previously mentioned normal ferroelectrics in that they exhibit a broad phase transition from the paraelectric to ferroelectric state, a strong frequency dependence of the dielectric constant (i.e., dielectric relaxation), and a weak residual polarization. Lead-based relaxor materials have complex, disordered perovskite structures.

Relaxor-type electrostrictive materials, such as those from the lead magnesium niobate–lead titanate, $Pb(Mg_{1/3}Nb_{2/3})O_3$–$PbTiO_3$ (or PMN-PT), solid solution, are highly suitable for actuator applications. This relaxor ferroelectric also exhibits an induced piezoelectric effect; that is, the electromechanical coupling factor k_t varies with the applied DC bias field. As the DC bias field increases, the coupling increases and saturates. Because this behavior is reproducible, these materials can be applied as ultrasonic transducers that are tunable by the bias field.[12] Recently, single-crystal relaxor ferroelectrics with the morphotropic phase boundary (MPB) composition have been developed, which show tremendous promise as ultrasonic transducers and electromechanical actuators. Single crystals of $Pb(Mg_{1/3}Nb_{2/3})O_3$ (PMN), $Pb(Zn_{1/3}Nb_{2/3})O_3$ (PZN) and binary systems of these materials combined with $PbTiO_3$ (PMN-PT and PZN-PT) exhibit extremely large electromechanical coupling factors.[13,14] Large coupling coefficients and large piezoelectric constants have been found for crystals from the morphotropic phase boundaries of these solid solutions. PZN-8%PT single crystals were found to possess a high k_{33} value of 0.94 for the (001) crystal cuts; this is very high compared to the k_{33} value of conventional PZT ceramics of around 0.70–0.80.

2.1.2.3 Composites

Piezocomposites comprising a piezoelectric ceramic and a polymer phase are promising materials because of their excellent and readily tailored properties. The geometry for two-phase composites can be classified according to the dimensional connectivity of each phase into 10 structures: 0-0, 0-1, 0-2, 0-3, 1-1, 1-2, 1-3, 2-2, 2-3, and 3-3.[15] A 1-3 piezocomposite, such as the PZT-rod/polymer composite, is a most promising candidate. The advantages of this composite are high coupling factors, low acoustic impedance, good match to water or human tissue, mechanical flexibility, broad bandwidth in combination with a low mechanical quality factor, and the possibility of making undiced arrays by structuring the electrodes. The thickness-mode electromechanical coupling of the composite can exceed the k_t (0.40–0.50) of the constituent ceramic, approaching almost the value of the rod-mode electromechanical coupling, k_{33} (0.70–0.80), of that ceramic.[16] Acoustic impedance is the square root of the product of its density and elastic stiffness. The acoustic match to tissue or water (1.5 Mrayls) of the typical piezoceramics (20–30 Mrayls) is significantly improved by forming a composite structure, that is, by replacing some of the heavy, stiff ceramic with a light, soft polymer. Piezoelectric composite materials are especially useful for underwater sonar and medical diagnostic and ultrasonic transducer applications.

2.1.2.4 Thin Films

Both zinc oxide (ZnO) and aluminum nitride (AlN) are simple binary compounds with a Wurtzite-type structure, which can be sputter-deposited as a *c*-axis-oriented thin film on a variety of substrates. ZnO has large piezoelectric coupling, and thin films of this material are widely used in bulk acoustic and surface acoustic wave devices. The fabrication of highly oriented (along *c*) ZnO films have been studied and developed extensively. The performance of ZnO devices is limited, however, because of their low piezoelectric coupling (20–30%). PZT thin films are expected to exhibit higher piezoelectric properties. At present the growth of PZT thin films is being carried out for use in microtransducers and microactuators.

2.2 MAGNETOSTRICTIVE MATERIALS

Magnetostriction was discovered by James Joule in the 19th century. Since then the effect has been investigated in magnetic metals, such as Fe, Ni, and Co, and oxides such as the ferrites. Research on giant magnetostrictive materials, including rare earth metals, commenced in the 1960s. The magnetic structure and magnetic anisotropy of rare earth metals were under intensive study at this time. In particular, giant magnetostriction of more than 1% was reported for Tb and Dy metal single crystals at cryogenic temperatures.[17,18] This level of induced strain is two to three orders of magnitude higher than the strains typically observed in Ni and the ferrites.

The research of the 1970s was focused on compounds consisting of rare earth (R) and transition metal (T) elements. The RT_2 alloys, exemplified by RFe_2 (known as the *cubic Laves phase*), were of particular interest. Exceptionally large magnetostriction of up to 0.2% was reported for $TbFe_2$ at room temperature.[19,20] This triggered a boom in research on this enhanced effect now known as *giant magnetostriction*. In 1974, $Tb_{0.3}Dy_{0.7}Fe_2$, which exhibits enormous magnetostriction under relatively small magnetic field strengths at room temperature, was produced.[21] In 1987, sophisticated crystal control technology was developed[22] and a magnetostriction jump phenomenon was discovered in [112]-oriented alloys.[23] (The mechanism for the jump is a sudden realignment of the spin with one of the crystallographic easy axes closer to the external field direction.) These technologies helped to stabilize the magnetostriction. Strains of 0.15% can be consistently achieved under an applied magnetic field of 500 Oe and compressive stress of 1 kgf/mm^2. The materials in the compositional range $Tb_{0.27-0.3}Dy_{0.7-0.73}Fe_{1.9-2.0}$ are called Terfenol-D (dysprosium modified terbium iron produced at "nol") and are currently commercially produced by several manufacturers.The saturated magnetostrictive strain for some magnetic materials at 300 K are summarized in Table 2.2.[24] Note that several materials in the table have a negative magnetostriction coefficient (that is, their length decreases under an applied magnetic field), whereas other materials, including Terfenol-D, have a positive magnetostriction coefficient.

TABLE 2.2
The Saturated Magnetostrictive Strain For Some Magnetic Materials at 300 K

Material	Induced Strain ($\Delta l/l$ 10^{-6})
Fe	−9
Ni	−35
Co	−60
(0.6)Co–(0.4)Fe	68
(0.6)Ni–(0.4)Fe	25
$TbFe_2$	1753
Terfenol-D	1600
$SmFe_2$	−1560
Metglass 2605SC	40

Source: From J. B. Restorff: *Encyclopedia of Applied Physics*, Vol. 9, p. 229, VCH, New York (1994).

There are three primary mechanical concerns in the design of Terfenol-D transducers: (1) optimization of the output strain, (2) mechanical impedance matching, and (3) strengthening against fracture.[25] A prestress in the range of 7–10 MPa is generally applied to the Terfenol-D actuator to enhance its magnetostrictive response. This level of stress is sufficient to align the magnetic moments parallel to an easy axis, which is almost perpendicular to an actual crystallographic axis. Actuator displacements are enhanced by a factor of three for samples subjected to the prestress treatment.[26] Mechanical impedance matching is required between a magnetostrictive actuator and its external load to ensure efficient energy transfer to the load. Terfenol-D has a low tensile strength, with a yield strength of 700 MPa in compression, but only 28 MPa in tension.[27]

Various composite structures have been investigated in recent years to more effectively address these mechanical problems and to optimize the magnetostrictive response. Terfenol-D:epoxy composites, for example, are automatically subject to the prestress treatment during the cure process.[28,29] Models of the system have been developed to investigate the effects of particulate volume fraction and thermal expansion of the polymer matrix on the strain-field-load characteristics of the composite. It was determined that a low-viscosity matrix facilitates the formation of particle chains and helps to inhibit the production of voids during the cure process. The models were verified with Terfenol-D:epoxy samples representing various particle loading conditions (10–40%) and polymer thermal expansion coefficients ($30–50 \times 10^{-6}$ degree^{-1}), Young's moduli (0.5–3 GPa), and viscosities (60–10,000 cps).

The magnetostrictive strain for Terfenol-D:epoxy composites is shown in Figure 2.5.[29] We see in this response the competing influences of the magnetostrictive force needed to overcome and strain the matrix and the prestress force, which increases with decreasing volume fraction of Terfenol-D. As the volume fraction increases from 10 to 20%, an overall increase in the force and strain occurs largely because of the higher concentration of magnetostrictive material in the composite. However, as the volume fraction increases beyond 20%, the decrease in the prestress that is applied to the magnetostrictive composite on curing ultimately limits the force and strain generated in the structure, and the net response is reduced in spite of the higher concentration of magnetostrictive material.

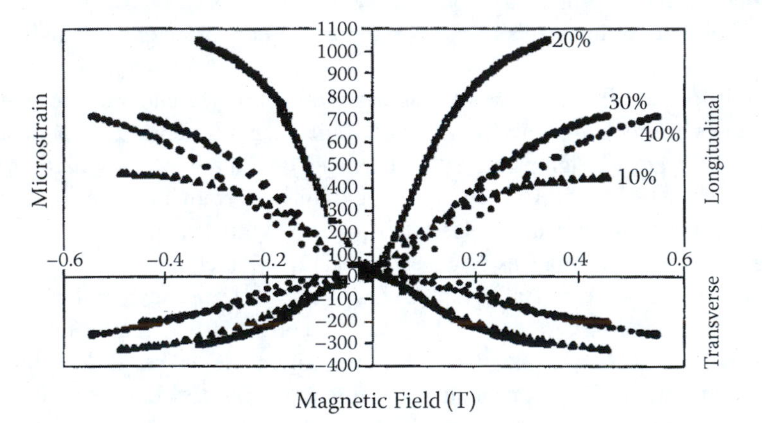

FIGURE 2.5 The longitudinal and transverse magnetostrictive strain as a function of applied magnetic field for Terfenol-D:epoxy composites with various volume fractions of Terfenol-D. (From T. A. Duenas and G. P. Carman: *Proc. Adaptive Struct. Mater. Syst.*, ASME, AD-**57**/MD-**83**, p. 63 [1998].)

Piezoelectric Materials	Magnetostrictive Materials

FIGURE 2.6 Various materials ready to use in the ATILA software. *New material's data can be easily uploaded by using "COPY" data key.*

2.3 MATHEMATICAL TREATMENT

2.3.1 TENSOR REPRESENTATION

In the solid-state theoretical treatment of the phenomenon of piezoelectricity, the strain x_{kl} is expressed in terms of the electric field, E_i, or electric polarization, P_i, as follows:

$$x_{kl} = d_{ikl} E_i = g_{ikl} P_i \tag{2.20}$$

where d_{ikl} and g_{ikl} are the piezoelectric coefficients. Represented in this way, we regard the quantities E_i and x_{kl} as first-rank and second-rank tensors, respectively, whereas d_{ikl} is considered a third-rank tensor. Generally speaking, if two physical properties are represented by tensors of p-rank and q-rank, *the quantity that combines the two properties in a linear relation is represented by a tensor of* $(p + q)$ *rank.*

The d_{ijk} tensor can be viewed as a three-dimensional array of coefficients comprising three "layers" of the following form:

$$\begin{array}{ll}
\text{First layer } (i = 1) & \begin{pmatrix} d_{111} & d_{112} & d_{113} \\ d_{121} & d_{122} & d_{123} \\ d_{131} & d_{132} & d_{133} \end{pmatrix} \\
\text{Second layer } (i = 2) & \begin{pmatrix} d_{211} & d_{212} & d_{213} \\ d_{221} & d_{222} & d_{223} \\ d_{231} & d_{232} & d_{233} \end{pmatrix} \\
\text{Third layer } (i = 3) & \begin{pmatrix} d_{311} & d_{312} & d_{313} \\ d_{321} & d_{322} & d_{323} \\ d_{331} & d_{332} & d_{333} \end{pmatrix}
\end{array} \tag{2.21}$$

2.3.2 CRYSTAL SYMMETRY AND TENSOR FORM

A physical property measured along two different directions must have the same value if these two directions are crystallographically equivalent. This consideration sometimes reduces the number of independent tensor coefficients representing a given physical property. Let us consider the third-rank piezoelectricity tensor as an example. The converse piezoelectric effect is expressed in tensor notation as

$$x_{kl} = d_{ikl} E_i \tag{2.22}$$

An electric field E, initially defined in terms of an (x, y, z) coordinate system, is redefined as E' in terms of another system rotated with respect to the first with coordinates (x', y', z') by means of a transformation matrix (a_{ij}), such that

$$E'_i = a_{ij} E_j \tag{2.23}$$

or

$$\begin{bmatrix} E'_1 \\ E'_2 \\ E'_3 \end{bmatrix} = \begin{pmatrix} a_{11} & a_{12} & a_{13} \\ a_{21} & a_{22} & a_{23} \\ a_{31} & a_{32} & a_{33} \end{pmatrix} \begin{bmatrix} E_1 \\ E_2 \\ E_3 \end{bmatrix} \tag{2.24}$$

The matrix (a_{ij}) is thus seen to be simply the array of direction cosines that allows us to transform the components of vector **E** referred to in the original coordinate axes to components referred to in the axes of the new coordinate system. The second-rank strain tensor is thus transformed in the following manner:

$$x'_{ij} = a_{ik} a_{jl} x_{kl} \tag{2.25}$$

or

$$\begin{bmatrix} x'_{11} & x'_{12} & x'_{13} \\ x'_{21} & x'_{22} & x'_{23} \\ x'_{31} & x'_{32} & x'_{33} \end{bmatrix} = \begin{pmatrix} a_{11} & a_{12} & a_{13} \\ a_{21} & a_{22} & a_{23} \\ a_{31} & a_{32} & a_{33} \end{pmatrix} \begin{bmatrix} x_{11} & x_{12} & x_{13} \\ x_{21} & x_{22} & x_{23} \\ x_{31} & x_{32} & x_{33} \end{bmatrix} \begin{pmatrix} a_{11} & a_{21} & a_{31} \\ a_{12} & a_{22} & a_{32} \\ a_{13} & a_{23} & a_{33} \end{pmatrix} \tag{2.26}$$

whereas the transformation of the third-rank piezoelectric tensor is expressed as

$$d'_{ijk} = a_{il}a_{jm}a_{kn}\, d_{lmn} \tag{2.27}$$

If the d_{lmn} tensor coefficients are symmetric with respect to m and n such that $d_{lmn} = d_{lmn}$, the following equivalences can be established:

$$\begin{array}{lll} d_{112} = d_{121} & d_{113} = d_{131} & d_{123} = d_{132} \\ d_{221} = d_{212} & d_{213} = d_{231} & d_{223} = d_{232} \\ d_{321} = d_{312} & d_{313} = d_{331} & d_{323} = d_{332} \end{array}$$

The number of independent coefficients is thus reduced from an original 27 (=3^3) to only 18, and the d_{lmn} tensor may then be represented by layers of the following form:

$$\begin{array}{ll} \text{First layer} & \begin{pmatrix} d_{111} & d_{112} & d_{131} \\ d_{112} & d_{122} & d_{123} \\ d_{131} & d_{123} & d_{133} \end{pmatrix} \\ \text{Second layer} & \begin{pmatrix} d_{211} & d_{212} & d_{231} \\ d_{212} & d_{222} & d_{223} \\ d_{231} & d_{223} & d_{233} \end{pmatrix} \\ \text{Third layer} & \begin{pmatrix} d_{311} & d_{312} & d_{313} \\ d_{312} & d_{322} & d_{323} \\ d_{313} & d_{323} & d_{333} \end{pmatrix} \end{array} \tag{2.28}$$

2.3.3 Matrix Notation

The reduction in tensor coefficients just carried out for the tensor quantity d_{ijk} makes it possible to render the three-dimensional array of coefficients in a more tractable two-dimensional matrix form. This is accomplished by abbreviating the suffix notation used to designate the tensor coefficients according to the following scheme:

Tensor notation	11	22	33	23,32	31,13	12,21
Matrix notation	1	2	3	4	5	6

The layers of tensor coefficients represented by equation 2.21 may now be rewritten as

$$\begin{array}{ll} \text{First layer} & \begin{pmatrix} d_{11} & (1/2)d_{16} & (1/2)d_{15} \\ (1/2)d_{16} & d_{12} & (1/2)d_{14} \\ (1/2)d_{15} & (1/2)d_{14} & d_{13} \end{pmatrix} \\ \text{Second layer} & \begin{pmatrix} d_{21} & (1/2)d_{26} & (1/2)d_{25} \\ (1/2)d_{26} & d_{22} & (1/2)d_{24} \\ (1/2)d_{25} & (1/2)d_{24} & (1/2)d_{23} \end{pmatrix} \\ \text{Third layer} & \begin{pmatrix} d_{31} & (1/2)d_{36} & (1/2)d_{35} \\ (1/2)d_{36} & d_{32} & (1/2)d_{34} \\ (1/2)d_{35} & (1/2)d_{34} & d_{33} \end{pmatrix} \end{array} \tag{2.29}$$

The last two suffixes in the tensor notation correspond to those of the strain components. Therefore, for the sake of consistency, we will also make similar substitutions in the notation for the strain components.

$$\begin{bmatrix} x_{11} & x_{12} & x_{13} \\ x_{21} & x_{22} & x_{23} \\ x_{31} & x_{32} & x_{33} \end{bmatrix} \rightarrow \begin{bmatrix} x_1 & (\tfrac{1}{2})x_6 & (\tfrac{1}{2})x_5 \\ (\tfrac{1}{2})x_6 & x_2 & (\tfrac{1}{2})x_4 \\ (\tfrac{1}{2})x_5 & (\tfrac{1}{2})x_4 & x_3 \end{bmatrix} \tag{2.30}$$

Note that here the number of independent coefficients for this second-rank tensor may also be reduced from nine (=3²) to six because it is a symmetric tensor and $x_{ij} = x_{ji}$. The factors of (1/2) in this and the piezoelectric layers of equation 2.29 are included to retain the general form that, similar to the corresponding tensor equation expressed by equation 2.22, includes no factors of 2; so we may write

$$x_j = d_{ij} E_i \qquad (i = 1, 2, 3\ ; j = 1, 2, \ldots, 6) \tag{2.31a}$$

or

$$\begin{bmatrix} x_1 \\ x_2 \\ x_3 \\ x_4 \\ x_5 \\ x_6 \end{bmatrix} = \begin{pmatrix} d_{11} & d_{21} & d_{31} \\ d_{12} & d_{22} & d_{32} \\ d_{13} & d_{23} & d_{33} \\ d_{14} & d_{24} & d_{34} \\ d_{15} & d_{25} & d_{35} \\ d_{16} & d_{26} & d_{36} \end{pmatrix} \begin{bmatrix} E_1 \\ E_2 \\ E_3 \end{bmatrix} \tag{2.31b}$$

When deriving a matrix expression for the direct piezoelectric effect in terms of the matrix form of the stress X_{ij}, the factors of (1/2) are not necessary, and the matrix may be represented as

$$\begin{bmatrix} X_{11} & X_{12} & X_{13} \\ X_{21} & X_{22} & X_{23} \\ X_{31} & X_{32} & X_{33} \end{bmatrix} \rightarrow \begin{bmatrix} X_1 & X_6 & X_5 \\ X_6 & X_2 & X_4 \\ X_5 & X_4 & X_3 \end{bmatrix} \tag{2.32}$$

so that

$$P_j = d_{ij} X_j \qquad (i = 1, 2, 3\ ; j = 1, 2, \ldots, 6) \tag{2.33a}$$

or

$$\begin{bmatrix} P_1 \\ P_2 \\ P_3 \end{bmatrix} = \begin{bmatrix} d_{11} & d_{12} & d_{13} & d_{14} & d_{15} & d_{16} \\ d_{21} & d_{22} & d_{23} & d_{24} & d_{25} & d_{26} \\ d_{31} & d_{32} & d_{33} & d_{34} & d_{35} & d_{36} \end{bmatrix} \begin{bmatrix} X_1 \\ X_2 \\ X_3 \\ X_4 \\ X_5 \\ X_6 \end{bmatrix} \tag{2.33b}$$

Although matrix notation has the advantage of being more compact and tractable than tensor notation, one must remember that the matrix coefficients, d_{ij}, do not transform like the tensor components, d_{ijk}.

Table 2.3 summarizes the matrices d_{ij} for all crystallographic point groups.[30] ATILA FEM software will treat these d_{ij} components for the calculation. Note that there are three independent components, d_{33}, d_{31}, and d_{15}, in a polycrystalline ceramic sample (equivalent to 4 mm or 6 mm) (refer to Figure 2.2).

TABLE 2.3
Piezoelectric Strain Coefficient (d) Matrices

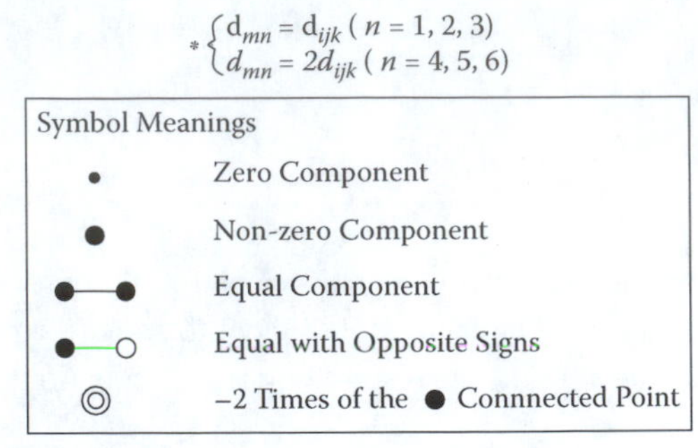

I Centro Symmetric Point Group

Point Group $\overline{1}$, $2/m$, mmm, $4/m$, $4/mmm$, $m3$, $m3m$, $\overline{3}$, $\overline{3}m$, $6/m$, $6/mmm$
All Components are Zero

II Noncentro Symmetric Point Group

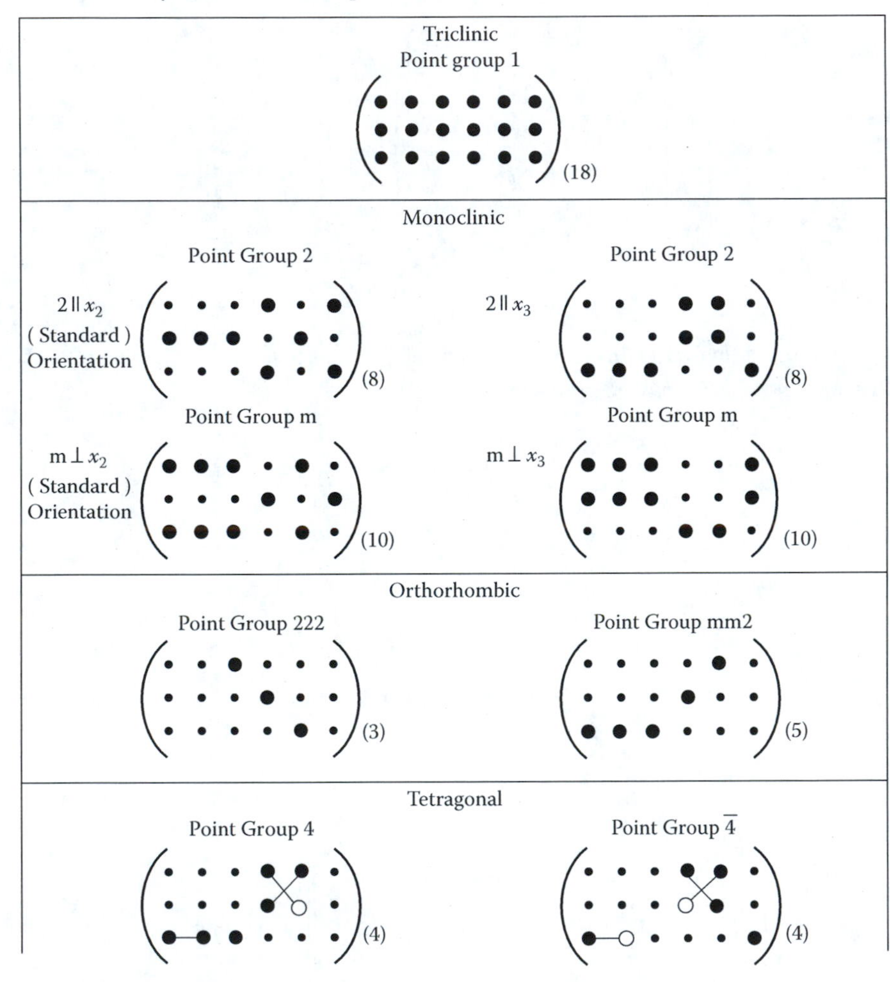

TABLE 2.3 (continued)
Piezoelectric Strain Coefficient (*d*) Matrices

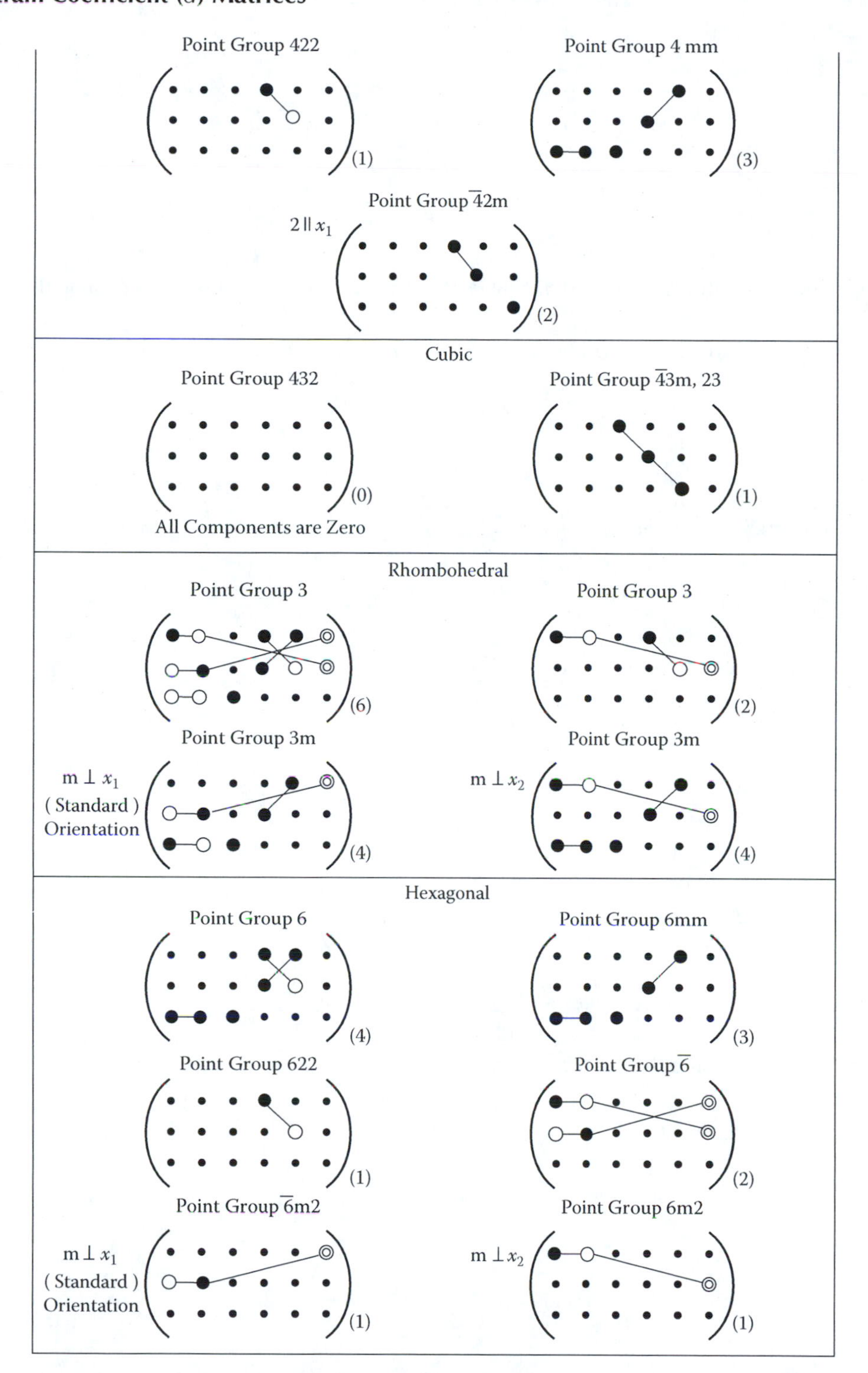

Source: From J. F. Nye: *Physical Properties of Crystals*, Oxford University Press, London, p. 123, 140 (1972). With permission.

EXAMPLE PROBLEM 2.1

Barium titanate ($BaTiO_3$) has a tetragonal crystal symmetry (point group 4 mm) at room temperature. The appropriate piezoelectric strain coefficient matrix is therefore of the form

$$d_{ij} = \begin{pmatrix} 0 & 0 & 0 & 0 & d_{15} & 0 \\ 0 & 0 & 0 & d_{15} & 0 & 0 \\ d_{31} & d_{31} & d_{33} & 0 & 0 & 0 \end{pmatrix}$$

(a) Determine the nature of the strain induced in the material when an electric field is applied along the crystallographic c axis.
(b) Determine the nature of the strain induced in the material when an electric field is applied along the crystallographic **a** axis.

SOLUTION

The matrix equation that applies in this case is

$$\begin{bmatrix} x_1 \\ x_2 \\ x_3 \\ x_4 \\ x_5 \\ x_6 \end{bmatrix} = \begin{pmatrix} 0 & 0 & d_{31} \\ 0 & 0 & d_{31} \\ 0 & 0 & d_{33} \\ 0 & d_{15} & 0 \\ d_{15} & 0 & 0 \\ 0 & 0 & 0 \end{pmatrix} \begin{bmatrix} E_1 \\ E_2 \\ E_3 \end{bmatrix} \tag{P2.1.1}$$

from which we may derive expressions for the induced strains

$$x_1 = x_2 = d_{31}E_3, x_3 = d_{33}\,E_3,\ x_4 = d_{15}\,E_2, x_5 = d_{15}\,E_1,\ x_6 = 0$$

so that the following determinations can be made:

(a) When E_3 is applied, elongation in the c direction ($x_3 = d_{33}E_3$, $d_{33} > 0$) and contraction in the a and b directions ($x_1 = x_2 = d_{31}E_3$, $d_{31} < 0$) occur.
(b) When E_1 is applied, a shear strain x_5 ($= 2x_{31}$) $= d_{15}E_1$ is induced. The case where $d_{15} > 0$ and $x_5 > 0$ is illustrated in Figure 2.7.

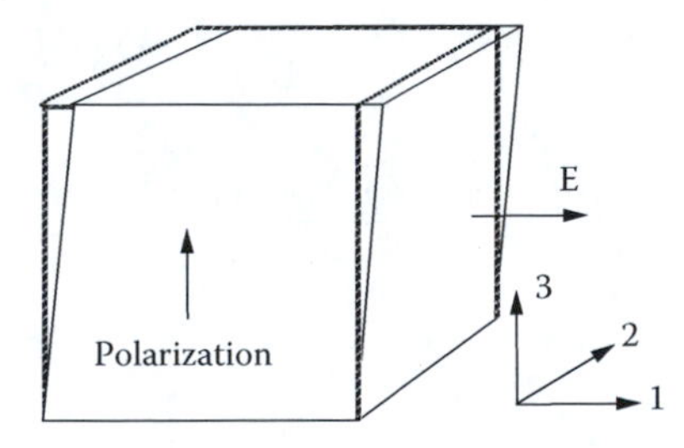

FIGURE 2.7 The shear strain induced through the piezoelectric effect for a $BaTiO_3$ single crystal with tetragonal point group symmetry 4 mm.

2.3.4 Hysteresis and Loss in Piezoelectrics

As the industrial demand for piezoelectric actuators has increased in recent years, fundamental research in this area has become more focused on issues related to reliability. The heat generated by these dielectric and mechanical loss mechanisms is a significant problem, especially for high-power applications.

The loss or hysteresis associated with a piezoelectric can be both detrimental and beneficial. When the material is used in a positioning device, hysteresis in the field-induced strain presents a serious problem in reproducibility of the motion. When the material is driven in resonance mode, as is the case with ultrasonic motor applications, the losses generate excessive heat in the device. Also, because the resonant strain amplification is proportional to the mechanical quality factor Q_m, low (intensive)-mechanical-loss materials are generally preferred for ultrasonic motors. On the other hand, a high mechanical loss, which corresponds to a low mechanical quality factor Q_m, is actually used for piezoelectric force sensors and acoustic transducers because it enables a broader range of operating frequencies. Systematic studies of the loss mechanisms in piezoelectrics operated under high-voltage and high-power conditions have been limited and, to date, little has been reported in the literature on the subject. A partial theoretical treatment of loss mechanisms can be found, such as that presented in the text by T. Ikeda[31]; however, these descriptions tend to neglect piezoelectric losses, which have become more important in the context of recent investigations. A phenomenological description of the loss mechanisms associated with piezoelectrics will be presented in this section.[32]

Three losses, dielectric, elastic, and piezoelectric, are separately installed in the ATILA FEM software so that the magnitudes of the electric impedance and the mechanical displacement in a piezoelectric component reflect differently according to the vibration mode. However, the simulation for heat generation in the device cannot be treated in the present software.

2.3.4.1 Dielectric Loss and Hysteresis

Let us consider first the dielectric loss and hysteresis associated with the electric displacement, D, (which is proportional and almost equal to the polarization, P) against the electric field (E curve pictured in Figure 2.8), neglecting the electromechanical coupling. When the electric displacement, D, (or polarization, P) traces a different curve with increasing applied electric field, E, from that when the applied electric field is decreasing, the material is said to be exhibiting *dielectric hysteresis*.

When the dielectric hysteresis is not very large, the observed variation in electric displacement, D, can be represented as if it had a slight phase lag with respect to the applied electric field. If the alternating electric field, E^*, associated with such a phase-lagging electric displacement has an angular frequency, $\omega = 2\pi f$, it can be expressed as follows:

$$E^* = E_o\, e^{j\omega t} \tag{2.34}$$

and the induced electric displacement, D^*, will oscillate at the same frequency but lag in phase by δ, such that

$$D^* = D_o\, e^{j(\omega t - \delta)} \tag{2.35}$$

If we express the relationship between D^* and E^* as

$$D^* = K^* \varepsilon_o E^* \tag{2.36}$$

where the *complex dielectric constant*, K^*, is

$$K^* = K' - j K'' \tag{2.37}$$

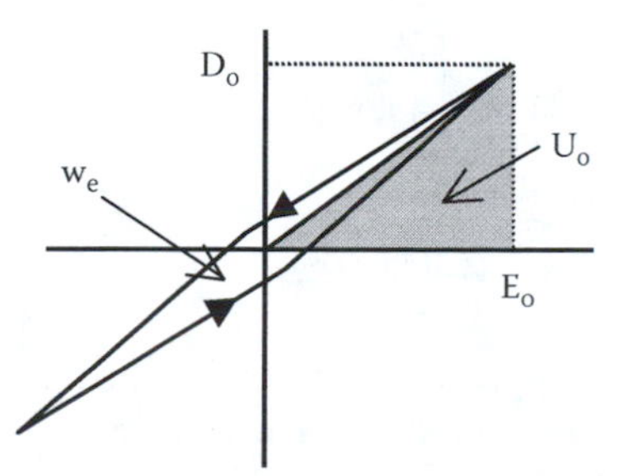

FIGURE 2.8 A plot of electric displacement, D, as a function of applied electric field, E, characteristic of dielectric hysteresis.

and where

$$K''/K' = \tan\delta \tag{2.38}$$

Note that the negative sign preceding the imaginary part of equation (2.37) is associated with the phase lag of the electric displacement and that $K'\varepsilon_o = (D_o/E_o)\cos\delta$ and $K''\varepsilon_o = (D_o/E_o)\sin\delta$. In terms of this complex description, the hysteresis loop should be elliptical in shape, which is not what is actually observed.

The integrated area inside the hysteresis loop, labeled w_e in Figure 2.6, is equivalent to the energy loss per cycle per unit volume of the dielectric. It is defined for an isotropic dielectric as

$$w_e = -\int D\,dE = -\int_0^{2\pi/\omega} D\frac{dE}{dt}\,dt \tag{2.39}$$

Substituting the real parts of the electric field E^* and electric displacement D^* into equation (2.40),

$$w_e = \int_0^{2\pi/\omega} D_o\cos(\omega t-\delta)\left[E_o\,\omega\sin(\omega t)\right]dt = E_oD_o\omega\sin(\delta)\int_0^{2\pi/\omega}\sin^2(\omega t)\,dt = \pi E_oD_o\sin(\delta) \tag{2.40}$$

so that

$$w_e = \pi K''\varepsilon_o E_o^2 = \pi K'\varepsilon_o E_o^2\tan\delta \tag{2.41}$$

When there is no phase lag (d = 0), the energy loss will be zero ($w_e = 0$), and the electrostatic energy stored in the dielectric will be recovered completely after a full cycle (100% efficiency). However, when there is a phase lag, an energy loss (or nonzero w_e) will occur for every cycle of the applied electric field, resulting in the generation of heat in the dielectric material. In this context, the quantity tan d is referred to as the *dissipation factor.*

The electrostatic energy stored during a half cycle of the applied electric field will be just $2U_e$, where U_e, the integrated area so labeled in Figure 2.7, represents the energy stored during a quarter cycle. This can be expressed as follows:

$$4U_e = \tfrac{1}{2}(2E_o)(2D_o\cos\delta) = 2(E_oD_o)\cos\delta = 2K'\varepsilon_o E_o^2 \tag{2.42}$$

Combining this equation with equation (2.41) yields an alternative expression for the dissipation factor:

$$\tan\delta = (1/2\pi)(w_e/U_e) \tag{2.43}$$

which highlights the significance of tan d in terms of the electrostatic energy efficiency.

2.3.4.2 General Considerations

Let us expand the treatment presented in the previous section to include loss mechanisms associated with piezoelectric materials. We will introduce the complex parameters K^{X*}, s^{E*}, and d^* to account for the hysteretic losses associated with the electric, elastic, and piezoelectric coupling energy as follows:

$$K^{X*} = K^X(1-j\tan\delta') \tag{2.44}$$

$$s^{E*} = s^E(1-j\tan\phi') \tag{2.45}$$

$$d^* = d(1-j\tan\theta') \tag{2.46}$$

where θ' is the phase delay of the strain under an applied electric field or the phase delay of the electric displacement under an applied stress. The angle δ' represents the phase delay of the electric displacement in response to an applied electric field under a constant stress (that is, zero stress) condition, and the angle ϕ' is the phase delay of the strain in response to an applied stress under a constant electric field (that is, a short-circuit) condition.

Samples of experimentally determined *D-E* (stress-free conditions), *x-X* (short-circuit conditions), *x-E* (stress-free conditions), and *D-X* (short-circuit conditions and accumulate the charges) hysteresis curves appear in Figure 2.9. These measurements are standard for the characterization of piezoelectric materials and simple to obtain.

The purely electrical and purely mechanical stored energies and hysteretic losses may be expressed in terms of the dielectric and mechanical hysteresis curves depicted in Figures 2.9a and 2.9b as follows:

$$U_e = (1/2)\, K^X \varepsilon_o E_o^2 \tag{2.47}$$

$$w_e = \pi K^X \varepsilon_o E_o^2 \tan \delta' \tag{2.48}$$

$$U_m = (1/2)\, s^E X_o^2 \tag{2.49}$$

$$w_m = \pi\, s^E X_o^2 \tan \phi' \tag{2.50}$$

according to the analysis carried out in the previous section.

The electromechanical loss, the X-E curve, is more complicated because the field-versus-strain domain cannot be used for directly calculating the energy. Let us consider the electromechanical energy, U_{em}, stored during a quarter cycle of the applied electric field cycle (that is, from $E = 0$ to $E = E_o$) in terms of our previous analysis:

$$U_{em} = -\int x\, dX = \frac{1}{2}\left[\frac{x_o^2}{s^E}\right] = \frac{1}{2}\left[\frac{(E\,d)^2}{s^E}\right] \tag{2.51}$$

Replacing d and s^E by $d^* = d(1\text{-}j\tan\theta')$ and $s^{E*} = s^E(1\text{-}j\tan\phi')$, we obtain

$$U_{em} = \frac{1}{2}\left[\frac{d^2}{s^E}\right] E_o^2 \tag{2.52}$$

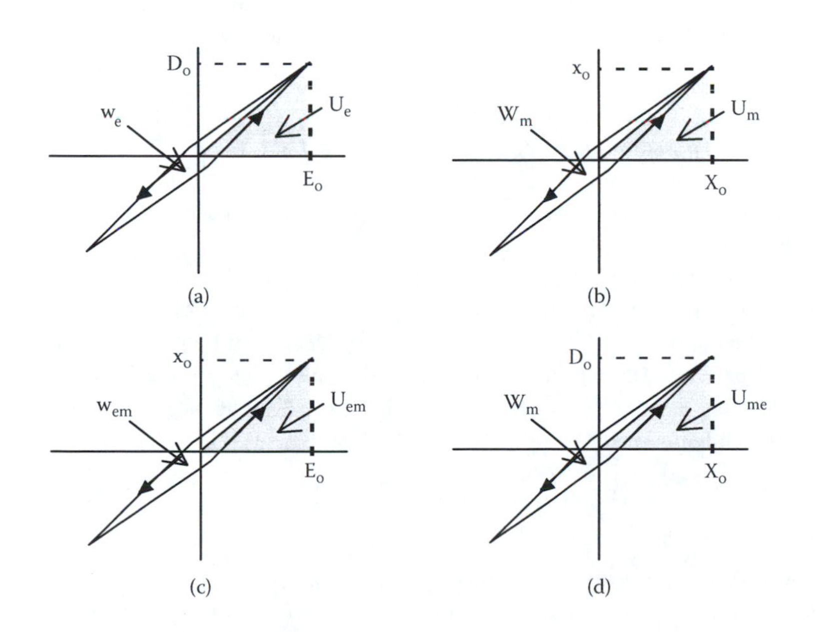

FIGURE 2.9 Samples of experimentally determined hysteresis curves: (a) *D-E* (stress-free conditions), (b) *x-X* (short-circuit conditions), (c) *x-E* (stress-free conditions), and (d) *D-X* (short-circuit conditions).

and

$$w_{em} = \pi \left[\frac{d^2}{s^E} \right] E_o^2 \; (2 \tan \theta' - \tan \phi') \tag{2.53}$$

So we see that the x-E hysteresis data analyzed in this manner does not lead to a relationship that involves only the piezoelectric loss, tan θ′; instead, we find that the observed loss involves contributions from both the piezoelectric loss, tan θ′, and the elastic loss, tan ϕ′.

Similarly, when we consider the D-X hysteresis data in this manner, the electromechanical energy U_{me} stored during a quarter cycle of the applied stress and the hysteresis loss, w_{me}, that occurs during a full stress cycle, are found to be

$$U_{me} = \frac{1}{2} \left[\frac{d^2}{K^X \varepsilon_o} \right] X_o^2 \tag{2.54}$$

and

$$w_{me} = \pi \left[\frac{d^2}{K^X \varepsilon_o} \right] X_o^2 \; (2 \tan \theta' - \tan \delta') \tag{2.55}$$

Hence, from the D-E and x-X data, we can obtain the contributions of the dielectric loss, tan δ′, and the mechanical loss, tan ϕ′, respectively, and we may evaluate finally the electromechanical loss, tan θ′.

These losses provide both heat generation and the vibration damping. The ATILA FEM software cannot simulate the temperature rise at present, but can simulate the admittance value and the vibration amplitude at the resonance frequency (refer to Figure 2.2.)

2.4 APPLICATIONS OF SMART TRANSDUCERS

2.4.1 Ultrasonic Transducers

Ultrasonic waves are now used in various fields. The sound source is made from piezoelectric ceramics as well as magnetostrictive materials. Piezoceramics are generally superior in efficiency and in size to magnetostrictive materials. In particular, hard piezoelectric materials with a high Q_m are preferable. A liquid medium is usually used for sound energy transfer. Ultrasonic washers, ultrasonic microphones for short-distance remote control and underwater detection, such as sonar and fish finding, and nondestructive testing are typical applications of piezoelectric materials. Ultrasonic scanning detectors are useful in medical electronics for clinical applications ranging from diagnosis to therapy and surgery.

2.4.1.1 Audible Sound

The audible range for humans differs individually. Average data are shown in Figure 2.10. The audible frequency ranges from 20 Hz to 20 kHz, whereas the audible intensity level ranges from 10^{-12} to 1 W/m^2 (this differs depending on the frequency).[33] This minimum audible sound power level (10^{-12} W/m^2) is often taken as the intensity standard, 0 dB, which corresponds to the sound pressure 2×10^{-5} Pa (or 0.0002 m bar).[33]

When we introduce a high-power ultrasound into liquid, we can create a ***cavitation*** effect, which is a vacuum pore. Humans cannot endure air sound intensity more than 200 m bar, which corresponds to only a 0.0002 deviation of the regular air pressure (1,000 hPa). On the contrary, when we irradiate 50 kHz ultrasound with a vibration amplitude of 0.16 mm into water (0.35×10^4 W/m^2 power density), the sound pressure exceeds 1 atm pressure (1,000 hPa) rather easily. As depicted in Figure 2.11, during the pressure-decreasing half cycle, vacuum pore (zero pressure cavity) is created. Once this vacuum pore is cyclically created at this high frequency, a strong shock wave is generated. This effect is called *cavitation*, and it is the principle of ultrasonic cleaning.

2.4.1.2 Sound Channel

The ultrasonic wave can travel more than 1,000 km in the sea.[33] Why does it propagate such a long distance without diverging? The principle is analogous to an optical fiber, in which a suitable distribution of the glass density traps the optical beam. The "sound channel" in the sea was discovered by Russian scientists. Figure 2.12 illustrates the principle of this sound channel. Underwater sound velocity increases with temperature rise and with pressure, in general. The water temperature decreases

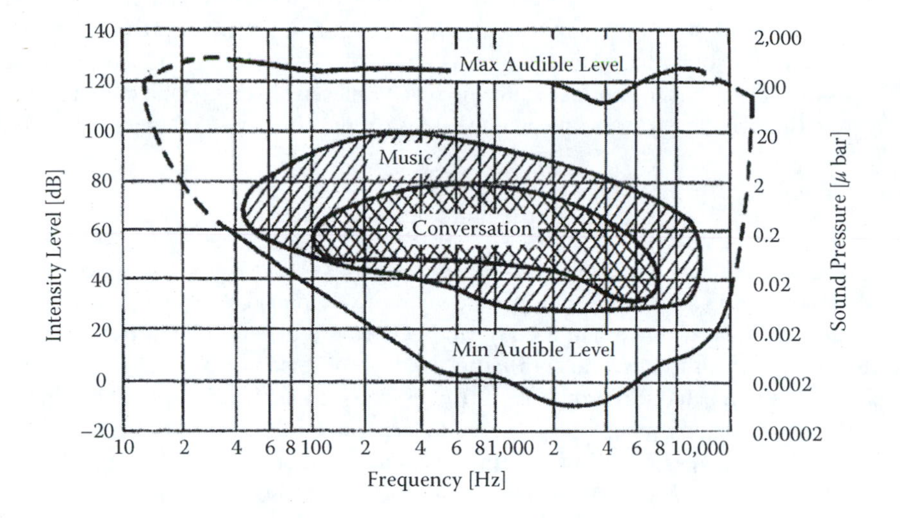

FIGURE 2.10 Audible frequency and intensity level for humans. Sound power density 10^{-12} W/m^2 is taken as the standard 0 dB. (From A. Kawabata: *Easy Ultrasonic Engineering*, Industrial Research Institute [1982]. With permission.)

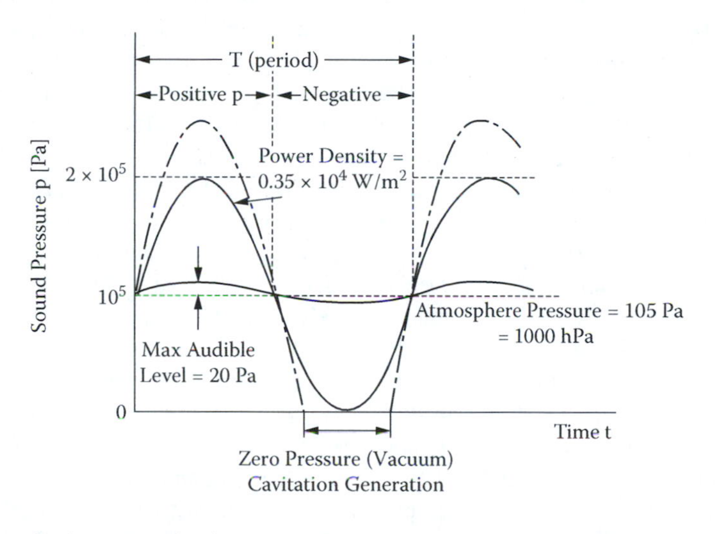

FIGURE 2.11 Explanation of *cavitation* generation in water. When the acoustic power density exceeds 0.35×10^4 W/m^2, zero pressure vacuum is generated. (From A. Kawabata: *Easy Ultrasonic Engineering*, Industrial Research Institute [1982]. With permission.)

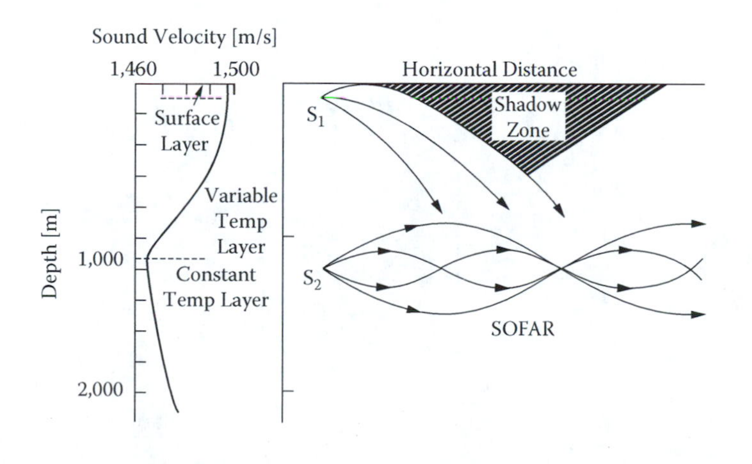

FIGURE 2.12 Principle of the sound channel (sound fixing and ranging: SOFAR) in the sea. The sound wave merges with the minimum sound velocity area. (From A. Kawabata: *Easy Ultrasonic Engineering*, Industrial Research Institute [1982]. With permission.)

from the sea surface down to 1,000 m, and then levels off. Accordingly, the sound velocity decreases gradually down to 1,000 m. In the level further below 1,000 m (constant temperature), because of the water pressure increase, the sound velocity increases again. Therefore, SOFAR (sound fixing and ranging) channel exists around 1,000 m below sea level.

It is possible for whales or submarines to communicate with each other using this sound channel. The sound propagates in the sea more than 20,000 km (half of the earth's surface) without diverging, similar to an optical fiber communication between the United States and Japan.

2.4.1.3 Medical Ultrasonic Imaging

One of the most important applications is based on ultrasonic echo field.[34,35] Ultrasonic transducers convert electrical energy into mechanical form when generating an acoustic pulse and convert mechanical energy into an electrical signal when detecting its echo. The transmitted waves propagate into a body, and echoes are generated that travel back to be received by the same transducer. These echoes vary in intensity according to the type of tissue or body structure, thereby creating images. An ultrasonic image represents the mechanical properties of the tissue, such as ***density*** and ***elasticity***. We can recognize anatomical structures in an ultrasonic image because the organ boundaries and fluid-to-tissue interfaces are easily discerned. The ultrasonic imaging process can also be performed in real time. This means that we can follow rapidly moving structures, such as the heart, without motion distortion. In addition, ultrasound is one of the safest diagnostic imaging techniques. It does not use ionizing radiation such as x-rays and thus is routinely used for fetal and obstetrical imaging. Useful areas for ultrasonic imaging include cardiac structures, the vascular systems, the fetus, and abdominal organs such as liver and kidney. Thus, by using ultrasound it is possible to see inside the human body without breaking the skin.

Figure 2.13 shows the basic ultrasonic transducer geometry. The transducer is mainly composed of matching, ***piezoelectric material,*** and ***backing layers***.[36] One or more matching layers are used to increase sound transmissions into tissues. The backing is added to the rear of the transducer to damp the acoustic backwave and to reduce the pulse duration. Piezoelectric materials are used to generate and detect ultrasound. In general, broadband transducers should be used for medical ultrasonic imaging. The broad-bandwidth response corresponds to a short pulse length, resulting in better axial resolution. Three factors are important in designing broad-bandwidth transducers: acoustic impedance matching, a high electromechanical coupling coefficient of the transducer, and electrical impedance matching. These pulse echo transducers operate based on thickness mode resonance of the piezoelectric thin plate. Further, a low planar mode coupling coefficient, k_p, is beneficial for limiting energies being expended in nonproductive lateral mode. A large dielectric constant is necessary to enable a good electrical impedance match to the system, especially with tiny piezoelectric sizes. Various types of transducers are used in ultrasonic imaging. Mechanical sector transducers consist of single, relatively large resonators and can provide images by mechanical scanning such as wobbling. Multiple-element array transducers permit discrete elements to be individually accessed by the imaging system and enable electronic focusing in the scanning plane to various adjustable penetration depths through the use of phase delays. There are two basic types of array transducers: linear and phased (or sector). A linear array is a collection of elements arranged in one direction, producing a rectangular display (see Figure 2.14). A curved linear (or convex) array is a modified linear array whose elements are arranged along an arc to permit an enlarged trapezoidal field of view. The elements of these linear-type array transducers are excited sequentially group by group with a sweep of the beam in one direction. These linear array transducers are used for radiological and obstetrical examinations. On the other hand, in a phased array transducer, the acoustic beam is steered by signals that are applied to the elements with delays, creating a sector display. This transducer is useful for cardiology applications where positioning between the ribs is necessary.

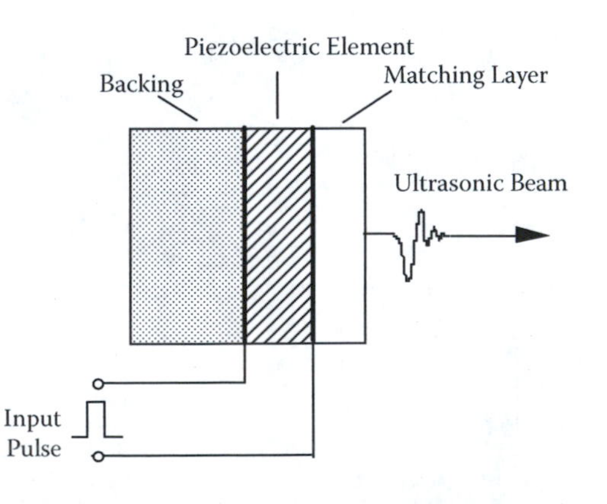

FIGURE 2.13 Basic transducer geometry for acoustic imaging applications.

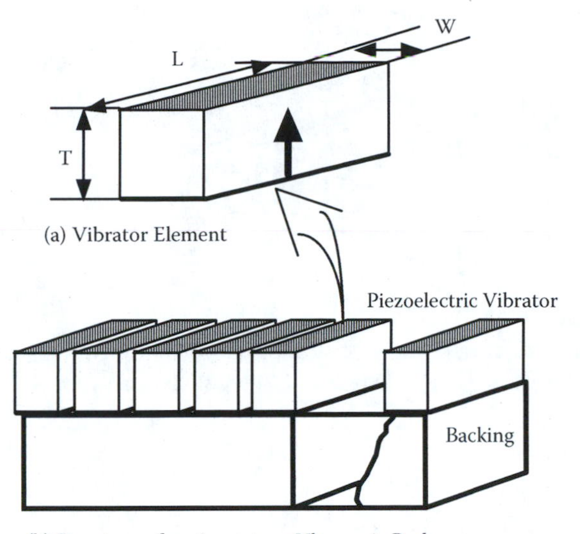

FIGURE 2.14 Linear array-type ultrasonic probe.

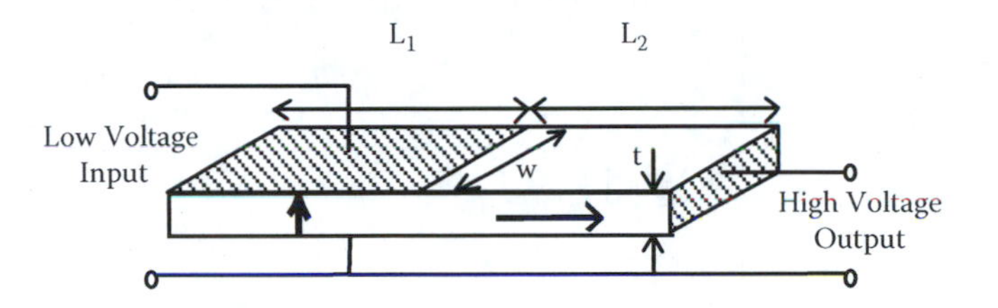

FIGURE 2.15 Piezoelectric transformer proposed by Rosen.

2.4.2 Piezoelectric Transformers

When input and output terminals are fabricated on a piezo device and input/output voltage is changed through the vibration energy transfer, the device is called a *piezoelectric transformer.* Piezoelectric transformers were used in color TVs because of their compact size in comparison with the conventional electromagnetic-coil-type transformers. Because serious problems were found initially in the mechanical strength (collapse occurred at the nodal point) and in heat generation, the development approach was the same as that used for fabricating ceramic actuators. Recent laptop computers with a liquid crystal display require a very thin, no-electromagnetic-noise transformer to start the glow of a fluorescent back-lamp. This application has recently accelerated the development of the piezo transformer.

Since the original piezo transformer was proposed by C. A. Rosen,[37] a variety of such transformers have been investigated. Figure 2.15 shows a fundamental structure where two differently-poled parts coexist in one piezoelectric plate. A standing wave with a wavelength equal to the sample length is excited, with a half wavelength existing on both the input (L_1) and output (L_2) parts. The voltage rise ratio r (*step-up ratio*) is given for the unloaded condition by

$$r = (4/\pi^2)\, k_{31} k_{33} Q_m (L_2/t)\left[2\sqrt{s_{33}{}^E / s_{11}{}^E} \,/\left(1+\sqrt{s_{33}{}^D / s_{11}{}^E}\right)\right]. \tag{2.56}$$

The ratio r is increased with an increase of (L_2/t), where t is the thickness.

NEC proposed a multilayer-type transformer (Figure 2.16) to increase the voltage rise ratio.[38] The multilayer structure also changes the input part impedance to match the input drive circuit impedance. More than 50 times step-up ratio can be obtained. Usage of the third-order longitudinal mode is another idea to distribute the stress concentration.

2.4.3 Piezoelectric Actuators

Piezoelectric/electrostrictive actuators may be classified into two categories based on the type of driving voltage applied to the device and the nature of the strain induced by the voltage (Figure 2.17): (1) rigid displacement devices for which the strain

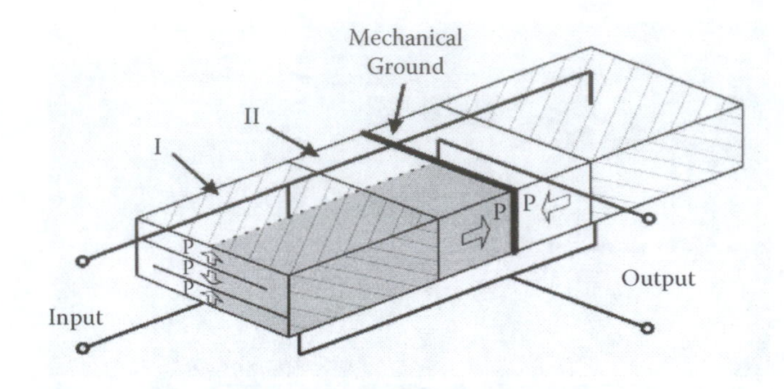

FIGURE 2.16 Multilayer-type transformer by NEC. (From S. Kawashima, O. Ohnishi, H. Hakamata, S. Tagami, A. Fukuoka, T. Inoue and S. Hirose: *Proc. IEEE Int Ustrasonic Symp. '94*, France [November 1994]. With permission.)

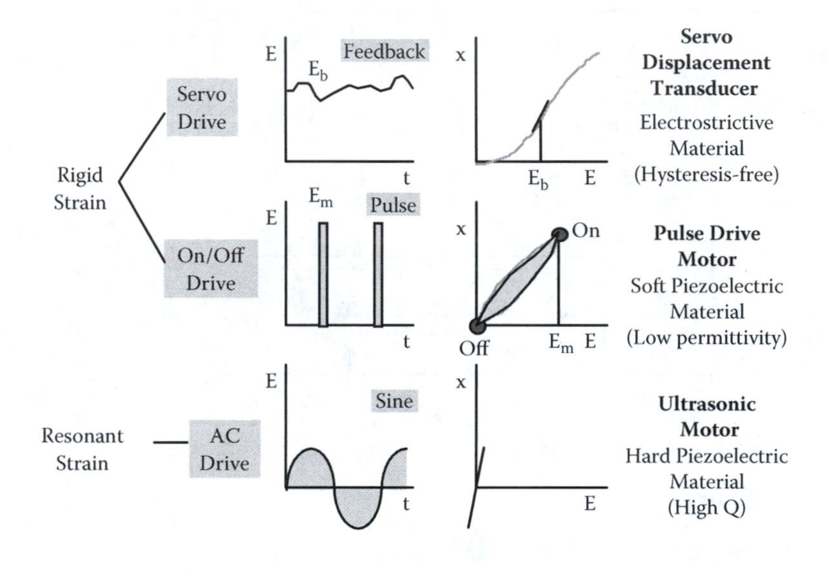

FIGURE 2.17 Classification of piezoelectric/electrostrictive actuators.

is induced unidirectionally along the direction of the applied DC field and (2) resonating displacement devices for which the alternating strain is excited by an AC field at the mechanical resonance frequency (ultrasonic motors). The first can be further divided into two types: servo displacement transducers (positioners), controlled by a feedback system through a position detection signal, and pulse drive motors, operated in a simple on/off switching mode, exemplified by inkjet printers.

The material requirements for these classes of devices are somewhat different, and certain compounds will be better suited to particular applications. The ultrasonic motor, for instance, requires a very hard piezoelectric with a high mechanical quality factor, Q_m, to suppress heat generation. Driving the motor at the antiresonance frequency, rather than at resonance, is also an intriguing technique to reduce the load on the piezoceramic and the power supply.[39] The servo displacement transducer suffers most from strain hysteresis and, therefore, a PMN electrostrictor (strain is induced in proportion to the square of the applied electric field—secondary electromechanical coupling) is used for this purpose. The pulse drive motor requires a low-permittivity material aimed at quick response with a certain power supply rather than a small hysteresis; so soft PZT piezoelectrics are preferred to the high-permittivity PMN for this application.

2.4.3.1 Servo Displacement Transducers

A typical example is found in a space truss structure proposed by the jet propulsion laboratory.[40] A stacked PMN actuator was installed at each truss nodal point and operated so that unnecessary mechanical vibration was suppressed immediately. A Hubble telescope has also been proposed using multilayer PMN electrostrictive actuators to control the phase of the incident light wave in the field of optical information processing (Figure 2.18).[41] The PMN electrostrictor provided superior adjustment of the telescope image because of negligible strain hysteresis.

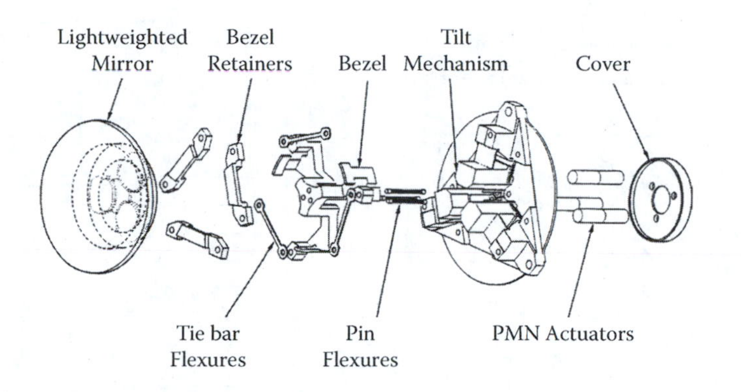

FIGURE 2.18 "Hubble" telescope using three PMN electrostrictive actuators for optical image correction.

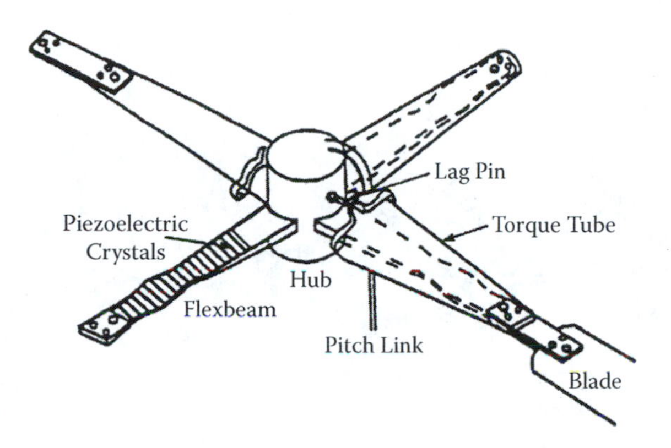

FIGURE 2.19 Bearingless rotor flexbeam with attached piezoelectric strips. A slight change in the blade angle provides for enhanced controllability.

The U.S. Army is interested in developing a rotor control system in helicopters. Figure 2.19 shows a bearingless rotor flexbeam with attached piezoelectric strips.[42] Various types of PZT-sandwiched beam structures have been investigated for such a flexbeam application and for active vibration control.[43]

Concerning home appliance applications, there is already a large market in VCR systems. The requirement for high-quality images has become very stringent for VCRs, especially when played in still, slow, or quick mode. As illustrated in Figure 2.20, when the tape is running at a speed different from the normal speed, the head trace deviates from the recording track depending on the velocity difference. Thus, the head traces on the guard band, generating *guard band noise*.[44] The autotracking scan system by Ampex operates with a piezoelectric actuator so that the head follows the recording track. The piezoelectric device generates no magnetic noise.

Bimorph structures are commonly used for this tracking actuator because of their large displacement. However, special care has been taken not to produce a *spacing angle* between the head and the tape, because a single bimorph exhibits deflection with slight rotation. Various designs have been proposed to produce a completely parallel motion.

2.4.3.2 Pulse Drive Motors

The dot matrix printer was the first widely commercialized product using ceramic actuators. Each character formed by such a printer is composed of a 24×24 dot matrix. A printing ribbon is subsequently impacted by a multiwire array. A sketch of the printer head appears in Figure 2.21a.[45] The printing element is composed of a multilayer piezoelectric device, in which 100 thin ceramic sheets 100 mm in thickness are stacked, together with a sophisticated magnification mechanism [Figure 2.21b]. The magnification unit is based on a monolithic hinge lever with a magnification of 30, resulting in an amplified displacement of 0.5 mm and an energy transfer efficiency greater than 50%.

A piezoelectric camera shutter was the largest production item in the late 1990s (Figure 2.22). A piece of piezoelectric bimorph can open and close the shutter in a millisecond through a mechanical wing mechanism.[46]

Toyota developed a piezo TEMS (Toyota electronic modulated suspension), which is responsive to each protrusion on the road in adjusting the damping condition, and installed it on a Celcio (equivalent to Lexus, internationally) in 1989.[47] In general,

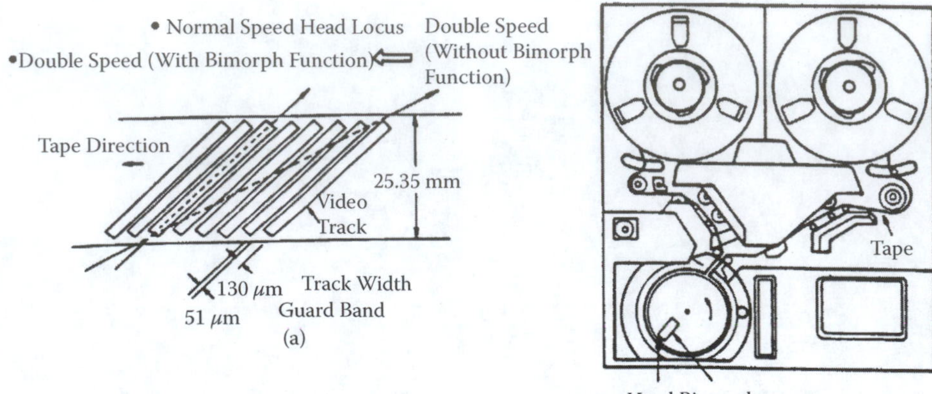

FIGURE 2.20 Locus of the video head and the function of the piezo actuator.

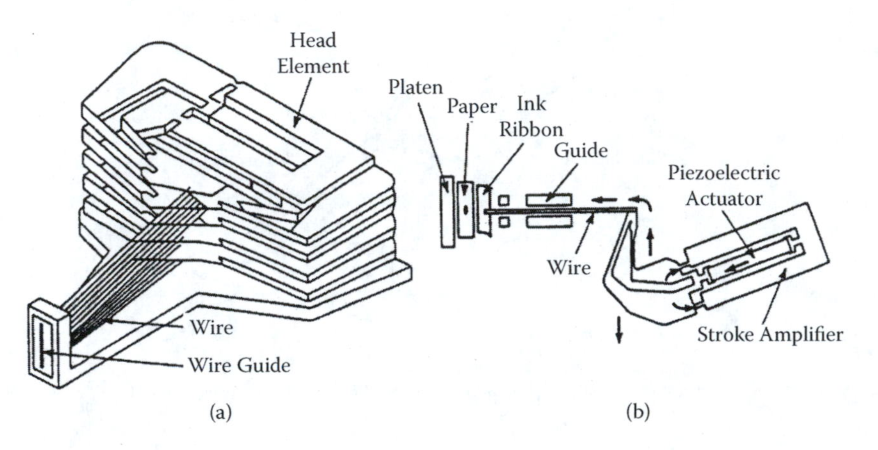

FIGURE 2.21 (a) Structure of a printer head and (b) a differential-type piezoelectric printer-head element. A sophisticated monolithic hinge-lever mechanism amplifies the actuator displacement by 30 times.

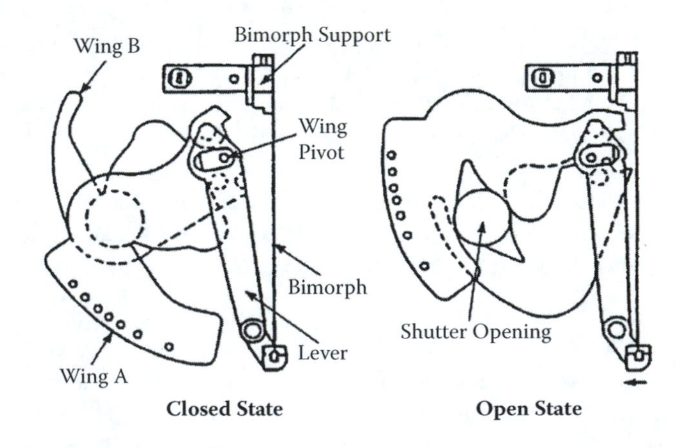

FIGURE 2.22 Camera shutter mechanism using a piezoelectric bimorph actuator.

as the damping force of a shock absorber in an automobile is increased (i.e., "hard" damper), the controllability and stability of the vehicle are improved. However, comfort is sacrificed because the road roughness is easily transferred to the passengers. The purpose of the electronically controlled shock absorber is to obtain both controllability and comfort simultaneously. Usually, the system is set to provide a low damping force ("soft") so as to improve comfort, and the damping force is changed to a high position according to the road condition and the car speed to improve the controllability. To respond to a road protrusion, a very high response of the sensor and actuator combination is required.

Figure 2.23 shows the structure of the electronically controlled shock absorber. The sensor is composed of five layers of 0.5-mm-thick PZT disks. The detecting speed of the road roughness is about 2 ms, and the resolution of the up-down deviation is 2 mm. The actuator is made of 88 layers of 0.5-mm-thick disks. Applying 500 V generates a displacement of about 50 mm, which is magnified 40 times through a piston and plunger pin combination. This stroke pushes the change valve of the damping force down and then opens the bypass oil route, leading to the decrease of the flow resistance (i.e., "soft").

Figure 2.24 illustrates the operation of the suspension system. The up-down acceleration and pitching rate were monitored when the vehicle was driven on a rough road. When the TEMS system was used (top figure), the up-down acceleration was suppressed to as small as the condition fixed at "soft," providing comfort. At the same time, the pitching rate was also suppressed to as small as the condition fixed at "hard," leading to better controllability. Figure 2.25 shows a walking piezo motor with four multilayer actuators.[48] The two shorter actuators function as clamps, and the longer two provide the movement by an inchworm mechanism.

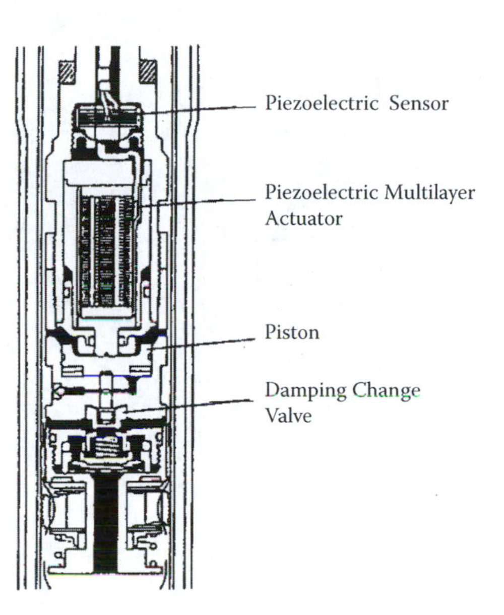

FIGURE 2.23 Electronic modulated suspension by Toyota. (From Y. Yokoya: *Electronic Ceramics*, **22**, No. 111, p. 55 [1991]. With permission.)

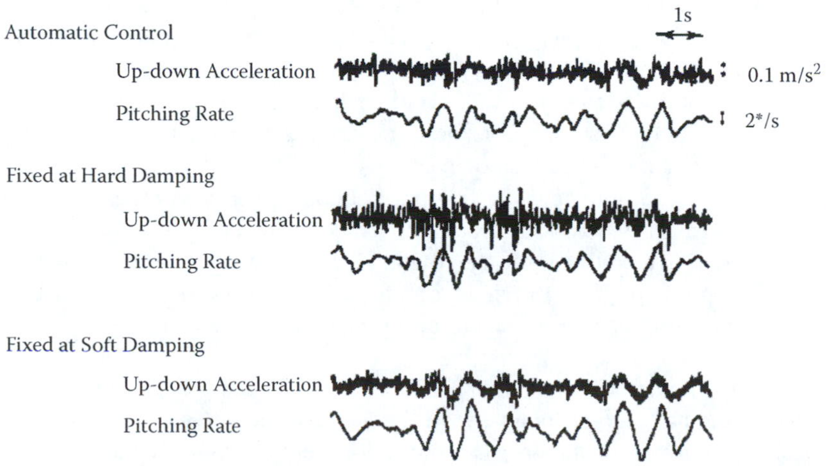

FIGURE 2.24 Function of the adaptive suspension system.

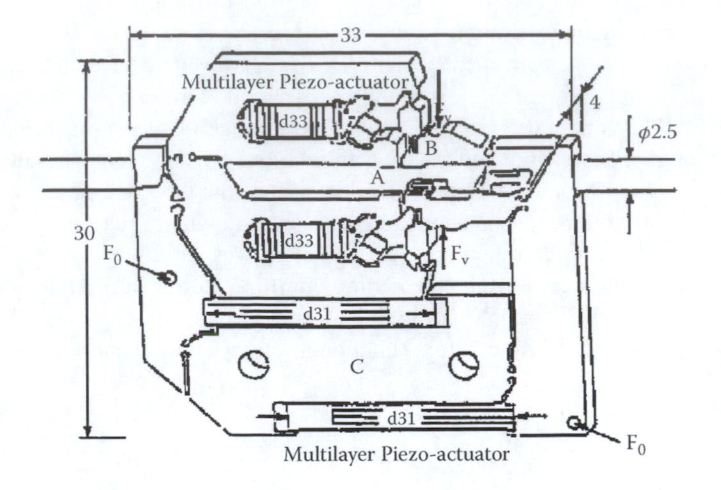

FIGURE 2.25 Walking piezo motor using an inchworm mechanism with four multilayer piezoelectric actuators by Philips.

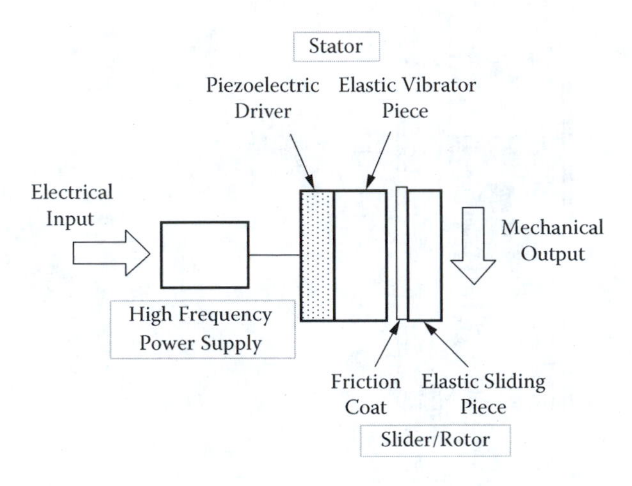

FIGURE 2.26 Fundamental construction of an ultrasonic motor.

2.4.3.3 Ultrasonic Motors

Electromagnetic motors were invented more than a hundred years ago. Although these motors still dominate the industry, a drastic improvement cannot be expected except through new discoveries in magnetic or superconducting materials. Regarding conventional electromagnetic motors, tiny motors smaller than 1 cm size are rather difficult to produce with sufficient energy efficiency. Therefore, a new class of motors using high-power ultrasonic energy, the ultrasonic motor, is gaining widespread attention. Ultrasonic motors made with piezoceramics whose efficiency is insensitive to size are superior in the minimotor area. Figure 2.26 shows the basic construction of most ultrasonic motors, which consist of a high-frequency power supply, a vibrator, and a slider. The vibrator is composed of a piezoelectric driving component and an elastic vibratory part, and the slider is composed of an elastic moving part and a friction coat.

Although there had been some earlier attempts, the first practical ultrasonic motor was proposed by H. V. Barth of IBM in 1973.[49] The rotor was pressed against two horns placed at different locations. By exciting one of the horns, the rotor was driven in one direction, and by exciting the other horn, the rotation direction was reversed. Various mechanisms based on virtually the same principle were proposed by V. V. Lavrinenko[50] and P. E. Vasiliev[51] in the former USSR. Because of difficulty in maintaining a constant vibration amplitude with temperature rise, and wear and tear, the motors were not of much practical use at that time.

In the 1980s, with increasing chip pattern density, the semiconductor industry began to demand much more precise and sophisticated positioners that would not generate magnetic field noise. This urgent need accelerated the development of ultrasonic motors. Another advantage of ultrasonic motors over conventional electromagnetic motors with expensive copper coils is the improved availability of piezoelectric ceramics at reasonable cost. Figure 2.27 shows the world's smallest ultrasonic motor with reasonable torque to be used in a cellular phone.[52] This motor comprises two inexpensive PZT rectangular plates bonded

FIGURE 2.27 (a) Photo of the world's smallest ultrasonic motor (1.5 mmf), which is composed of a metal tube and two rectangular PZT plates.[52] (b) Zoom/focus mechanism for a cellular phone camera application with two of the metal tube motors (courtesy of Samsung Electromechanics).

on a metal tube. Hula-hoop motion is excited on a metal tube, which rotates ferrules. A targeted price of about 40 cents per unit including its high-frequency drive circuit is not unrealizable in the near future. Let us summarize the merits and demerits of the ultrasonic motor in comparison with electromagnetic motors:

Merits

1. Low speed and high torque—direct drive
2. Quick response, wide velocity range, hard brake, and no backlash—excellent controllability, fine position resolution
3. High power/weight ratio and high efficiency
4. Quiet drive
5. Compact size and light weight
6. Simple structure and easy production process
7. Negligible effect from external magnetic or radioactive fields, and also no generation of these fields

Demerits

1. Necessity for a high-frequency power supply
2. Less durability because of frictional drive
3. Drooping torque versus speed characteristics

The detailed structures and operating principles are provided in chapter 3.

REFERENCES

1. B. Jaffe, W. Cook, and H. Jaffe: *Piezoelectric Ceramics*, London: Academic Press (1971).
2. W. G. Cady: *Piezoelectricity*, New York: McGraw-Hill, Revised Edition by Dover Publications (1964).
3. M. E. Lines and A. M. Glass: *Principles and Applications of Ferroelectric Materials*, Oxford: Clarendon Press (1977).
4. K. Uchino: *Piezoelectric Actuators and Ultrasonic Motors*, Kluwer Academic Publishers, MA (1996).
5. T. Ikeda: *Fundamentals of Piezoelectric Materials Science*, Ohm Publishing, Tokyo (1984).
6. L. E. Kinsler, A. R. Frey, A. B. Coppens, and J. V. Sanders: *Fundamentals of Acoustics*, John Wiley and Sons, New York (1982).
7. Y. Ito and K. Uchino: Piezoelectricity, *Wiley Encyclopedia of Electrical and Electronics Engineering*, Vol. **16**, p. 479, John Wiley and Sons, New York (1999).
8. W. A. Smith: *Proc. SPIE—The International Society for Optical Engineering* 1733 (1992).
9. H. Takeuchi, S. Jyomura, E. Yamamoto, and Y. Ito: *J. Acoust. Soc. Am.*, **74**, 1114 (1982).
10. Y. Yamashita, K. Yokoyama, H. Honda, and T. Takahashi: *Jpn. J. Appl. Phys.*, **20**, Suppl. 20–4, 183 (1981).
11. Y. Ito, H. Takeuchi, S. Jyomura, K. Nagatsuma, and S. Ashida: *Appl. Phys. Lett.*, **35**, 595 (1979).
12. H. Takeuchi, H. Masuzawa, C. Nakaya, and Y. Ito: *Proc. IEEE 1990 Ultrasonics Symp.*, 697 (1990).
13. J. Kuwata, K. Uchino, and S. Nomura: *Jpn. J. Appl. Phys.*, **21**, 1298 (1982).
14. T. R. Shrout, Z. P. Chang, N. Kim, and S. Markgraf: *Ferroelectr. Lett.*, **12**, 63 (1990).
15. R. E. Newnham, D. P. Skinner, and L. E. Cross: *Materials Research Bulletin*, **13**, 525 (1978).
16. W. A. Smith: *Proc. 1989 IEEE Ultrasonic Symp.*, 755 (1989).

17. S. Legvold, J. Alstad, and J. Rhyne: *Phys. Rev. Lett.* **10**, 509 (1963).
18. A. E. Clark, B. F. DeSavage, and R. Bozorth: *Phys. Rev.* **138**, A216 (1965).
19. N. C. Koon, A. Schindler, and F. Carter: *Phys. Lett.* **A37**, 413 (1971).
20. A. E. Clark, H. Belson: *AIP Conf. Proc.*, No. 5, 1498 (1972).
21. A. E. Clark: *Ferroelectric Materials*, Vol. **1**, p. 531, Eds. K. H. J. Buschow and E. P. Wohifarth, North-Holland, Amsterdam (1980).
22. J. D. Verhoven, E. D. Gibson, O. D. McMasters, and H. H. Baker: *Metal Trans. A*, **18A**, No. 2, 223 (1987).
23. A. E. Clark, J. D. Verhoven, O. D. McMasters, and E. D. Gibson: *IEEE Trans. Mag.* MAG-22, p. 973 (1986).
24. J. B. Restorff: *Encyclopedia of Applied Physics*, Vol. **9**, p. 229, VCH Publ., New York (1994).
25. A. B. Flatau, M. J. Dapino, and F. T. Calkins: Chap. 5.26 Magnetostrictive Composites in *Comprehensive Composite Materials*, Elsevier Science, Oxford (2000).
26. R. Kellogg and A. B. Flatau: *Proc. SPIE*, **3668**(19), 184 (1999).
27. J. L. Butler: *Application Manual for the Design Etrema Terfenol-D Magnetostrictive Transducers*, Edge Technologies, Ames, IA (1988).
28. G. P. Carman, K. S. Cheung, and D. Wang: *J. Intelligent Mater. Syst. Struct.*, **6**, 691 (1995).
29. T. A. Duenas and G. P. Carman: *Proc. Adaptive Struct. Mater. Syst.*, ASME, AD-**57**/MD-**83**, p. 63 (1998).
30. J. F. Nye: *Physical Properties of Crystals*, Oxford University Press, London, p. 123, 140 (1972).
31. T. Ikeda: *Fundamentals of Piezoelectric Materials Science*, Ohm Publication Company, Tokyo, p. 83 (1984).
32. K. Uchino and S. Hirose: *IEEE-UFFC Trans.* **48**, 307–321 (2001).
33. A. Kawabata: *Easy Ultrasonic Engineering*, Industrial Research Institute (1982).
34. B. A. Auld: *Acoustic Fields and Waves in Solids*, 2nd ed., Melbourne: Robert E. Krieger (1990).
35. G. S. Kino: *Acoustic Waves: Device Imaging and Analog Signal Processing*, Englewood Cliffs, NJ: Prentice-Hall (1987).
36. C. S. Desilets, J. D. Fraser, and G. S. Kino: *IEEE Trans. Sonics Ultrason*ics, SU-25, 115 (1978).
37. C. A. Rosen: *Proc. Electronic Component Symp.*, p. 205 (1957).
38. S. Kawashima, O. Ohnishi, H. Hakamata, S. Tagami, A. Fukuoka, T. Inoue, and S. Hirose: *Proc. IEEE Int Ustrasonic Symp. '94*, France (November 1994).
39. S. Hirose, S. Takahashi, K. Uchino, M. Aoyagi, and Y. Tomikawa: *Proc. Mater. Smart Systems, Mater. Res. Soc.* Vol. **360**, p. 15 (1995).
40. J. T. Dorsey, T. R. Sutter, and K. C. Wu: *Proc. 3rd Int. Conf. Adaptive Structures*, p. 352 (1992).
41. B. Wada: JPL Document D-10659, p. 23 (1993).
42. F. K. Straub: *Smart Mater. Struct.*, **5**, 1 (1996).
43. P. C. Chen and I. Chopra: *Smart Mater. Struct.*, **5**, 35 (1996).
44. A. Ohgoshi and S. Nishigaki: Ceramic Data Book '81, p. 35 Inst. Industrial Manufacturing Technology, Tokyo (1981).
45. T. Yano, E. Sato, I. Fukui, and S. Hori: *Proc. Int. Symp. Soc. Information Display*, p. 180 (1989).
46. Y. Tanaka: *Handbook on New Actuators for Precision Control*, Fuji Technosystem, p. 764 (1994).
47. Y. Yokoya: *Electronic Ceramics*, **22**, No. 111, p. 55 (1991).
48. M. P. Koster: *Proc. 4th Int. Conf. New Actuators*, Germany, p. 144 (1994).
49. H. V. Barth: *IBM Technical Disclosure Bull.* **16**, 2263 (1973).
50. V. V. Lavrinenko, S. S. Vishnevski, and I. K. Kartashev: Izvestiya Vysshikh Uchebnykh Zavedenii, *Radioelektronica* **13**, 57 (1976).
51. P. E. Vasiliev et al.: UK Patent Application GB 2020857 A (1979).
52. S. Cagatay, B. Koc, and K. Uchino: *IEEE Trans.—UFFC*, **50**(7), 782–786 (2003).

3 Structures of Smart Transducers

Piezoelectric actuator and transducer designs are commonly fabricated with a bimorph, multilayer, or flextensional structure. On the contrary, magnetostrictors require a bias mechanical stress mechanism in addition to a magnetic coil and shield. Ultrasonic motor designs are also described in this chapter. Most of the designs in this chapter are computer-simulated in Part II, "How to Use ATILA."

3.1 DESIGN CLASSIFICATION

A classification of electromechanical ceramic actuators based on structure type is presented in Figure 3.1. Simple devices directly use the longitudinally or transversely induced strain. The simple disk and multilayer types make use of longitudinal strain, and the cylinder types (the simple cylinder, separated cylinder, and honeycomb designs) utilize the transverse strain. Complex devices do not use the induced strain directly but rather use a magnified displacement, produced through a spatial magnification mechanism (demonstrated by the unimorph, bimorph, moonie, and hinge lever designs) or through a sequential drive mechanism (inchworm). Among the designs shown in Figure 3.1, the multilayer and bimorph types are the most commonly used structures. Although the multilayer type produces only relatively modest displacements (10 μm), it offers a respectable generative force (100 kgf), a quick response speed (10 μs), long lifetime (10^{11} cycles), and a high electromechanical coupling factor k_{33} (70%). The bimorph type provides large displacements (500 μm) but can only offer a relatively low generative force (100 gf), a much slower response speed (1 ms), a shorter lifetime (10^8 cycles), and a rather low electromechanical coupling factor k_{eff} (10%). We will examine each structure more closely in the sections that follow.

3.2 MULTILAYERS

Ferroelectric ceramic multilayer devices have been investigated intensively for capacitor and actuator applications because they have low driving voltages and they are highly suitable for miniaturization and integration onto hybrid structures. Miniaturization and hybridization are key concepts in the development of modern micromechatronic systems. The goal for multilayer actuators is to eventually incorporate layers thinner than 10 μm, which is thicker than the current standard for multilayer capacitors. The typical layer thickness for multilayer actuators at this time is about 60 μm. A multilayer structure typically exhibits a field-induced strain of 0.1% along its length l (for example, *a 1 cm sample will exhibit a 10 μm displacement*) and has a fundamental resonance frequency given by

$$f_r = \frac{1}{2l\sqrt{\rho s_{33}^D}} \tag{3.1}$$

where ρ is the density and s_{33}^D is the elastic compliance (for example, *a 1 cm sample will have a 100 kHz resonance frequency*). New multilayer configurations or heterostructures comprising electromechanical materials and modified electrode patterns are anticipated that will be incorporated into evermore sophisticated smart systems. There are two general methods for fabricating multilayer ceramic devices: the ***cut-and-bond method*** and the ***tape-casting method***.

The tape-casting method, in which ceramic green sheets with printed electrodes are laminated and cofired with compatible internal electrodes, is far more conducive to mass production than the cut-and-bond method and produces devices with much thinner layers, so that low drive voltages may be employed.

The multilayer structure essentially comprises alternating ferroelectric and conducting layers that are ***cofired*** to produce a dense, cohesive unit, as shown in Figure 3.2. A ferroelectric layer sandwiched between a pair of electrodes constitutes a single displacement element. Hundreds of these units may be connected in parallel to the potential difference supplied by the external electrodes, which are connected to the many interleaved internal electrodes of the stack as shown in Figure 3.2. A flowchart for the manufacturing process is shown in Figure 3.3. ***Green sheets*** are prepared in two steps. First, the ceramic powder is

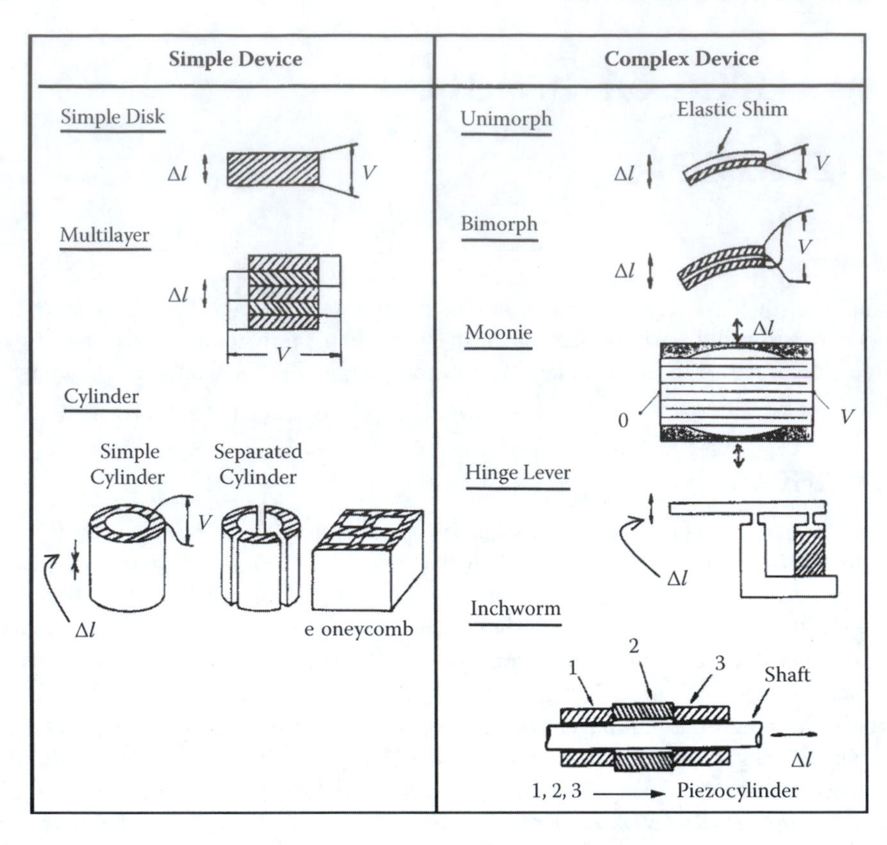

FIGURE 3.1 A classification of electromechanical ceramic actuators based on structure type.

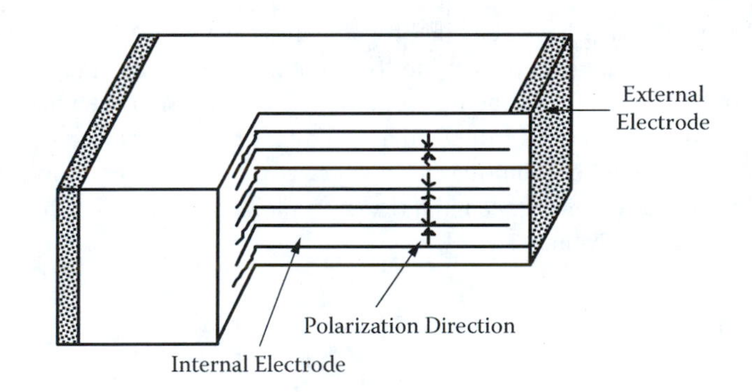

FIGURE 3.2 The structure of a multilayer actuator.

combined with an appropriate liquid solution to form a *slip*. The *slip* mixture generally includes the ceramic powder and a liquid comprising a ***solvent***, a ***deflocculant***, a ***binder***, and a ***plasticizer***. During the second part of the process, the slip is cast into a film under a special straight blade, called a ***doctor blade***, whose distance above the carrier determines the film thickness. Once dried, the resulting film, called a ***green sheet***, has the elastic flexibility of synthetic leather. The volume fraction of the ceramic in the now polymerized matrix at this point is about 50%. The green sheet is cut into an appropriate size, and internal electrodes are printed using silver, palladium, or platinum ink. Several tens to hundreds of these layers are then laminated and pressed using a hot press. After the stacks are cut into small chips, the green bodies are sintered at around 1200°C in a furnace, with special care taken to control the initial binder evaporation at 500°C. The sintered chips are polished and externally electroded; lead wires are attached and, finally, the chips are coated with a waterproof spray.

A cross-sectional view of a conventional ***interdigital electrode*** configuration is shown in Figure 3.4a.[1] The area of the internal electrode is slightly smaller than the cross-sectional area of the device. Note that every two layers of the internal electrodes extend to one side of the device and connect with the external electrode on that side so that all active layers of the device are effectively connected in parallel. The small segments in each layer that are not addressed by the internal electrodes

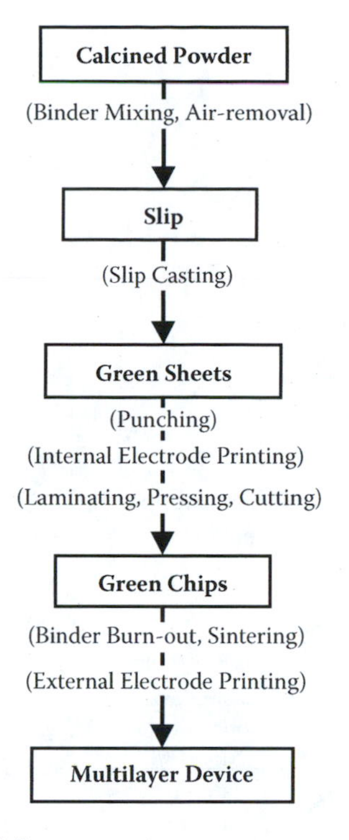

FIGURE 3.3 A flowchart for the fabrication of a multilayer ceramic actuator.

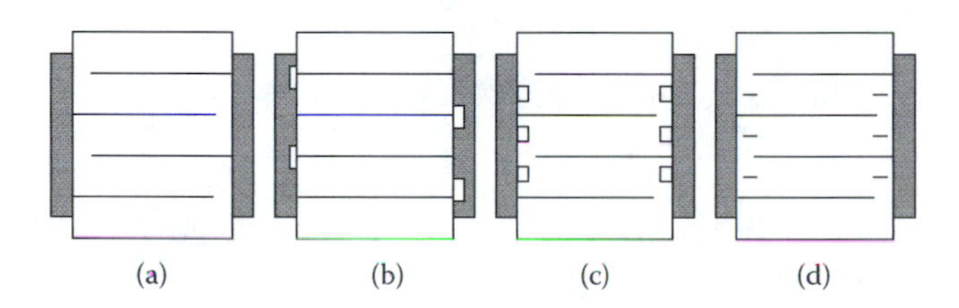

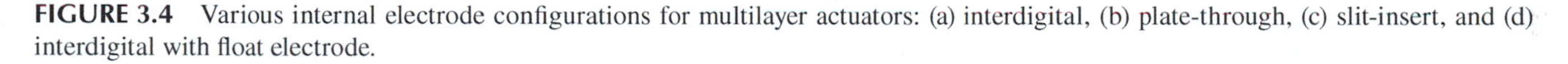

FIGURE 3.4 Various internal electrode configurations for multilayer actuators: (a) interdigital, (b) plate-through, (c) slit-insert, and (d) interdigital with float electrode.

remain inactive, thereby restricting the overall generative displacement and leading to detrimental stress concentrations in the device. A multilayer structure is represented in Figure 3.5a, and the strain distribution measured in a test device is shown in Figure 3.5b. The derivative of the displacement distribution provides an estimate of the stress concentration in the device.[2] The internal stress distribution was also predicted using the finite element method. The results of this analysis are summarized in Figure 3.6. The maximum tensile and compressive stresses are 1×10^8 N/m^2 and 1.2×10^8 N/m^2, respectively, which are very close to the critical strength of the ceramic.[2]

Crack propagation has been investigated in a variety of multilayer systems.[3] A crack pattern commonly observed in lead nickel niobate-lead zinc nidoate (PNNZT) piezoelectric actuators under bipolar drive is shown in Figure 3.7a. It occurs much as predicted in the theoretical treatment of these systems.[2] The crack originates at the edge of an internal electrode and propagates toward the adjacent electrode. Delamination between the electrode and the ceramic occurs simultaneously, leading to a Y-shaped crack. An interesting difference was observed in the case of the antiferroelectric PNZST system shown in Figure 3.7b. Here, a Y-shaped crack is again produced, but it originates in the ceramic between the electrodes.[4] This is probably due to the combination of the two distinct induced strains in the ceramic, the anisotropic piezoelectric strain and the more isotropic strain associated with the antiferroelectric response to the applied field.

A modified electrode configuration, called the *plate-through design* (see Figure 3.4b), was developed by NEC as a solution to this particular mechanical problem.[5] The electrode in this modified configuration extends over the entire surface of the ceramic so that the stress concentration cannot develop. This modification requires that an insulating tab terminate every two

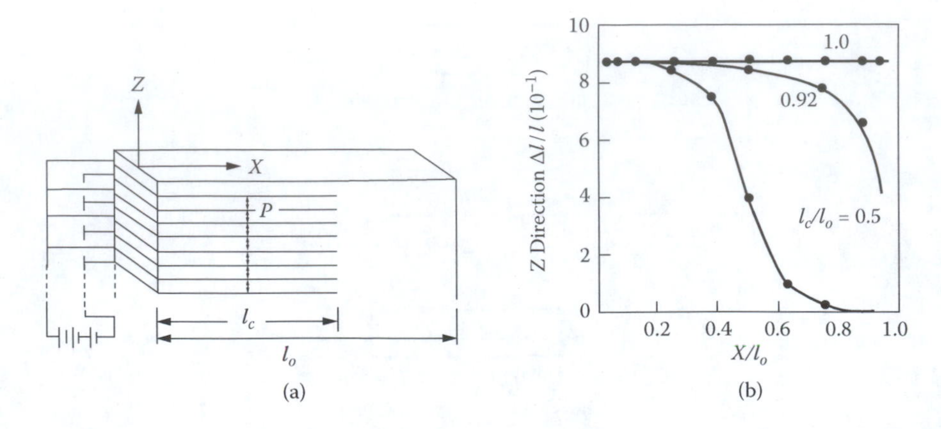

FIGURE 3.5 A multilayer structure: (a) a schematic depiction and (b) the induced strain distribution measured in a test device. (From S. Takahashi, A. Ochi, M. Yonezawa, T. Yano, T. Hamatsuki, and I. Fukui: *Jpn. J. Appl. Phys.* **22**, Suppl. 22-2, 157 [1983]. With permission.)

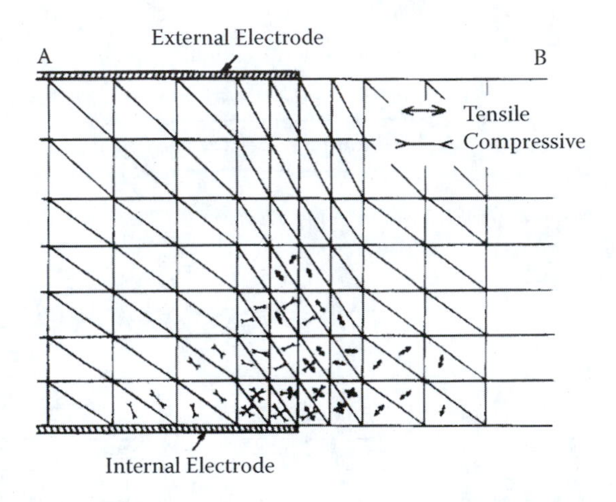

FIGURE 3.6 The internal stress distribution for a multilayer actuator as predicted by finite element analysis. (From S. Takahashi, A. Ochi, M. Yonezawa, T. Yano, T. Hamatsuki, and I. Fukui: *Jpn. J. Appl. Phys.* **22**, Suppl. 22-2, 157 [1983]. With permission.)

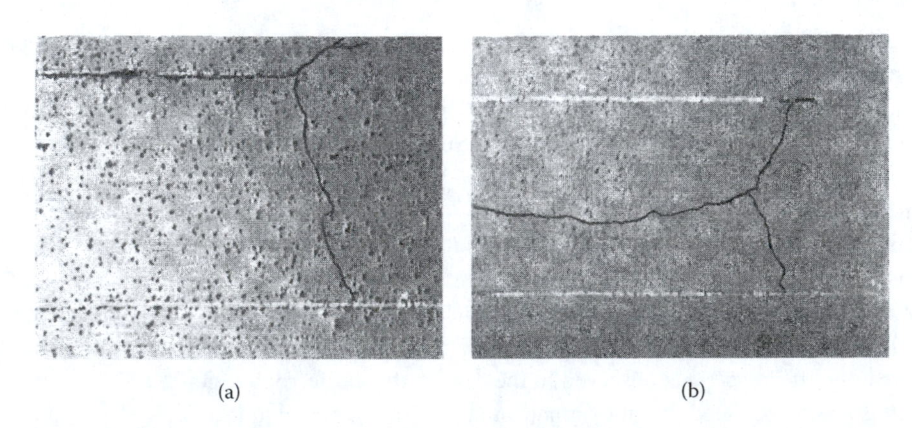

FIGURE 3.7 Crack generation in multilayer ceramic actuators under bipolar drive: (a) piezoelectric PNNZT, and (b) antiferroelectric PNZST. (From A. Furuta and K. Uchino: *Ferroelectrics* **160**, 277 [1994]. With permission.)

electrode layers on the sides of the device where the external electrodes are painted. Key issues in producing reliable devices with this alternative electrode design are concerned with the precise application of the insulating terminations and improvement of the adhesion between the ceramic and internal electrode layers. The designers of NEC addressed the first of these issues by developing an electrophoretic technique for applying glass terminators to the device. The problem of adhesion was resolved by making use of a special electrode paste containing powders of both the Ag-Pd electrode material and the ceramic phase. The displacement curve for a (0.65)PMN-(0.35)PT multilayer actuator with 99 layers of 100-μm-thick sheets ($2 \times 3 \times 10$ mm^3) is shown in Figure 3.8a.[6] We see from these data that an 8.7 μm displacement is generated by an applied voltage of 100 V, accompanied by a slight hysteresis. This curve is consistent with the typical response of a disk device. The transient response of the induced displacement after the application of a rectangular voltage is shown in Figure 3.8b. Rising and falling responses as quick as 10 μs are observed.

The ***slit-insert design*** and the ***interdigital with float electrode*** are shown in Figure 3.4c and 3.4d, respectively. The induced stress concentration is relieved in the slit-insert design, whereas the electric field concentration is avoided with the interdigital with float electrode configuration.[7]

Another common crack pattern is the vertical crack, which will occasionally occur in the layer just adjacent to the top or bottom inactive layer. The inactive layer serves as an interface between the device and the object to which the actuator is attached. A crack occurring in the layer adjacent to the inactive bottom layer of a multilayer device is pictured in Figure 3.9. It originates from a transverse tensile stress produced in this layer due to clamping from the adjacent thick inactive layer. One solution that has been proposed for this problem is to prepole the top and bottom layers of the device.

A multilayer actuator incorporating a new interdigital internal electrode configuration has been developed by NEC-Tokin.[8] In contrast to devices with the conventional interdigital electrode configuration, for the modified design, line electrodes are printed on the piezoelectric green sheets, which are stacked so that alternate electrode lines are displaced by one-half pitch. This

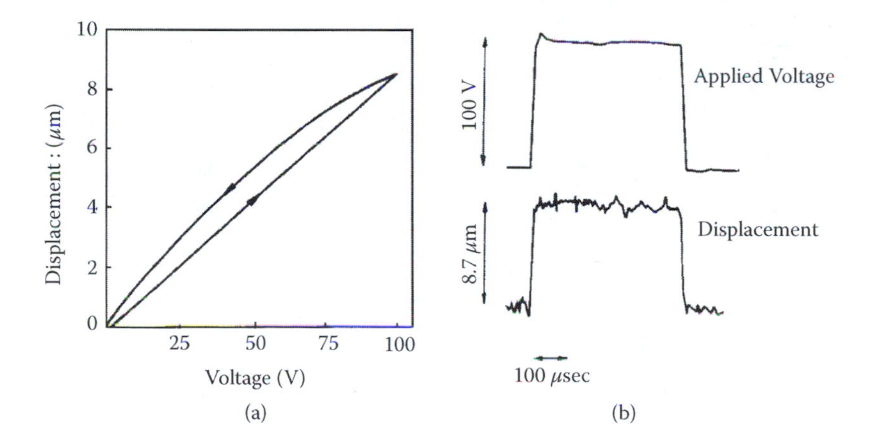

FIGURE 3.8 The response of a (0.65)PMN–(0.35)PT multilayer actuator: (a) the displacement as a function of applied voltage and (b) the displacement response to a step voltage. (From S. Takahashi: *Sensor Technology* **3**, No. 12, 31 [1983]. With permission.)

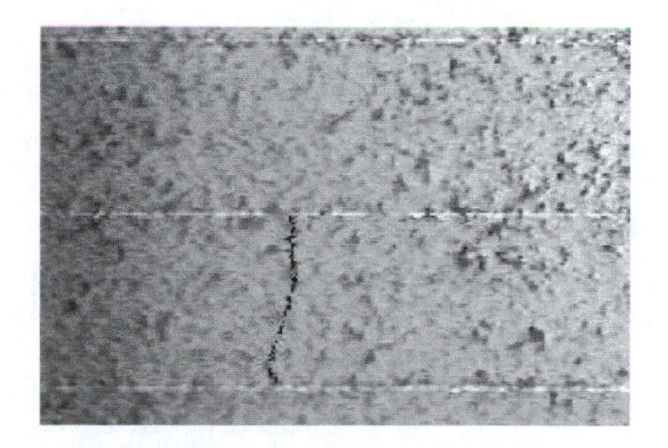

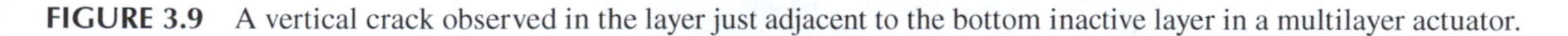

FIGURE 3.9 A vertical crack observed in the layer just adjacent to the bottom inactive layer in a multilayer actuator.

actuator produces displacements normal to the stacking direction through the longitudinal piezoelectric effect. Long ceramic actuators up to 74 mm in length have been manufactured, which can generate longitudinal displacements up to 55 μm.

A three-dimensional positioning actuator with a stacked structure has been proposed by PI Ceramic, in which both transverse and shear strains are induced to generate displacements.[9] As shown in Figure 3.10a, this actuator consists of three parts: the top 10-mm-long Z-stack generates the displacement along the *z* direction, whereas the second and the bottom 10-mm-long X and Y stacks provide the *x* and *y* displacements through shear deformation, as illustrated in Figure 3.10b. The device can produce 10 μm displacements in all three directions when 500 V is applied to the 1-mm-thick layers.

Various failure detection techniques have been proposed for implementation in smart actuator devices to essentially monitor their own "health."[10] One such "intelligent" actuator system that utilizes acoustic emission [AE] detection is shown in Figure 3.11. The actuator is controlled by two feedback mechanisms: position feedback, which can compensate for positional drift and hysteresis, and breakdown detection feedback, which can shut down the actuator system safely in the event of an imminent failure. AE from a piezoelectric actuator driven by a cyclic electric field is a good indicator of mechanical failure. The emissions are most pronounced when a crack propagates in the ceramic at the maximum speed. A portion of this smart piezoelectric actuator is therefore dedicated to sensing and responding to acoustic emissions. The AE rate in a piezoelectric device can increase by three orders of magnitude just prior to complete failure. During the operation of a typical multilayer

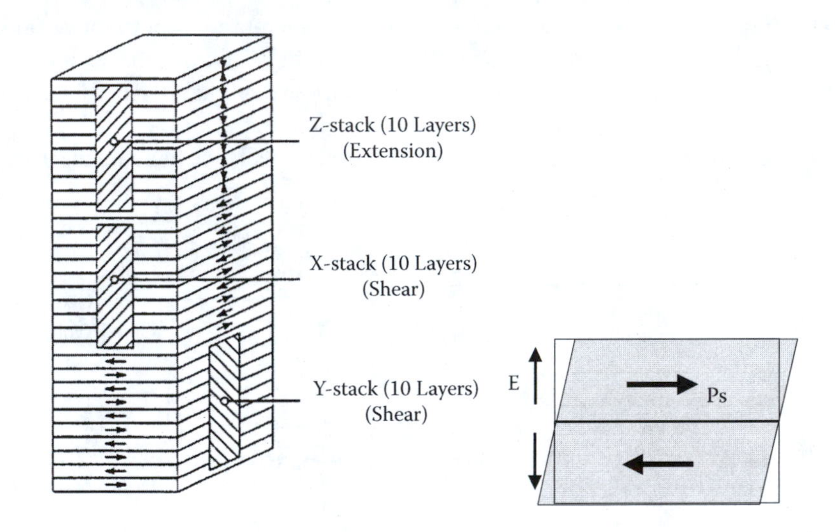

FIGURE 3.10 A three-dimensional positioning actuator with a stacked structure proposed by PI Ceramic: (a) a schematic diagram of the structure and (b) an illustration of the shear deformation. (From A. Banner and F. Moller: *Proc. 4th. Int. Conf. New Actuators*, AXON Tech. Consult. GmbH, p. 128 [1995]. With permission.)

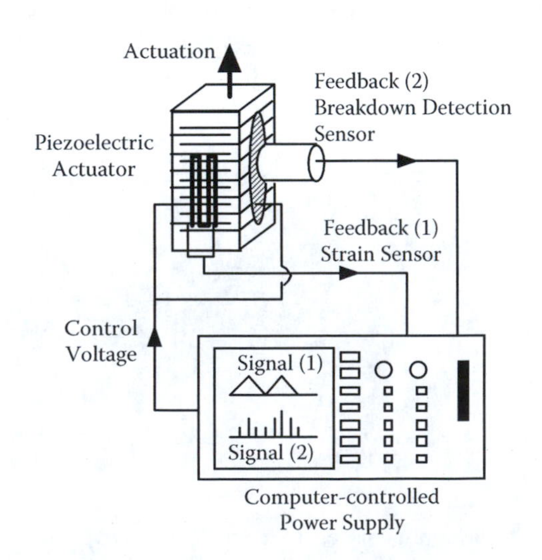

FIGURE 3.11 An intelligent actuator system with both position and breakdown detection feedback mechanisms.

piezoelectric actuator, the AE-sensing portion of the device will monitor the emissions and respond to any dramatic increase in the emission rate by initiating a complete shutdown of the system.

Another recent development for device failure self-monitoring is based on a strain-gauge-type electrode configuration, as pictured in Figure 3.12.[11] Both the electric-field-induced strain and the occurrence of cracks in the ceramic can be detected by closely monitoring the resistance of a strain-gauge-shaped electrode embedded in a ceramic actuator. The resistance of such a smart device is plotted as a function of applied electric field in Figure 3.13. The field-induced strain of a "healthy" device is represented by the series of curves depicted in Figure 3.13a. Each curve corresponds to a distinct number of drive cycles. A sudden decrease in the resistance, as shown in Figure 3.13b, is a typical symptom of device failure.

The effect of aging is manifested clearly by the gradual increase of the resistance with the number of drive cycles, as shown for a smart device in Figure 3.13a. Ceramic aging is an extremely important factor to consider in the design of a reliable actuator device, although there have been relatively few investigations done to better understand and control it. Aging is associated with two types of degradation: depoling and mechanical failure. Creep and zero-point drift in the actuator displacement are caused by depoling of the ceramic. The strain response is also seriously impaired when the device is operated under conditions of very high electric field, elevated temperature, high humidity, and high mechanical stress. The lifetime of a multilayer piezoelectric actuator operated under a DC bias voltage can be described by the empirical relationship as follows:

$$t_{DC} = A\,E^{-\eta}\,\exp(W_{DC}/kT) \tag{3.2}$$

where η has a value of about 3 and W_{DC} is an activation energy ranging from 0.99 to 1.04 eV.[12]

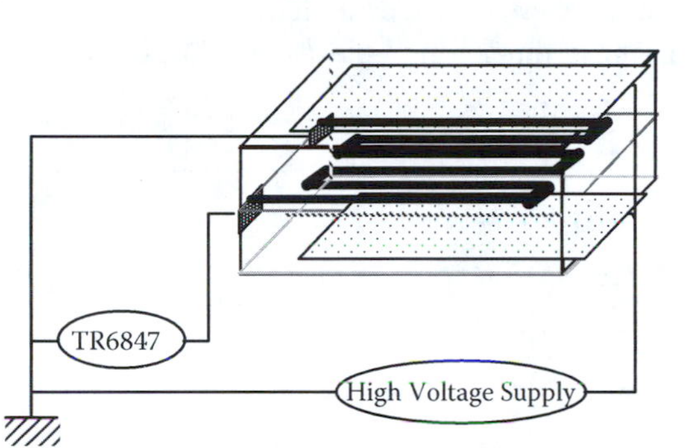

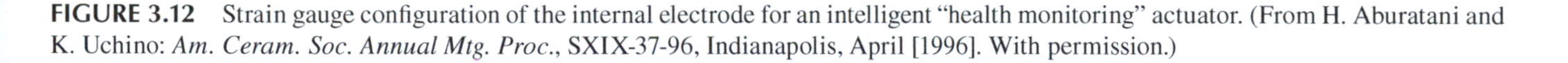

FIGURE 3.12 Strain gauge configuration of the internal electrode for an intelligent "health monitoring" actuator. (From H. Aburatani and K. Uchino: *Am. Ceram. Soc. Annual Mtg. Proc.*, SXIX-37-96, Indianapolis, April [1996]. With permission.)

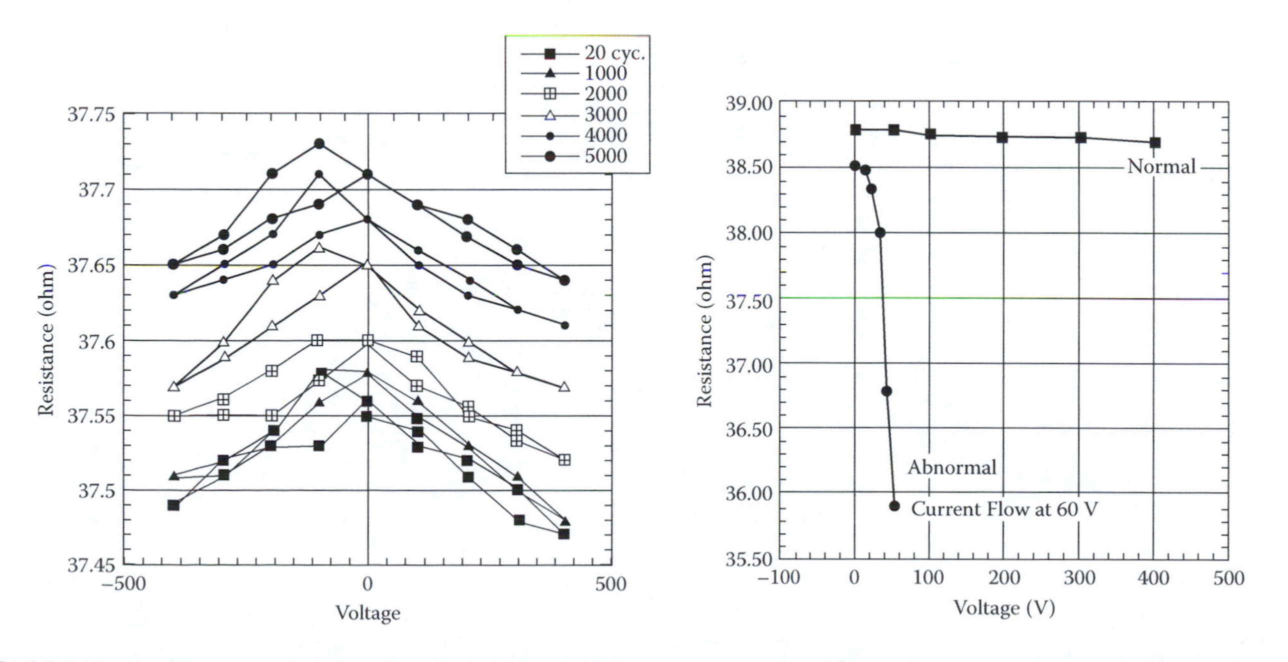

FIGURE 3.13 Resistance as a function of applied electric field for a smart actuator with a strain-gauge-type internal electrode for self-monitoring of potential failure: (a) the electric-field-induced strain response of a "healthy" device and (b) the response of a failing device. (From H. Aburatani and K. Uchino: *Am. Ceram. Soc. Annual Mtg. Proc.*, SXIX-37-96, Indianapolis, April [1996]. With permission.)

Investigations have been conducted on the heat generation from multilayer piezoelectric ceramic actuators of various sizes.[13] The temperature rise, ΔT, monitored in actuators driven at 3 kV/mm and 300 Hz, is plotted as a function of the quantity, V_e/A, in Figure 3.14, where V_e is the effective actuator volume (corresponding to the electroded portion of the device) and A is its surface area. The linear relationship observed is expected for a ratio of device volume, V_e, to its surface area, A. We see from this trend that a configuration with a small V_e/A will be the most conducive to suppressing heating within the device. Flat and cylindrical shapes, for example, are preferable to cube and solid rod structures, respectively.

3.3 UNIMORPH/BIMORPH

Unimorph and ***bimorph*** devices are simple structures comprising ceramic and inactive elastic plates bonded surface to surface. Unimorph devices have one plate, and bimorph structures have two ceramic plates bonded onto an ***elastic shim***. We will focus on the bimorph structure here.

The bending deformation in a bimorph occurs because the two piezoelectric plates are bonded together, and each plate produces its own extension or contraction under the applied electric field. This effect is also employed in piezoelectric speakers. The induced voltage associated with the bending deformation of a bimorph has been used in accelerometers. This is a very popular and widely used structure mainly because it is easily fabricated (the two ceramic plates are just bonded with an appropriate resin), and the devices readily produce a large displacement. The drawbacks of this design include a low response speed (1 kHz) and low generative force because of the bending mode. A metallic sheet shim is occasionally used between the two piezoceramic plates to increase the reliability of the bimorph structure, as illustrated in Figure 3.15. When this type of shim is used, the structure will maintain its integrity even if the ceramic fractures. The bimorph is also generally tapered to increase the response frequency while maintaining optimum tip displacement. Anisotropic elastic shims, made from such materials as

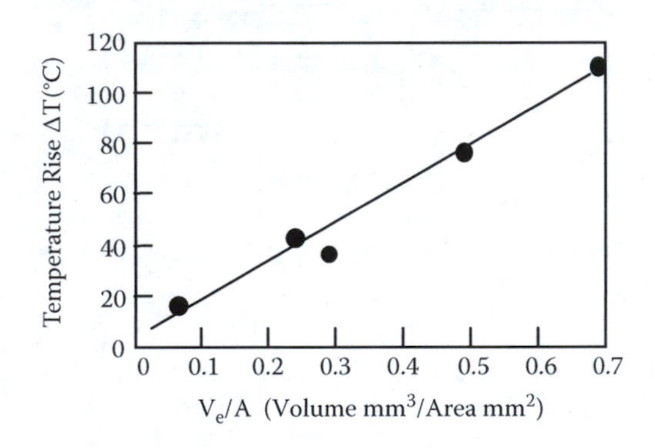

FIGURE 3.14 The temperature rise, ΔT, monitored in actuators driven at 3 kV/mm and 300 Hz, plotted as a function of the quantity, V_e/A, where V_e is the effective actuator volume (corresponding to the electroded portion of the device) and A is its surface area. (From J. Zheng, S. Takahashi, S. Yoshikawa, K. Uchino, and J. W. C. de Vries: *J. Am. Ceram. Soc.* **79**, 3193 [1996]. With permission.)

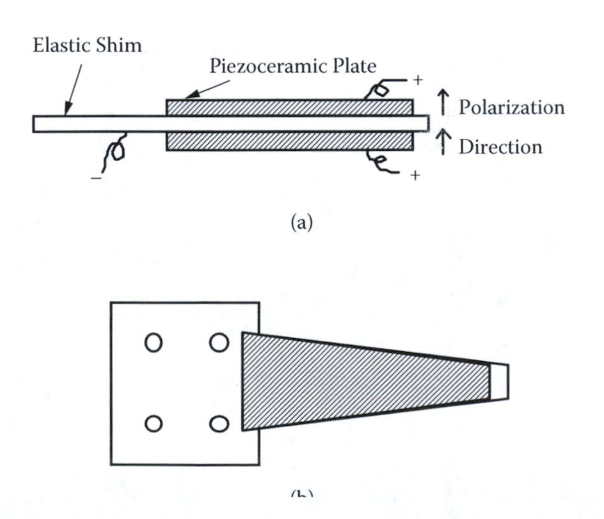

FIGURE 3.15 The basic structure of a piezoelectric bimorph: (a) side view of the device and (b) top view.

oriented carbon-fiber-reinforced plastics, have been used to enhance the displacement magnification rate by a factor of 1.5 as compared to the displacement of a similar device with an isotropic shim.[14]

Many studies have been conducted on these devices, which have produced equations describing the tip displacement and the resonance frequency. Two shimless bimorph designs are illustrated in Figure 3.16. Two poled piezoceramic plates of equal thickness and length are bonded together with either their polarization directions opposing each other or parallel to each other. When the devices are operated under an applied voltage, *V*, with one end clamped (the *cantilever condition*), the tip displacement, δ, is given by

$$\delta = (3/2)\, d_{31}\, (l^2/t^2)\, V \tag{3.3a}$$

$$\delta = 3\, d_{31}\, (l^2/t^2)\, V \tag{3.3b}$$

where d_{31} is the piezoelectric strain coefficient of the ceramic, t is the combined thickness of the two ceramic plates, and l is the length of the bimorph.

Equation 3.3a applies to the antiparallel polarization condition and equation 3.3b to the parallel polarization condition. Note that the difference between the two cases arises from the difference in electrode gap. The separation between the electrodes is equal to the combined thickness of the two plates for the antiparallel polarization case (Figure 3.16a) and half that thickness for the parallel polarization case (Figure 3.16b). The fundamental resonance frequency in both cases is determined by the combined thickness of the two plates, t, according to the following equation:

$$f_o = 0.161 \left[\frac{t}{l^2 \sqrt{\rho\, s_{11}^E}} \right] \tag{3.4}$$

where ρ is the mass density of the ceramic and s_{11}^E is its elastic compliance.[15]

Example Problem 3.1

Using a PZT-based ceramic with a piezoelectric strain coefficient of $d_{31} = (-300)$ pC/N, design a shimless bimorph with a total length of 30 mm (where 5 mm is used for cantilever clamping), which can produce a tip displacement of 40 μm under an applied voltage of 20 V. Calculate the response speed of this bimorph. The mass density of the ceramic is $\rho = 7.9$ g/cm^3, and its elastic compliance is $s_{11}^E = 16 \times 10^{-12}$ m^2/N.

Solution

When the device is to be operated under low voltages, the antiparallel polarization type of device pictured in Figure 3.16a is preferred over the parallel polarization type pictured in Figure 3.16b because it produces a larger displacement under these conditions. Substituting the length of $l = 25$ mm into equation 3.3b, we obtain the combined piezoelectric plate thickness, t:

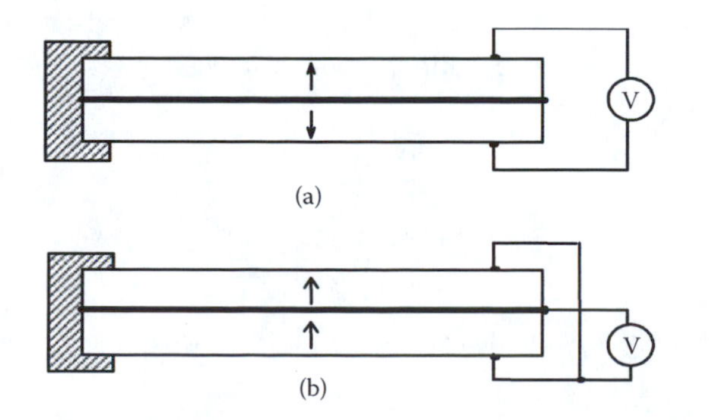

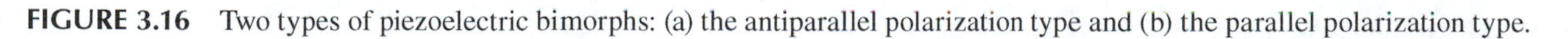

FIGURE 3.16 Two types of piezoelectric bimorphs: (a) the antiparallel polarization type and (b) the parallel polarization type.

$$t = l\sqrt{3(d_{31}V/\delta)}$$

$$= 25 \times 10^{-3}\,(m)\sqrt{3[300 \times 10^{-12}(C/N)\ 20\,(V)/40 \times 10^{-6}\,(m)]} \quad \text{(P3.1.1)}$$

$$\rightarrow t = 530\ \mu\text{m}.$$

The ceramic is cut into plates 265 μm in thickness, 30 mm in length, and 4–6 mm in width. The plates are electroded and poled and then bonded together in pairs. The width of the bimorph is usually chosen such that $w/l < 1/5$ to optimize the magnitude of the bending displacement.

The response time is estimated by the resonance period. We can determine the fundamental resonance frequency of the structure from the following equation:

$$f_o = 0.161\left[\frac{t}{l^2\sqrt{\rho s_{11}^E}}\right] \quad \text{(P3.1.2)}$$

$$= 0.161\left[\frac{530 \times 10^{-6}\,(m)}{[25 \times 10^{-3}\,(m)]^2\sqrt{7.9 \times 10^3\,(kg/m^3)16 \times 10^{-12}\,(m^2/N)}}\right] = 378\,(Hz)$$

$$\rightarrow\rightarrow \textit{Response Time} \approx \frac{1}{f_o} = \frac{1}{378\,(s^{-1})} = 2.6\,(\text{ms})$$

Example Problem 3.2

A unimorph bending actuator can be fabricated by bonding a piezoceramic plate to a metallic shim.[16] The tip deflection, δ, of the unimorph supported in a cantilever configuration is given by

$$\delta = \frac{d_{31}\,E\,l^2\,Y_c\,t_c}{(Y_m[t_o^2 - (t_o - t_m)^2] + Y_c[(t_o + t_c)^2 - t_o^2])} \quad \text{(P3.2.1)}$$

Here E is the electric field applied to the piezoelectric ceramic; d_{31}, the piezoelectric strain coefficient; l, the length of the unimorph; Y, Young's modulus for the ceramic or the metal; and t, the thickness of each material. The subscripts "c" and "m" denote the ceramic and the metal, respectively. The quantity t_o is the distance between the *strain-free neutral plane* and the bonding surface and is defined as follows:

$$t_o = \frac{t_c\,t_m^2\,(3t_c + 4t_m)Y_m + t_c^4\,Y_c}{6\,t_c\,t_m\,(t_c + t_m)Y_m} \quad \text{(P3.2.2)}$$

Assuming $Y_c = Y_m$, calculate the optimum (t_m/t_c) ratio that will maximize the deflection, δ, under the following conditions:

(a) A fixed ceramic thickness, t_c
(b) A fixed total thickness, $t_c + t_m$

Solution

Setting $Y_c = Y_m$, the equations (P3.2.1) and (P.3.2.2) become

$$\delta = \frac{d_{31} E l^2 t_c}{([t_o^2 - (t_o - t_m)^2] + [(t_o + t_c)^2 - t_o^2])} \tag{P3.2.3}$$

$$t_o = \frac{t_c t_m^2 (3t_c + 4t_m) + t_c^4}{6 t_c t_m (t_c + t_m)} \tag{P3.2.4}$$

Substituting t_o as it is expressed in equation (3.2.4) into equation (P3.2.3) yields

$$\delta = \frac{d_{31} E l^2 3 t_m t_c}{(t_m + t_c)^3} \tag{P3.2.5}$$

(a) The function $f(t_m) = (t_m t_c)/(t_m + t_c)^3$ must be maximized for a fixed ceramic thickness t_c.

$$\frac{d f(t_m)}{d t_m} = \frac{(t_c - 2t_m) t_c}{(t_m + t_c)^4} = 0 \tag{P3.2.6a}$$

Thus, the metal plate thickness should be $t_m = t_c/2$ and $t_o = t_c/2$. (b) In the latter case for a fixed total thickness, $t_{tot} = t_c + t_m$., equation (P3.2.6a) becomes

$$\frac{d f(t_m)}{d t_m} = \frac{(t_{tot} - 2t_m)}{t_{tot}^3} = 0 \tag{P3.2.6b}$$

Thus, it is determined that both the metal and ceramic plate thickness should be the same: $t_m = t_c = t_{tot}/2$ and $t_o = t_{tot}/3$.

The generative displacement of a bimorph is decreased when it is supported at both ends, as shown in Figure 3.17a. When the special shim design pictured in Figure 3.17b is used, however, the center displacement of the bimorph is enhanced by a factor of four as compared to the displacement of a conventionally supported device.[17]

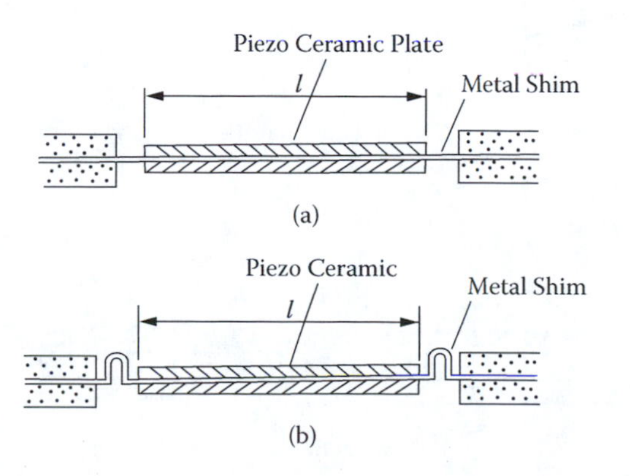

FIGURE 3.17 Double support mechanisms for a bimorph: (a) conventional design and (b) special shim design. (From C. Tanuma, Y. Suda, S. Yoshida, and K. Yokoyama: *Jpn. J. Appl. Phys.* **22**, Suppl. 22-2, 154 [1983]. With permission.)

The bimorph displacement inevitably includes some rotational motion. A special mechanism generally must be employed to eliminate this rotational component of the induced motion. Several such mechanisms are pictured in Figure 3.18.

A twin structure developed by SONY is shown in Figure 3.18a.[18] It is employed for video tracking control. The design incorporates a flexible head support installed at the tips of two parallel bimorphs. The complex bimorph design pictured in Figure 3.18b has been proposed by Ampex. It has electrodes on the top and the bottom of the device that are independently addressed such that any deviation from perfectly parallel motion is compensated for and effectively eliminated. The proposed design also includes a sensor electrode that detects the voltage generated in proportion to the magnitude of the bending displacement. The ring-shaped bimorph design pictured in Figure 3.18c has been developed by Matsushita Electric for use in a video tracking control system. The design is compact and provides a large displacement normal to the ring. It also remains firmly supported in operation because of the dual clamp configuration incorporated in the design.

The "Thunder" actuator developed by Face International is essentially a unimorph, but has a curved shape, and a tensile stress is maintained on the piezoceramic plate.[19]

This actuator is very reliable when operated under both high-field and high-stress conditions. The device is fabricated at high temperatures, and its curved shape is due to the thermal expansion mismatch between the bonded PZT and metal plates.

3.4 MOONIE/CYMBAL

A composite actuator structure called the ***moonie*** has been developed to amplify the pressure sensitivity and the small displacements induced in a piezoelectric ceramic.[20] The moonie has characteristics intermediate between the conventional multilayer and the bimorph actuators; it exhibits an order-of-magnitude larger displacement (100 μm) than the multilayer and a much larger generative force (10 kgf) with a quicker response (100 μs) than the bimorph. This device consists of a thin multilayer ceramic element and two metal plates with a narrow moon-shaped cavity bonded together, as pictured in Figure 3.19a. A moonie with dimensions 5×5×2.5 mm can generate a 20 μm displacement under an applied voltage of 60 V, which is eight times as large as the generative displacement of a multilayer of similar dimensions.[21] A displacement twice that of the moonie can be obtained with the ***cymbal*** design, illustrated in Figure 3.19b. The generative displacement of this device is quite uniform, showing negligible variation for points extending out from the center of the end cap.[21] Another advantage the cymbal has over the moonie is its relatively simple fabrication. The end caps for this device are made in a single-step punching process that is both more simple and more reproducible than the process involved in making the end caps for the moonie structure.

A "ring-morph" has been developed that is made with a metal shim and two PZT rings rather than PZT disks.[22] The bending displacement is amplified by more than 30% when an appropriate inner ring diameter is chosen.

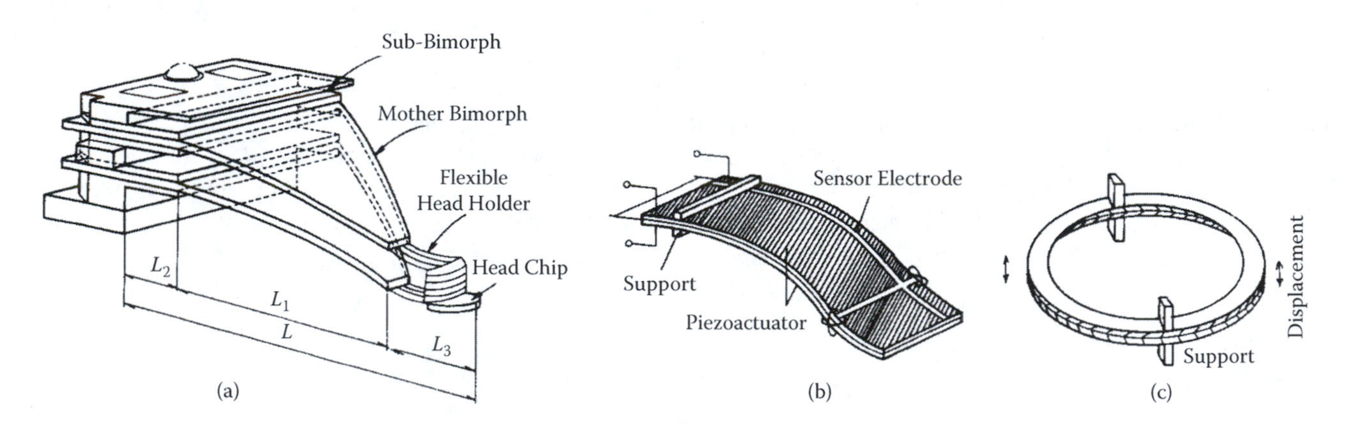

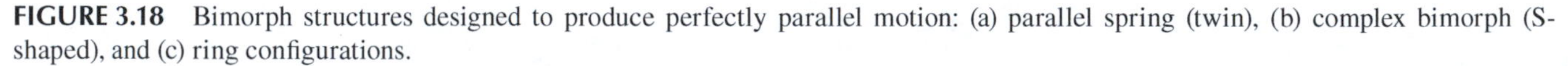

FIGURE 3.18 Bimorph structures designed to produce perfectly parallel motion: (a) parallel spring (twin), (b) complex bimorph (S-shaped), and (c) ring configurations.

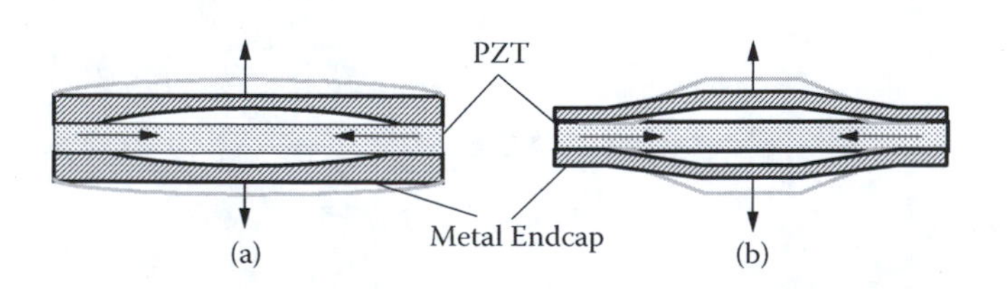

FIGURE 3.19 Flextensional structures: (a) the moonie (from H. Goto, K. Imanaka, and K. Uchino: *Ultrasonic Techno* **5**, 48 [1992]. With permission) and (b) the cymbal (from A. Dogan: Ph. D. Thesis, Penn State University [1994]. With permission).

3.5 DISPLACEMENT AMPLIFICATION MECHANISMS

In addition to the flextensional structures, such as bimorphs and cymbals, there are structures that make use of several displacement amplification mechanisms.

A ***monolithic hinge structure*** is made from a monolithic ceramic body by cutting indented regions in the monolith, as shown in Figure 3.20a. This effectively creates a lever mechanism that may function to either amplify or reduce the displacement. It was initially designed to reduce the displacement (including backlash) produced by a stepper motor and thus increase its positioning accuracy. Recently, however, monolithic hinge levers have been combined with piezoelectric actuators to amplify their displacement.

If the indented region of the hinge can be made sufficiently thin to promote optimum bending while maintaining extensional rigidity (ideal case), a mechanical amplification factor for the lever mechanism close to the apparent geometric lever length ratio is expected. The actual amplification, however, is generally less than this ideal value. If we consider the dynamic response of the device, a somewhat larger hinge thickness (identified as 2b in Figure 3.20a) producing an actual amplification of approximately half the apparent geometric ratio is found to be optimum for achieving maximum generative force and response speed. The characteristic response (that is, in terms of the displacement, generative force, and response speed) of an actuator incorporating a hinge lever mechanism is generally intermediate between that of the multilayer and bimorph devices.

When a piezoelectric actuator is to be utilized in a large mechanical control system, the ***oil pressure displacement amplification mechanism*** shown in Figure 3.21 is often implemented. It is a mechanism especially well suited for oil pressure servo systems. The output displacement Δl_2 is effectively amplified by a 1:2 area ratio according to

$$\Delta l_2 = (d_1/d_2)^2\, \Delta l_1 \tag{3.5}$$

Although the response is slow because of the viscosity of the oil, the overall loss associated with the amplified displacement is very small.

An ***inchworm*** is a linear motor that advances in small steps over time. An example of an inchworm device is the one designed by Burleigh Instruments, pictured in Figure 3.22.[23] It comprises three piezoelectric tubes. Two of the tubes (labeled

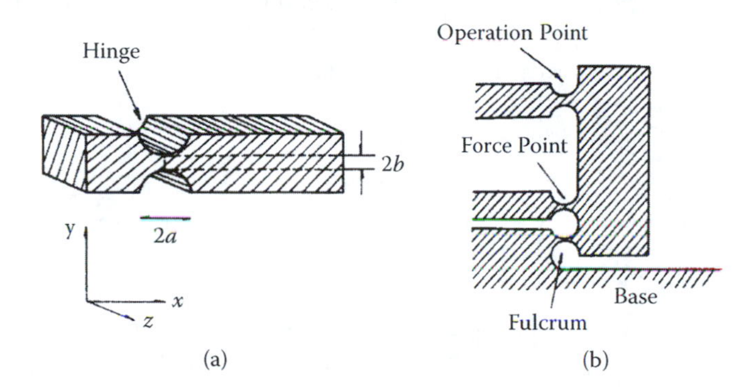

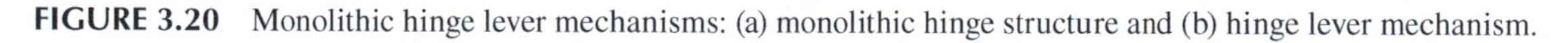

FIGURE 3.20 Monolithic hinge lever mechanisms: (a) monolithic hinge structure and (b) hinge lever mechanism.

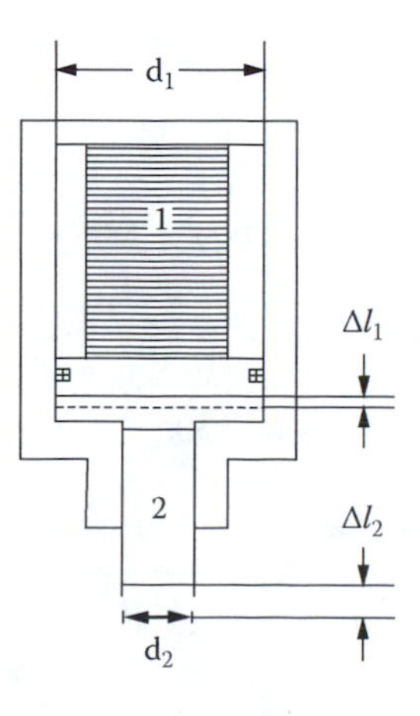

FIGURE 3.21 An oil pressure displacement amplification mechanism.

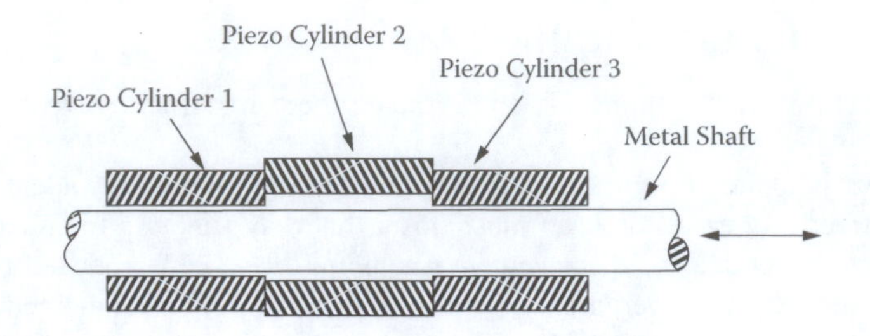

FIGURE 3.22 An inchworm structure. (From Burleigh Instruments Inc., East Rochester, NY, Product Catalog. With permission.)

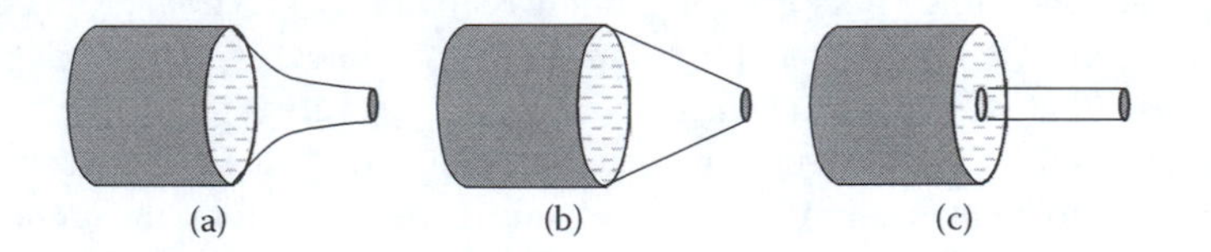

FIGURE 3.23 Three example configurations of the horn: (a) Exponential taper: highest efficiency, but costly in manufacturing; (b) straight taper: cheaper in manufacturing, with a reasonable efficiency; (c) step type: cheapest, but some mechanical energy is bounced back.

1 and 3 in the figure) function to either clamp or release the metal shaft, whereas the remaining tube (labeled 2 in the figure) actually moves the inchworm along the shaft. When three separate drive voltages of differing phases are applied to the three tubes, the inchworm can move forward and backward along the metal shaft. Although this motor is rather slow (0.2 mm/s), the high resolution per step (1 nm) is a very attractive feature of the device.

So far we have discussed the displacement amplification mechanism under an off-resonance mode. For a resonance mode, the most popular mechanism is a ***horn***. Figure 3.23 illustrates some horn configurations. The axial displacement is increased in inverse proportion to the radius of the vibrator. These three samples exhibit the following features:

1. Exponential taper: highest efficiency, but costly in manufacturing
2. Straight taper: cheaper in manufacturing, with a reasonable efficiency
3. Step type: cheapest, but some mechanical energy is bounced back

3.6 ULTRASONIC MOTOR

3.6.1 Classification of Ultrasonic Motors

From a customer's point of view, there are rotary and linear-type motors. If we categorize them according to the vibrator shape, there are rod, π-shaped, ring (square), and cylinder types. Two categories are being investigated for ultrasonic motors from a vibration characteristic viewpoint: a standing-wave type and a propagating-wave type. The reader should now review the wave formulas. The standing wave is expressed by

$$u_s(x,t) = A \cos kx \cos \omega t, \tag{3.6}$$

whereas the propagating wave is expressed as

$$u_p(x,t) = A \cos (kx - \omega t). \tag{3.7}$$

Using a trigonometric relation, equation (3.7) can be transformed as

$$u_p(x,t) = A \cos kx \cos \omega t + A \cos (kx - \pi/2) \,.\, \cos (\omega t - \pi/2). \tag{3.8}$$

This leads to an important result: ***a propagating wave can be generated by superposing two standing waves with a phase difference of 90° between them, both in time and in space***. This principle is necessary to generate a propagating wave on a limited volume/size substance because only standing waves can be excited stably in a solid medium of finite size.

The standing-wave type is sometimes referred to as a vibratory-coupler type or a "woodpecker" type, in which a vibratory piece is connected to a piezoelectric driver and the tip portion generates a flat, elliptical movement. Figure 3.24 shows a simple

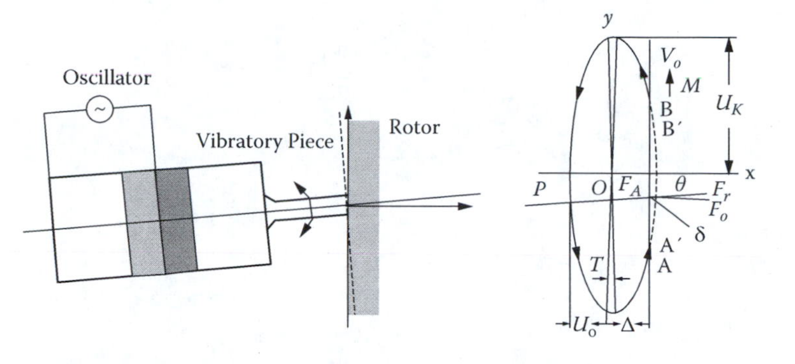

FIGURE 3.24 (a) Vibratory coupler type motor and (b) its tip locus.

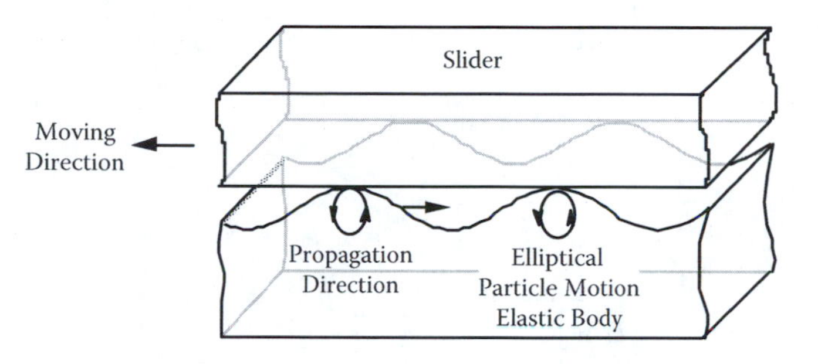

FIGURE 3.25 Principle of the propagating-wave-type motor.

model proposed by T. Sashida.[24] A vibratory piece is attached to a rotor or a slider with a slight cant angle. When a vibration is excited at the piezoelectric vibrator, the vibratory piece generates bending because of restriction by the rotor so that the tip moves along the rotor face between A → B and freely between B → A. If the vibratory piece and the piezo vibrator are tuned properly, they form a resonating structure, which is an elliptical locus. Therefore, only the duration A → B provides a unidirectional force to the rotor through friction and, therefore, an intermittent rotational torque or thrust. However, because of the inertia of the rotor, the rotation speed ripple is not observed to be large. The standing-wave type, in general, is low in cost (one vibration source) and has high efficiency (up to 98% theoretically), but lacks control in both the clockwise and counterclockwise directions.

By comparison, the propagating-wave type (a surface wave or "surfing" type) combines two standing waves with a 90° phase difference both in time and in space. The principle is shown in Figure 3.25. A surface particle of the elastic body draws an elliptical locus because of the coupling of longitudinal and transverse waves. This type requires, in general, two vibration sources to generate one propagating wave, leading to low efficiency (not more than 50%), but it is controllable in both rotational directions.

3.6.2 Standing Wave Motors

T. Sashida developed a rotary-type motor similar to the fundamental structure.[24] Four vibratory pieces were installed on the edge face of a cylindrical vibrator and pressed onto the rotor. This is one of the prototypes that triggered the present active development of ultrasonic motors. A rotation speed of 1500 rpm, torque of 0.08 N.m, and an output of 12 W (efficiency 40%) are obtained under an input of 30 W at 35 kHz. This type of ultrasonic motor can provide a speed much higher than the inchworm types because of its high operating frequency and amplified vibration displacement at the resonance frequency.

Hitachi Maxel significantly improved the torque and efficiency by replacing Sashida's vibratory pieces with a torsional coupler (Figure 3.26) and by increasing the pressing force with a bolt.[25] The torsion coupler looks like an old-fashioned TV channel knob, consisting of two legs that transform the longitudinal vibration generated by the Langevin vibrator to a bending mode of the knob disk and a vibratory extruder. Note that this extruder is aligned with a certain cant angle to the legs that transforms the bending to a torsional vibration. This transverse moment coupled with the bending up-down motion leads to an elliptical rotation on the tip portion, as illustrated in Figure 3.26b. The optimum pressing force to get the maximum thrust is obtained when the ellipse locus is deformed roughly by half. A motor 30 mm×60 mm in size and with a 20–30° in cant angle between leg and vibratory piece can generate torques as high as 1.3 N.m with an efficiency of 80%. However, this type provides

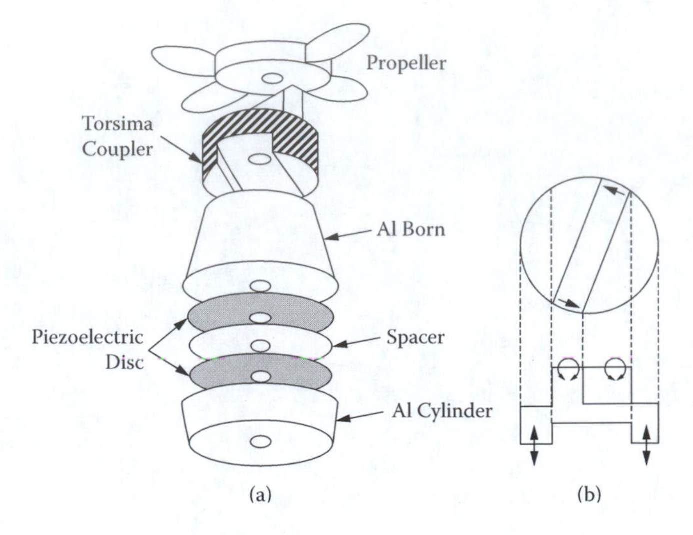

FIGURE 3.26 (a) Torsion coupler ultrasonic motor and (b) the motion of the torsional coupler.

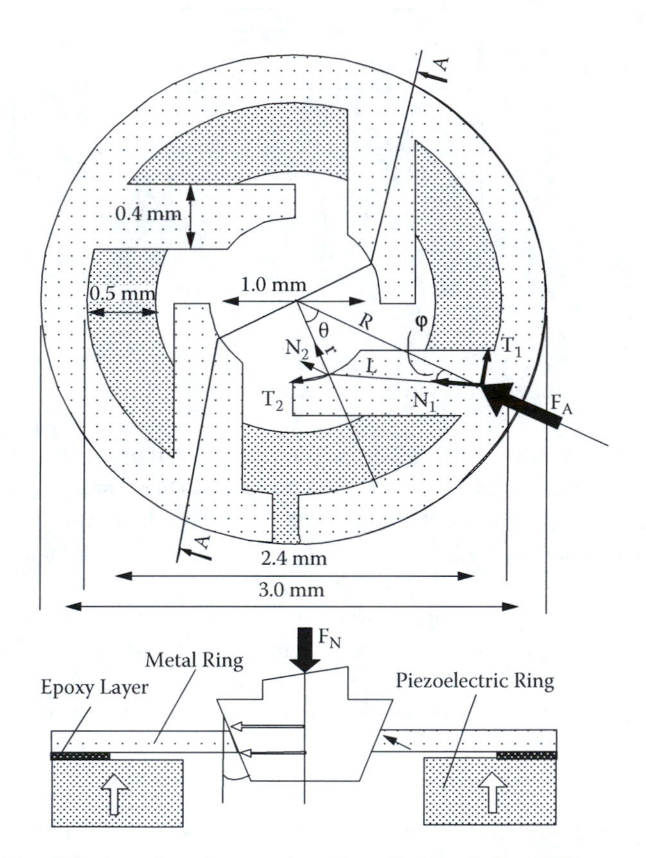

FIGURE 3.27 "Windmill" motor with a disk-shaped torsion coupler. (From B. Koc, P. Bouchilloux, and K. Uchino: *IEEE Trans. UFFC*, **47**, 836 (2000). With permission.)

only unidirectional rotation. Note that even though the drive of the motor is intermittent, the output rotation becomes very smooth because of the inertia of the rotor.

A compact ultrasonic rotary motor, as tiny as 3 mm in diameter, has been developed at the Pennsylvania State University. As shown in Figure 3.27, the stator consists of a piezoelectric ring and a metal ring with windmill-shaped arms bonded together so as to generate a coupling of the radial and arm-bending vibrations.[26] Because the number of components is reduced and the fabrication process is much simplified, the fabrication price is decreased remarkably, and a disposable design becomes feasible. When driven at 160 kHz, a maximum revolution of 600 rpm and a maximum torque of 1 mN.m were obtained for a motor with a diameter of 11 mm.

A "windmill" motor design with basically a flat and wide configuration was followed by a thin and long configuration—a metal hollow tube motor. It is composed of a metal tube and two rectangular PZT plates bonded on it (see Figure 3.28).[27,28] When one of the PZT plates is driven by a sinusoidal voltage, the metal tube exhibits a "wobbling" motion, similar to a hula-hoop or Chinese dish spinning.

A no-load speed of 1,800 rpm and output torque of more than 1 mNm were obtained at 80 V for both directions for a motor 2.4 mm ϕ in diameter and 12 mm in length. This significantly high torque was obtained because of the dual stator configuration and the high pressing force between the stator and rotors made of metal.

NEC-Tokin developed a piezoelectric ceramic cylinder for a torsional vibrator.[29] Using an interdigital-type electrode pattern printed with a 45° cant angle on the cylinder surface, torsion vibration was generated, which is applicable for a simple ultrasonic motor.

Ueha proposed a two-vibration-mode coupled-type motor (Figure 3.29); that is, a torsional Langevin vibrator was combined with three multilayer actuators to generate larger longitudinal and transverse surface displacements of the stator as well as to control their phase difference.[30] The phase change can change the rotation direction.

Uchino invented a π-shaped linear motor.[31] This linear motor is equipped with a multilayer piezoelectric actuator and fork-shaped metallic legs, as shown in Figure 3.30. Because there is a slight difference in the mechanical resonance frequency between the two legs, the phase difference between the bending vibrations of both legs can be controlled by changing the drive frequency. The walking slider moves in a way similar to a horse using its fore and hind legs when trotting. A test motor, $20 \times 20 \times 5$ mm^3 in dimension, exhibits a maximum speed of 20 cm/s and a maximum thrust of 0.2 kgf, with a maximum efficiency of 20%, when driven at 98 kHz at 6 V (actual power = 0.7 W). This motor has been employed in a precision X-Y stage.

Tomikawa's rectangular plate motor is also intriguing.[32] As shown in Figure 3.31, a combination of the two modes of vibration forms an elliptical displacement. The two modes chosen were the first longitudinal mode (L_1 mode) and the eighth bending

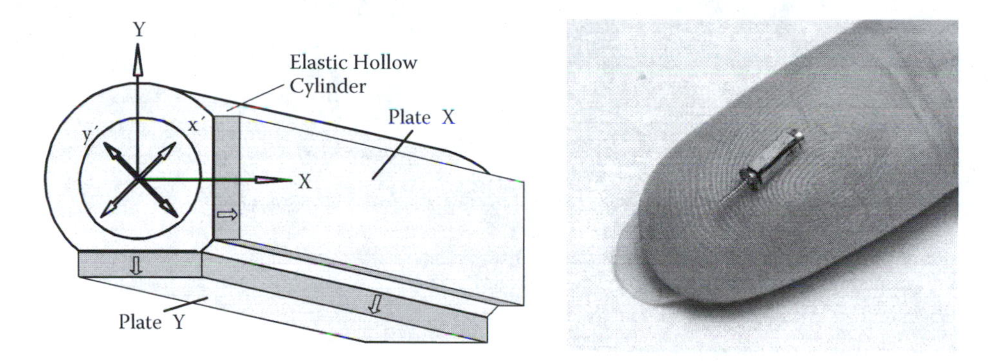

FIGURE 3.28 "Metal tube" motor using a metal tube and two rectangular PZT plates. (a) Schematic structure and (b) photo of the world's smallest motor (1.5 mmϕ). (From S. Cagatay, B. Koc, and K. Uchino, *IEEE Trans. UFFC*, 50(7), 782–786 [2003]. With permission.)

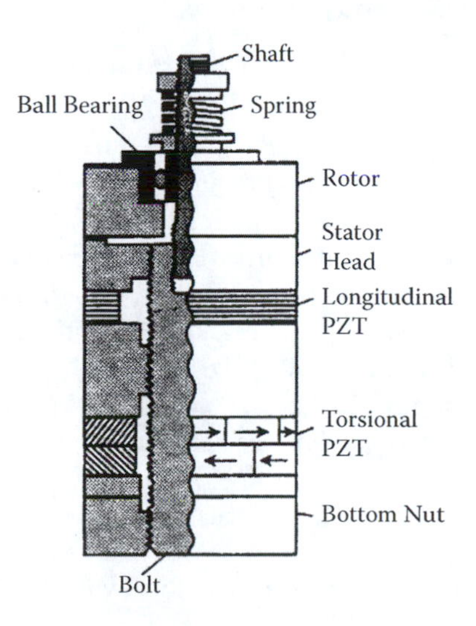

FIGURE 3.29 Two-vibration-mode coupled-type motor. (From K. Nakamura, M. Kurosawa, and S. Ueha: *Proc. Jpn. Acoustic Soc.*, No.1-1-18, p.917 [October 1993]. With permission.)

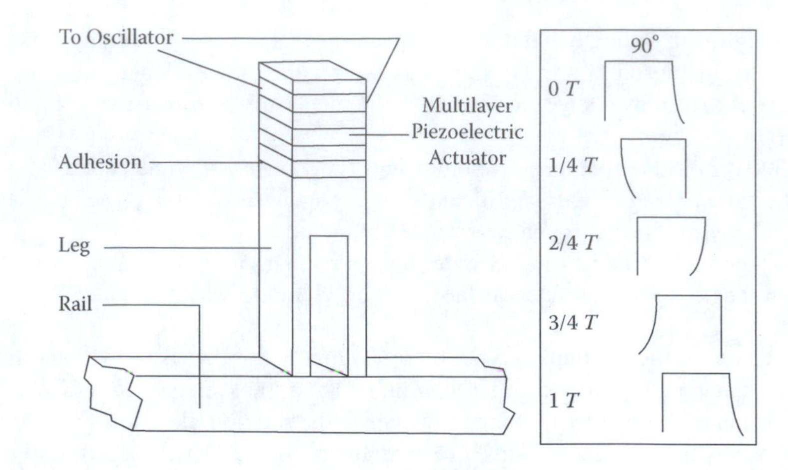

FIGURE 3.30 π-shaped linear ultrasonic motor: (a) construction and (b) walking principle. Note the 90° phase difference of two legs similar to that associated with human walking. (From K. Uchino, K. Kato, and M. Tohda: *Ferroelectrics* **87**, 331 [1988]. With permission.)

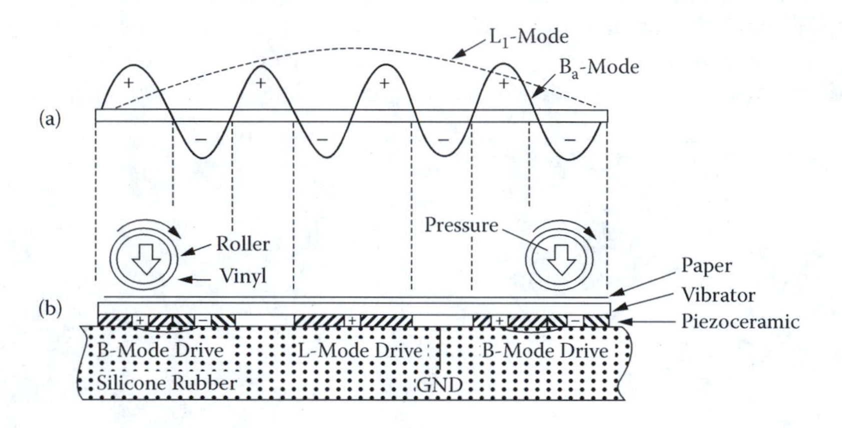

FIGURE 3.31 L_1 and B_8 double-mode vibrator motor. (From Y. Tomikawa, T. Nishituka, T. Ogasawara, and T. Takano: *Sensors Mater.*, **1**, 359 (1989). With permission.)

mode (B_8), whose resonance frequencies were almost the same. By applying voltages with a phase difference of 90° to the L-mode and B-mode drive electrodes, elliptical motion in the same direction can be obtained at both ends of this plate, leading to rotation of the rollers in contact with these points. Anticipated applications are paper or card senders.

3.6.3 Traveling Wave Motors

Figure 3.32 shows the famous Sashida motor,[33] which triggered the present ultrasonic motor boom. By means of the traveling elastic wave induced by a thin piezoelectric ring, a ring-type slider in contact with the "rippled" surface of the elastic body bonded onto the piezoelectric is driven in both directions by exchanging the sine and cosine voltage inputs. Another advantage is its thin design, which makes it suitable for installation in cameras as an automatic focusing device. Eighty percent of the exchange lenses in Canon's EOS camera series have already been replaced by the ultrasonic motor mechanism.

The PZT piezoelectric ring is divided into 16 positively and negatively poled regions and two asymmetric electrode gap regions so as to generate a ninth mode propagating wave at 44 kHz. A prototype was composed of a brass ring 60 mm in outer diameter, 45 mm in inner diameter, and 2.5 mm in thickness, bonded onto a PZT ceramic ring 0.5 mm in thickness with divided electrodes on the back. The rotor was made of polymer coated with hard rubber or polyurethane. Figure 3.33 shows Sashida's motor characteristics.

Canon utilized the "surfing" motor for a camera's automatic-focusing mechanism, installing the ring motor compactly in the lens frame. It is noteworthy that the stator elastic ring has many teeth that can magnify the transverse elliptical displacement and improve the speed. The lens position can be shifted back and forth with a screw mechanism. The advantages of this motor over a conventional electromagnetic motor are the following:

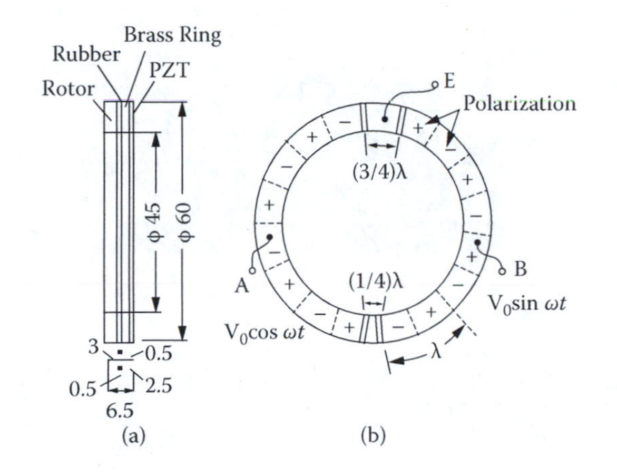

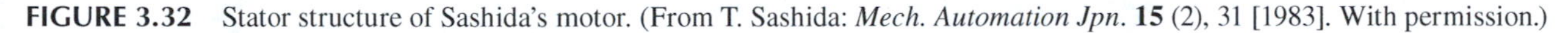
FIGURE 3.32 Stator structure of Sashida's motor. (From T. Sashida: *Mech. Automation Jpn.* **15** (2), 31 [1983]. With permission.)

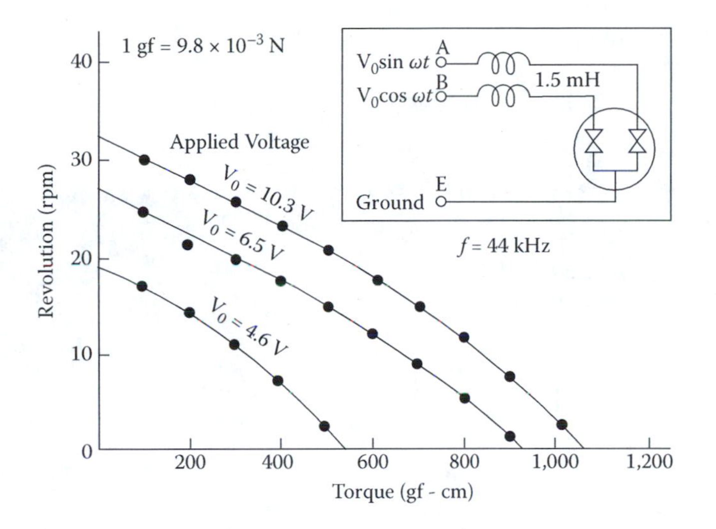

FIGURE 3.33 Motor characteristics of Sashida's motor. (From T. Sashida: *Mech. Automation Jpn.* **15** (2), 31 [1983]. With permission.)

1. Silent drive because of the ultrasonic frequency drive and no gear mechanism (i.e., more suitable for video cameras with microphones)
2. Thin motor design and no speed reduction mechanism such as gears, leading to space saving
3. Energy saving

A general problem encountered with these traveling-wave type motors is the support of the stator. In the case of a standing-wave motor, the nodal points or lines are generally supported; this affects the resonance vibration minimally. A traveling wave, however, does not have such steady nodal points or lines. Thus, special considerations are necessary. In Figure 3.32 the stator is basically supported very gently along the axial direction on felt so as not to suppress the bending vibration. It is important to note that the stop pins that latch onto the stator teeth only provide high rigidity against the rotation.

Matsushita Electric proposed a nodal line support method using a higher-order vibration mode.[34] A statorwide ring is supported at the nodal circular line, and "teeth" are arranged on the maximum amplitude circle to get larger revolution.

Seiko Instruments miniaturized the ultrasonic motor to dimensions as tiny as 10 mm in diameter using basically the same principle.[35] Figure 3.34 shows the construction of one of these small motors with a 10 mm diameter and a 4.5 mm thickness. A driving voltage of 3 V and a current of 60 mA produces 6,000 rev/min (no load) with a torque of 0.1 mN.m. AlliedSignal developed ultrasonic motors similar to Shinsei's, which are utilized as mechanical switches for launching missiles.[36]

Simple disk structures are preferable to unimorphs for motors because the bending action of the unimorph cannot generate sufficient mechanical power, and its electromechanical coupling factor is generally less than 10%.[37,38] Therefore, instead of the unimorph structure, a simple disk is generally used in these motors. Excitation of axial-asymmetric modes can produce a rotation of the inner and outer circumferences, resulting in "hula hoop" action.

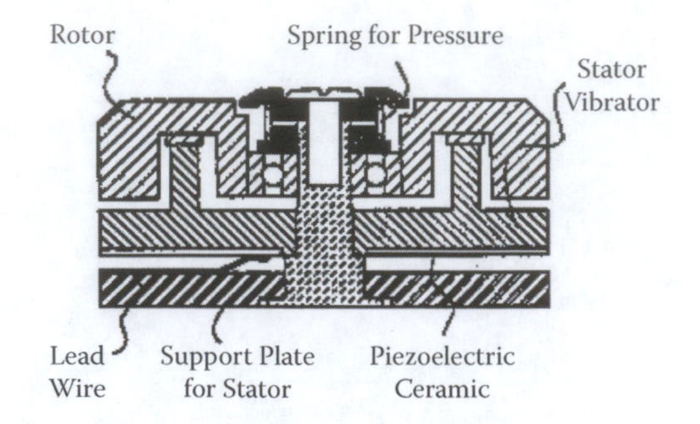

FIGURE 3.34 Construction of Seiko's motor. (From M. Kasuga, T. Satoh, N. Tsukada, T. Yamazaki, F. Ogawa, M. Suzuki, I. Horikoshi, and T. Itoh: *J. Soc. Precision Eng.* **57**, 63 [1991]. With permission.)

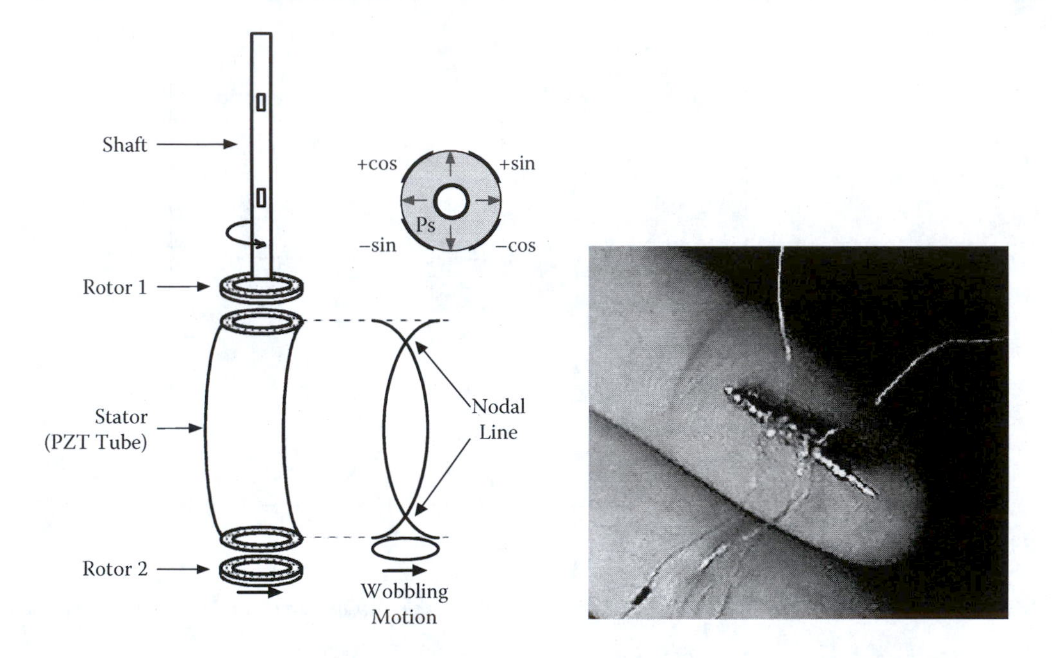

FIGURE 3.35 The PZT tube motor developed by researchers at Penn State and the Institute of Materials Research and Engineering in Singapore (from S. Dong, S. P. Lim, K. H. Lee, J. Zhang, L. C. Lim, and K. Uchino: Piezoelectric ultrasonic micromotors with 1.5 mm diameter, *IEEE UFFC Trans.*, **50**(4), 361–367 [2003]. With permission): (a) schematic illustration and (b) a photo.

An interesting design referred to as the *spinning plate* motor was proposed by NEC-Tokin.[39] Penn State and the Institute of Materials Research and Engineering (IMRE) miniaturized the size down to 1.5 mm in diameter.[40] Figure 3.35 shows its principle of operation and a photograph. A combination rotary/bending vibration is excited in a PZT rod when sine and cosine voltages are applied to the divided electrodes. The inner surface of a cuplike rotor is brought in contact with the "spinning" rod to produce the rotation.

In summary, the standing-wave type, in general, is low in cost (one vibration source) and has high efficiency (up to 98% theoretically), but lack of control in both the clockwise and counterclockwise directions is a problem. In comparison, the propagating-wave type combines two standing waves with a 90° phase difference both in time and space. This type requires, in general, two vibration sources to generate one propagating wave, leading to low efficiency (not more than 50%), but it is controllable in both rotational directions by exchanging the driving sine and cosine voltage sources.

3.7 UNDERWATER TRANSDUCER

3.7.1 Langevin Transducer

Langevin succeeded in transmitting ultrasonic pulses into the Mediterranean Sea, South of France in 1917. We can learn most of the practical development approaches from this original transducer design (Figure 3.36). First, 40 kHz was chosen for the

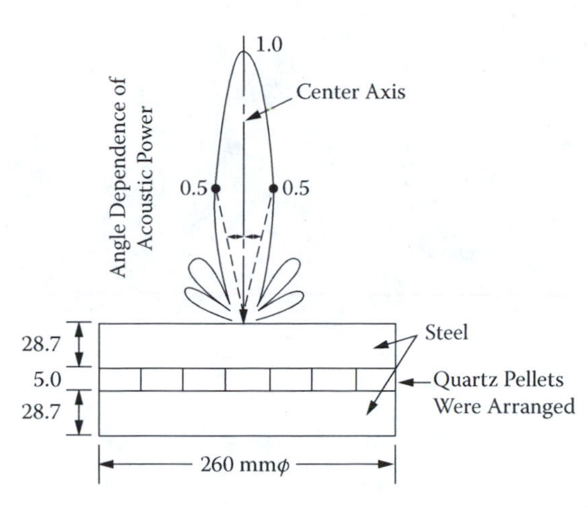

FIGURE 3.36 Original design of the Langevin underwater transducer and its acoustic power directivity.

sound wave frequency. Increasing the frequency (shorter wavelength) leads to a better monitoring resolution of the objective; however, it also leads to a rapid decrease in the reachable distance. Note that quartz single crystals were the only available piezoelectric material in the early 20th century. Because the sound velocity in quartz is about 5 km/s, 40 kHz corresponds to the wavelength of 12.5 cm in quartz. If we use mechanical resonance in the piezoelectric material, a 12.5/2 (6.25-cm-thick) quartz single crystal piece is required. However, in that period, it was not possible to produce such a large single crystal.

To overcome this dilemma, Langevin invented a new transducer construction: small quartz crystals arranged in a mosaic were sandwiched by two steel plates. Because the sound velocity in steel is in a similar range to quartz, taking 6.25 cm as the total thickness, he succeeded in setting the thickness resonance frequency around 40 kHz. This sandwich structure is called ***Langevin type***, which is popularly utilized nowadays. Note that quartz is located at the center, which corresponds to the nodal plane of the thickness vibration mode, where the maximum stress/strain (or the minimum displacement) is generated.

Further, to provide a sharp directivity for the sound wave, Langevin used a sound radiation surface with a diameter of 26 cm (more than double the wavelength). Because the half-maximum-power angle ϕ can be evaluated as

$$\phi = 30 \times (\lambda/2a) \text{ [degree]}, \tag{3.9}$$

where λ is the wavelength in the transmission medium (not in steel) and a is the radiation surface radius, if λ = 1,500 [m/s]/40 [kHz] = 3.75 cm and a = 13 cm, we obtain ϕ = 4.3° for this original design. He succeeded practically in detecting a U-boat 3,000 m away. Also, Langevin also observed many bubbles generated during his experiments, which seems to be due to the "cavitation" effect.

3.7.2 Acoustic Lens and Horn

A dolphin is one of the best animal models to explain ideal acoustic transducer systems. The dolphin and the whale do not have vocal chords, but they have a breathing hole on the head, as shown in Figure 3.37. By breathing strongly through this hole, a dolphin generates a sound similar to a human whistle. Although this breathing hole is a point sound source, due to additional physical structures, the sound beam can be focused rather sharply toward the front direction (Figure 3.37 bottom). The cranium shape seems to be a parabola antenna that reflects the radial propagating sound into a rather parallel frontward beam. Further, "melon" made of soft tissue such as paraffin (with a sound velocity lower than water) behaves as a convex acoustic lens, and focuses the sound beam more sharply. It is noteworthy that if we use a glass lens with a sound velocity higher than water, a convex acoustic lens diverges the sound beam (similar to an optical concave lens). This is the initial sound-transmitting process.

While receiving the returned sound signal reflected from some object, if the strong original transmitting signal returns directly to the receiving organ (ear), a serious blackout problem occurs. To overcome this problem, the dolphin has an acoustic impedance mismatch layer (air), which shuts out the noise like a double-glass window.

The dolphin's receiver is its bottom jaws, through which the sound is transferred to the ear. Note that the highly sensitive ear is isolated from the cranial bones to protect it from direct sound penetration. This situation is rather different from the human ear, which is connected directly to the bones (actually, we have a headset that uses direct bone transmission of sound for a hearing aid). Because the dolphin has completely separate right and left jaws, the two ears can detect the right and left sound signals independently, similar to a 3D stereo system.

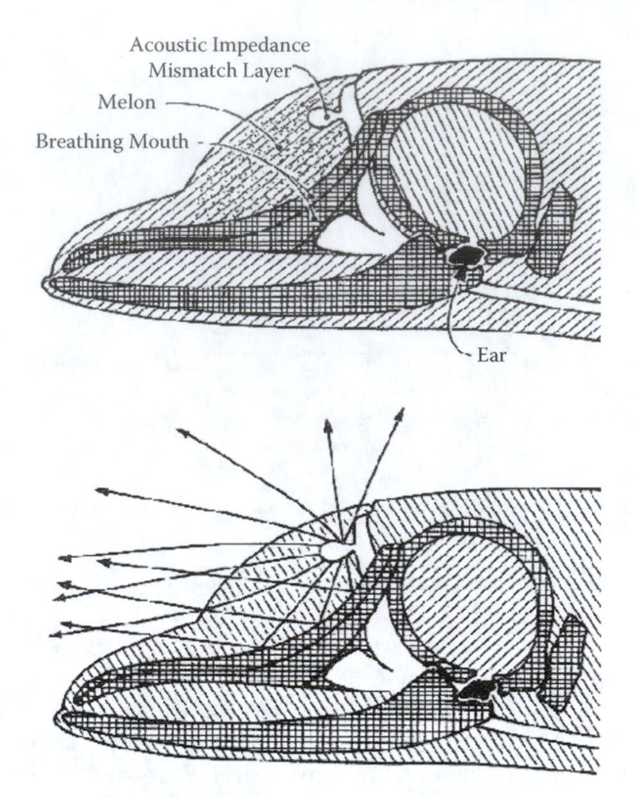

FIGURE 3.37 Dolphin head: mesh portion = bone, hatch portion = soft tissue, white portion = air, black portion = ear. Top figure exhibits the biological names, and the bottom illustrates the acoustic beam radiation situation.

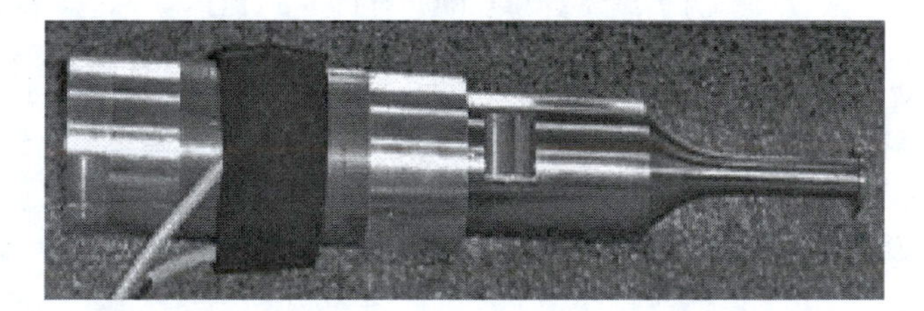

FIGURE 3.38 Langevin transducer with a horn.

The original steel portion of the Langevin transducer can be modified with a ***horn*** concept to increase the displacement amplitude. By tapering the steel tip portion as shown in Figure 3.38, we can significantly amplify the displacement level, which can be utilized for ultrasonic cutters and cavitation instruments.

When we use a concave radiation surface for a high-frequency wave (GHz) at the tip of the horn, as illustrated in Figure 3.39, we can focus the acoustic beam into a 0.1 mm range, which can be used in an acoustic microscope. There are two acoustic microscope types: reflection type and transmission type. This acoustic microscope is helpful for detecting a cancer portion with a harder elasticity than a normal portion in patient tissue.

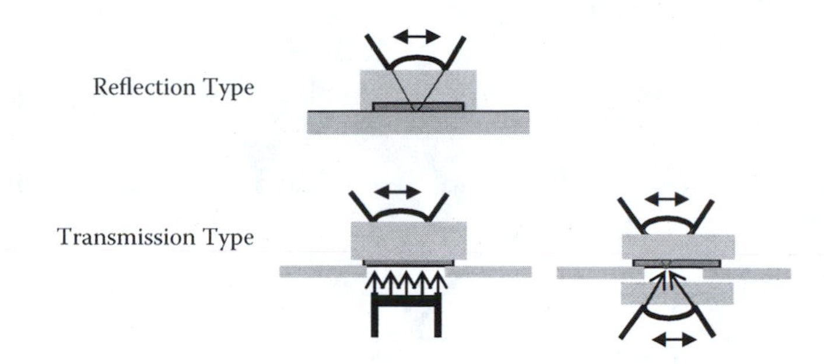

FIGURE 3.39 Classification of mechanical scan acoustic microscopes (SAM).

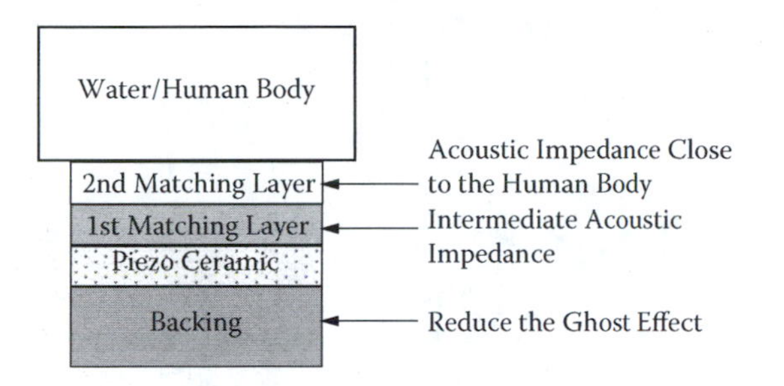

FIGURE 3.40 Acoustic transducer design for medical imaging applications.

3.7.3 Acoustic Impedance Matching

If we increase the driving frequency to 70 times the fish-finder frequency (50 kHz), that is, 3.5 MHz, we can increase the resolution by a factor of 70, and this can be used to detect the infant situation in the human body. However, the practical application is not so easy. Because the piezoelectric ceramic (such as PZT) has a very large acoustic impedance in comparison with the human body (similar to water), most of the high-power ultrasonic sound power is reflected at the interface, and the transmitted sound power is very low. To transmit the sound energy effectively, we need to suitably design the acoustic impedance-matching layer. When the acoustic impedances of PZT and water are Z_1 and Z_2, the recommended matching impedance of the matching layer is

$$\sqrt{(Z_1 Z_2)}\ .$$

As shown in Figure 3.40, we sometimes use multiple matching layers. Also, similar to a dolphin, the backing layer is used for reducing the ghost effect. Soft rubber is popularly used as a backing material.

The ATILA software has the capability of showing the transmitted voltage response (TVR) and the sound beam pattern, as shown in Figure 3.41.

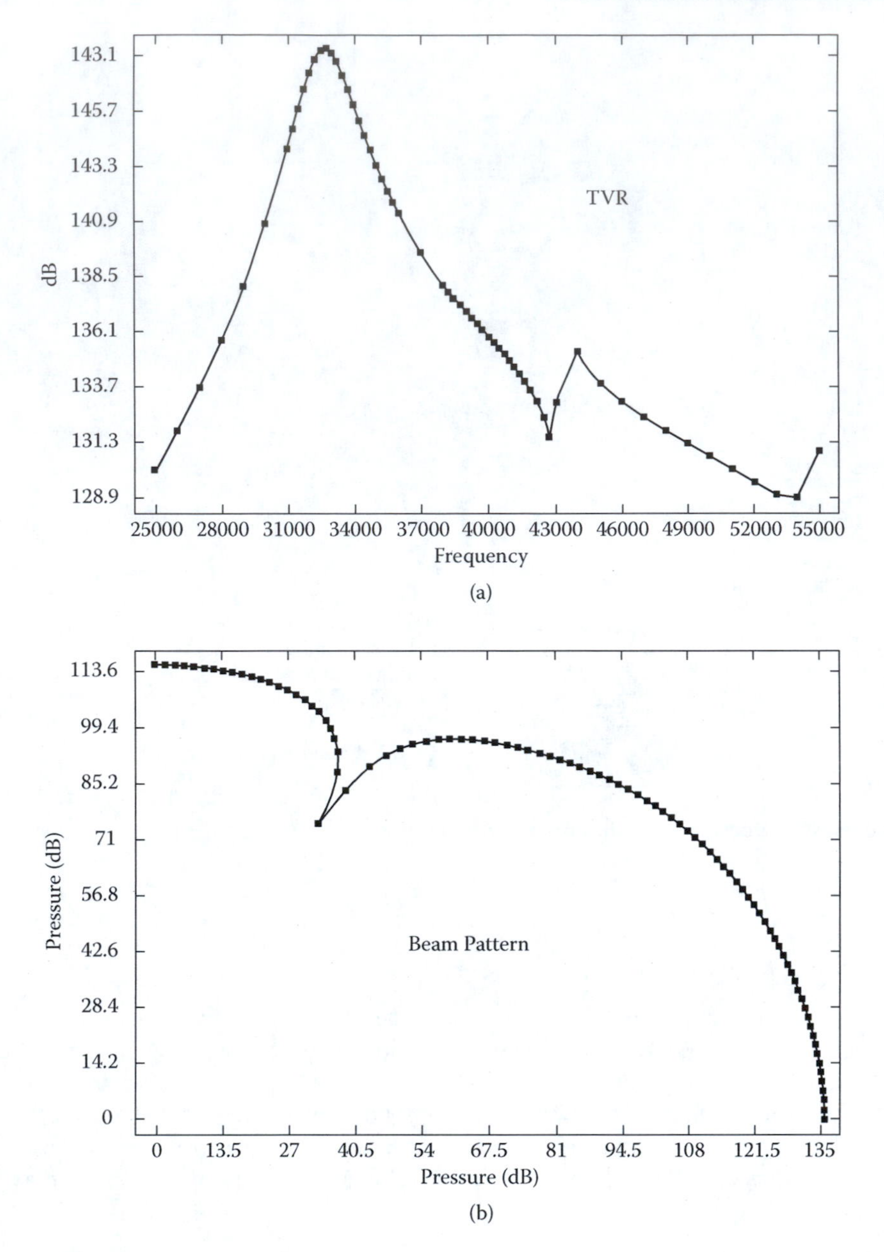

FIGURE 3.41 (a) Transmitted voltage response (TVR) as a function of frequency, and (b) the beam pattern as a function of the angle for a Tonpilz transducer under water.

REFERENCES

1. K. Uchino, S. Nomura, L. E. Cross, R. E. Newnham, and S. J. Jang: *J. Mater. Sci.* **16**, 569 (1981).
2. S. Takahashi, A. Ochi, M. Yonezawa, T. Yano, T. Hamatsuki, and I. Fukui: *Jpn. J. Appl. Phys.* **22**, Suppl. 22-2, 157 (1983).
3. A. Furuta and K. Uchino: *J. Am. Ceram. Soc.* **76**, 1615 (1993).
4. A. Furuta and K. Uchino: *Ferroelectrics* **160**, 277 (1994).
5. S. Takahashi: *Fabrication and Application of Piezoelectric Materials* (Ed. Shiosaki), chap. 14 Actuators, CMC Pub. (1984).
6. S. Takahashi: *Sensor Technology* **3**, No. 12, 31 (1983).
7. H. Aburatani, K. Uchino, A. Furuta, and Y. Fuda: *Proc. 9th Int'l Symp. Appl. Ferroelectrics* p. 750 (1995).
8. J. Ohashi, Y. Fuda, and T. Ohno: *Jpn. J. Appl. Phys.* **32**, 2412 (1993).
9. A. Banner and F. Moller: *Proc. 4th. Int. Conf. New Actuators*, AXON Tech. Consult. GmbH, p. 128 (1995).
10. K. Uchino and H. Aburatani: *Proc. 2nd Int. Conf. Intelligent Mater.*, p. 1248 (1994).
11. H. Aburatani and K. Uchino: *Am. Ceram. Soc. Annual Mtg. Proc.*, SXIX-37-96, Indianapolis, April (1996).
12. K. Nagata: *Proc. 49th Solid State Actuator Study Committee*, JTTAS, Japan (1995).
13. J. Zheng, S. Takahashi, S. Yoshikawa, K. Uchino, and J. W. C. de Vries: *J. Am. Ceram. Soc.* **79**, 3193 (1996).
14. T. Kitamura, Y. Kodera, K. Miyahara, and H. Tamura: *Jpn. J. Appl. Phys.* **20**, Suppl. 20-4, 97 (1981).
15. K. Nagai and T. Konno Edit.: *Electromechanical Vibrators and Their Applications*, Corona Pub., Tokyo, (1974).
16. K. Abe, K. Uchino, and S. Nomura: *Jpn. J. Appl. Phys.* **21**, L408 (1982).
17. C. Tanuma, Y. Suda, S. Yoshida, and K. Yokoyama: *Jpn. J. Appl. Phys.* **22**, Suppl. 22-2, 154 (1983).
18. Okamoto et al.: *Broadcast Technology* No. 7, p. 144 (1982).
19. www.faceco.com
20. H. Goto, K. Imanaka, and K. Uchino: *Ultrasonic Techno* **5**, 48 (1992).
21. A. Dogan: "Metal-Ceramic Composites: Cymbal." Ph. D. Thesis, Penn State University (1994).
22. S. Dong, X.-H. Du, P. Bouchilloux and K. Uchino, *J. Electroceramics* **8**, 155–161 (2002).
23. Burleigh Instruments Inc., East Rochester, NY, Catalog, 1980.
24. T. Sashida: *Oyo Butsuri* **51**, 713 (1982).
25. A. Kumada: *Jpn. J. Appl. Phys.*, **24**, Suppl. 24-2, 739 (1985).
26. B. Koc, P. Bouchilloux, and K. Uchino: *IEEE Trans. UFFC*, **47**, 836 (2000).
27. B. Koc, S. Cagatay, and K. Uchino, *IEEE Ultrasonic, Ferroelectric, Frequency Control Trans.*, **49**(4), 495–500 (2002).
28. S. Cagatay, B. Koc, and K. Uchino, *IEEE Trans. UFFC*, **50**(7), 782–786 (2003).
29. Y. Fuda and T. Yoshida: *Ferroelectrics*, **160**, 323 (1994).
30. K. Nakamura, M. Kurosawa, and S. Ueha: *Proc. Jpn. Acoustic Soc.*, No.1-1-18, p.917 (October 1993).
31. K. Uchino, K. Kato, and M. Tohda: *Ferroelectrics* **87**, 331 (1988).
32. Y. Tomikawa, T. Nishituka, T. Ogasawara, and T. Takano: *Sensors Mater.* **1**, 359 (1989).
33. T. Sashida: *Mech. Automation Jpn.* **15** (2), 31 (1983).
34. K. Ise: *J. Acoust. Soc. Jpn.* **43**, 184 (1987).
35. M. Kasuga, T. Satoh, N. Tsukada, T. Yamazaki, F. Ogawa, M. Suzuki, I. Horikoshi, and T. Itoh: *J. Soc. Precision Eng.* **57**, 63 (1991).
36. J. Cummings and D. Stutts: *Am. Ceram. Soc. Trans. Design for Manufacturability of Ceramic Components*, p. 147 (1994).
37. A. Kumada: *Ultrasonic Technology* **1** (2), 51 (1989).
38. Y. Tomikawa and T. Takano: *Nikkei Mechanical*, Suppl., p. 194 (1990).
39. T. Yoshida: *Proc. 2nd Memorial Symp. Solid Actuators of Japan: Ultra-Precise Positioning Techniques and Solid Actuators for Them*, p. 1 (1989).
40. S. Dong, S. P. Lim, K. H. Lee, J. Zhang, L. C. Lim, and K. Uchino: Piezoelectric ultrasonic micromotors with 1.5 mm diameter, *IEEE UFFC Trans.*, **50**(4), 361–367 (2003).

4 Drive/Control Techniques of Smart Transducers

There are three general methods for actuator drive/control that are most commonly employed: DC drive, pulse drive, and AC drive. These methods are typically used for displacement transducers, pulse drive motors and ultrasonic motors, respectively. Displacement transducers are usually controlled in closed-loop mode. Open-loop control can also be employed, but only when strain hysteresis is negligible and temperature fluctuation during operation is very small. Closed-loop control is a feedback method in which the electric-field-induced displacement of a ceramic actuator is monitored, deviation from the desired displacement is detected, and an electric signal proportional to this deviation is fed back to the ceramic actuator through an amplifier to effectively correct the deviation. Feedback is generally used for these devices to alleviate problems associated with the nonlinearity and hysteresis commonly encountered in piezoelectric materials. The pulse drive motor is typically operated in open-loop mode, but special care must be taken to suppress overshoot and mechanical ringing that can occur after the pulse voltage is applied. The AC voltage applied to ultrasonic motors is not very large, but the drive frequency must be precisely matched with the mechanical resonance frequency of the device for optimum performance. Heat generation, which is a potentially significant problem with this design, can be effectively minimized with the proper selection of operating parameters.

The dynamical drive of a piezoelectric device is a primary focus in this chapter, relating with the FEM calculation in terms of the modal analysis and frequency dependence of a piezo component.

4.1 CLASSIFICATION OF PIEZOELECTRIC ACTUATORS

Piezoelectric and electrostrictive actuators may be classified into two major categories, based on the type of drive voltage applied to the device and the nature of the strain induced by the voltage, as depicted in Figure 4.1. The categories are ***rigid displacement devices***, for which the strain is induced unidirectionally, aligned with the applied DC field, and ***resonant displacement devices***, for which an alternating strain is excited by an AC field at the mechanical resonance frequency (***ultrasonic motors***). The first category can be further divided into two general types: ***servo displacement transducers*** (***positioners***), which are controlled by a feedback system through a position detection signal, and ***pulse drive motors***, which are operated in a simple on/off switching mode. The drive/control techniques presented in this chapter will be discussed in terms of this classification scheme.

The response of the resonant displacement devices is not directly proportional to the applied voltage, but is dependent on the drive frequency. Although the positioning accuracy of this class of devices is not as high as that of the rigid displacement devices, ultrasonic motors are able to produce very rapid motion owing to their high-frequency operation. Servo displacement transducers, which are controlled by a feedback voltage superimposed on a DC bias, are used as positioners for optical and precision machinery systems. In contrast, a pulse drive motor generates only on/off strains, which are suitable for the impact elements of dot matrix or inkjet printers. An impact actuator can be created with a pulse-driven piezoelectric element. A simple flight actuator design is depicted schematically in Figure 4.2. The 2 mm steel ball can be launched as high as 20 mm by this device when a rapid 5 μm displacement is induced in the multilayer actuator.[1]

The material requirements for each class of devices will be different, and certain compositions will be better suited for particular applications. The servo displacement transducer suffers most from strain hysteresis and, therefore, a $Pb(Mg_{1/3}Nb_{2/3})O_3$-$PbTiO_3$ [PMN] electrostrictive material is preferred for this application. It should be noted that even when a feedback system is employed, the presence of a pronounced strain hysteresis general results in a much slower response speed. The pulse drive motor, for which a quick response rather than a small hysteresis is desired, requires a low-permittivity material. Soft $Pb(Zr,Ti)O_3$ [PZT] piezoelectrics are preferred over the high-permittivity PMN for this application. The ultrasonic motor, on the other hand, requires a very hard piezoelectric with a high mechanical quality factor, Q_m, to maximize the AC strain and to minimize heat generation. Note that the figure of merit for the resonant strain is characterized by $dElQ_m$ (d: the piezoelectric strain coefficient, E: the applied electric field, l: the sample length, and Q_m: the mechanical quality factor). Although hard lead zirconate titanate (PZT) materials have smaller d coefficients, they also have significantly larger Q_m values, thus providing the high resonant strains needed for these devices.

4.2 PULSE DRIVE

When a pulsed electric field is applied to a piezoelectric actuator, a mechanical vibration is excited, the characteristics of which depend on the pulse profile. Displacement overshoot and ringing are frequently observed. Quick and precise positioning is

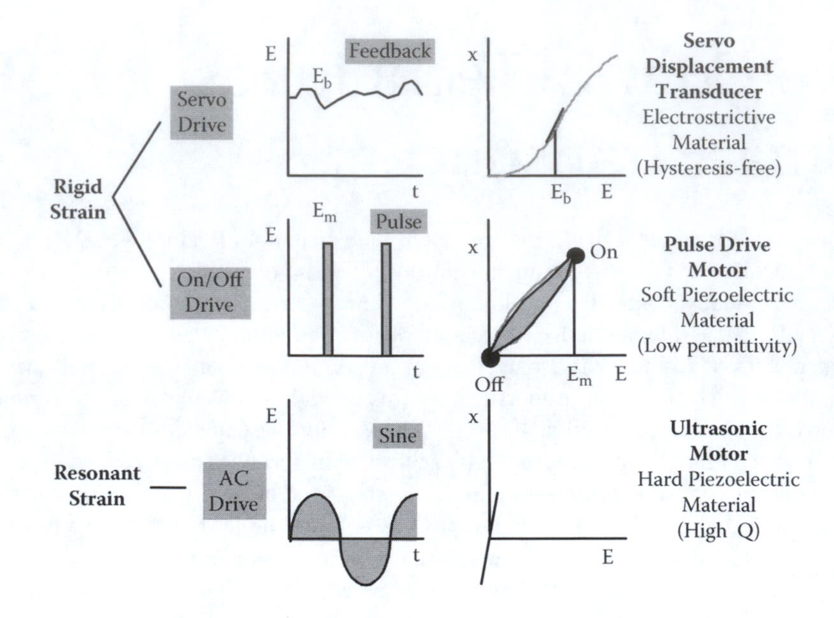

FIGURE 4.1 Classification of piezoelectric/electrostrictive actuators according to the type of drive voltage and the nature of the induced strain.

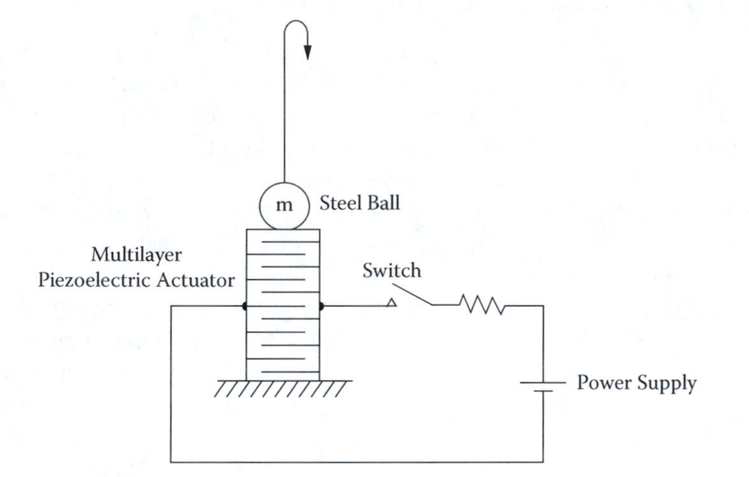

FIGURE 4.2 A simple flight actuator design.

difficult to achieve and, moreover, can lead to the destruction of the actuator because of the large tensile stress associated with overshoot. We will examine more closely the transient response of a piezoelectric device driven by a pulsed electric field in this section.

4.2.1 The Piezoelectric Equations

If the applied electric field, E, and the stress, X, are small, the strain, x, and the electric displacement, D, induced in a piezoelectric can be represented by the following equations:

$$x_i = s_{ij}^{E} X_j + d_{mi} E_m \tag{4.1}$$

$$D_m = d_{mi} X_i + \varepsilon_o K_{mk}^{X} E_k \tag{4.2}$$

where ($i, j = 1, 2, \ldots, 6$; $m, k = 1, 2, 3$). These are the *piezoelectric equations*. According to the tensor theory presented in chapter 2, we know that for the lowest symmetry trigonal crystal there are 21 independent s_{ij}^{E} coefficients, 18 d_{mi} coefficients, and 6 K_{mk}^{X} coefficients. We also recognized in that discussion that the number of independent coefficients decreases with increasing crystallographic symmetry. When considering polycrystalline ceramic specimens, the poling direction is typically designated as the Z axis. A poled ceramic is isotropic with respect to this Z axis and has a Curie group designation $C\infty_v$ (∞m). There are 10 nonzero matrix elements (s_{11}^E, s_{12}^E, s_{13}^E, s_{33}^E, s_{44}^E, d_{31}, d_{33}, d_{15}, K_{11}^X, and K_{33}^X) that apply in this case.

When considering the response of an electrostrictive material under DC bias (E_b) the strain is given by:

$$x_i = M_{mi}(E_b + E_m)^2 = M_{mi}E_b^2 + 2M_{mi}E_bE_m + \delta E_m^2 \quad (4.3)$$

If the alternative definition for d_{mi},

$$d_{mi} = 2M_{mi}E_b, \quad (4.4)$$

is substituted in equations (4.1) and (4.2), we can think in terms of an ***induced piezoelectric effect*** when analyzing the response of an electrostrictive material.

4.2.2 The Longitudinal Vibration Modes

4.2.2.1 The Longitudinal Vibration Mode Based on the Transverse Piezoelectric Effect

Let us consider a longitudinal mechanical vibration in the simple piezoelectric ceramic plate with thickness b, width w, and length l ($b \ll w \ll l$) based on the transverse piezoelectric effect d_{31}, pictured in Figure 4.3 (and as case (a) in Table 4.1). If the polarization is in the z direction and the x-y planes are the planes of the electrodes, the extensional vibration in the x direction is represented by the following dynamic equation:

$$\rho\frac{\partial^2 u}{\partial t^2} = F = \frac{\partial X_{11}}{\partial x} + \frac{\partial X_{12}}{\partial y} + \frac{\partial X_{13}}{\partial z} \quad (4.5)$$

where u is the displacement in the x direction of a small volume element in the ceramic plate. The relationship between the stress, the electric field (only E_z exists), and the induced strain is described by the following set of equations:

$$x_1 = s_{11}^E X_1 + s_{12}^E X_2 + s_{13}^E X_3 + d_{31}E_z$$

$$x_2 = s_{12}^E X_1 + s_{11}^E X_2 + s_{13}^E X_3 + d_{31}E_z$$

$$x_3 = s_{13}^E X_1 + s_{13}^E X_2 + s_{33}^E X_3 + d_{33}E_z \quad (4.6)$$

$$x_4 = s_{44}^E X_4$$

$$x_5 = s_{44}^E X_5$$

$$x_6 = 2(s_{11}^E - s_{12}^E)X_6$$

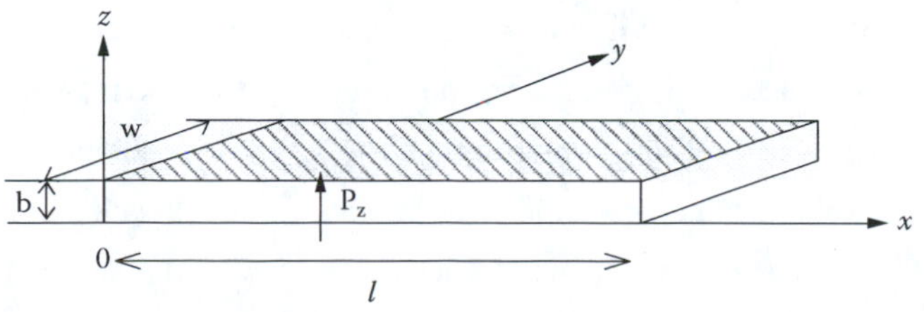

FIGURE 4.3 Transverse vibration of a rectangular piezoelectric plate.

TABLE 4.1
Characteristics of various resonators with different modes of vibration

	Factor	Boundary Conditions	Resonator Shape	Definition
a	k_{31}	$X_1 \neq 0, X_2 = X_3 = 0$ $x_1 \neq 0, x_2 \neq 0, x_3 \neq 0$	3 1	$\frac{d_{31}}{\sqrt{s_{11}^E \varepsilon_0 K_{33}^X}}$
b	k_{33}	$X_1 = X_2 = 0, X_3 \neq 0$ $x_1 = x_2 \neq 0, x_3 \neq 0$	3 Fundamental Mode	$\frac{d_{33}}{\sqrt{s_{13}^E \varepsilon_0 K_{33}^x}}$
c	k_p	$X_1 = X_2 \neq 0, X_3 = 0$ $x_1 = x_2 \neq 0, x_3 \neq 0$	3 Fundamental Mode	$k_{31}\sqrt{\frac{2}{1-\sigma}}$
d	k_t	$X_1 = X_2 \neq 0, X_3 \neq 0$ $x_1 = x_2 = 0, x_3 \neq 0$	3 Thickness Mode	$k_{33}\sqrt{\frac{\varepsilon_0 K_{33}^x}{c_{33}^D}}$
e	k_p'	$X_1 = X_2 \neq 0, X_3 \neq 0$ $x_1 = x_2 \neq 0, x_3 = 0$	3 Radial Mode	$\frac{k_p - A\, k_{33}}{\sqrt{1-A^2}\sqrt{1-k_{33}^2}}$
f	k_{31}'	$X_1 \neq 0, X_2 \neq 0, X_3 = 0$ $x_1 \neq 0, x_2 = 0, x_3 \neq 0$	3 2 Width Mode	$\frac{k_{31}}{\sqrt{1-k_{31}^2}}\sqrt{\frac{1+\sigma}{1-\sigma}}$
g	k_{31}''	$X_1 \neq 0, X_2 = 0, X_3 \neq 0$ $x_1 \neq 0, x_2 \neq 0, x_3 \neq 0$	3 2 1 Width Mode	$\frac{k_{31} - B\, k_{33}}{\sqrt{1-k_{33}^2}}$
h	k_{33}'''	$X_1 \neq 0, X_2 = 0, X_3 \neq 0$ $x_1 \neq 0, x_2 = 0, x_3 = 0$	1 2 3 Thickness Mode	$\sqrt{\frac{\frac{(k_p - A\, K_{33})^2}{1-A^2} - (k_{31} - B\, k_{33})^2}{1 - k_{33}^2 - (k_{31} - B\, k_{33})^2}}$
i	k_{33}'	$X_1 \neq 0, X_2 = 0, X_3 \neq 0$ $x_1 = 0, x_2 \neq 0, x_3 \neq 0$	1 3 Width Mode	$\frac{k_{33} - B\, k_{31}}{\sqrt{(1-B^2)(1-k_{31}^2)}}$
j	$k_{24}=k_{15}$	$X_1 = X_2 = X_3 = 0, X_4 \neq 0$ $x_1 = x_2 = x_3 = 0, x_4 \neq 0$		$\frac{d_{15}}{\sqrt{\varepsilon_0 K_{11}^X s_{44}^E}}$

$$\textit{Here}: \quad A = \frac{\sqrt{2}\, s_{13}^E}{\sqrt{s_{33}^E\,(s_{11}^E + s_{12}^E)}}, \quad B = \frac{s_{13}^E}{\sqrt{s_{11}^E\, s_{33}^E}}$$

It is important to note at this point that when an AC electric field of increasing frequency is applied to this piezoelectric plate, length, width, and thickness extensional resonance vibrations are excited. If we consider a typical PZT plate with dimensions 100 mm × 10 mm × 1 mm, these resonance frequencies correspond roughly to 10 kHz, 100 kHz, and 1 MHz, respectively. We will consider here the fundamental mode for this configuration, the length extensional mode. When the frequency of the applied field is well below 10 kHz, the induced displacement follows the AC field cycle, and the displacement magnitude is given by $d_{31}E_3l$. As we approach the fundamental resonance frequency, a delay in the length displacement with respect to the applied field begins to develop, and the amplitude of the displacement becomes enhanced. At frequencies above 10 kHz, the length displacement no longer follows the applied field, and the amplitude of the displacement is significantly reduced.

When a very long, thin plate is driven in the vicinity of this fundamental resonance, X_2 and X_3 may be considered zero throughout the plate. Because shear stress will not be generated by the applied electric field E_z, only the following single equation applies:

$$X_1 = x_1/s_{11}^E - (d_{31}/s_{11}^E)\, E_z \tag{4.7}$$

Substituting equation (4.7) into equation (4.5), and assuming that $x_1 = \partial u/\partial x$ and $\partial E_z/\partial x = 0$ (because each electrode is at the same potential), we obtain the following dynamic equation:

$$\rho \frac{\partial^2 u}{\partial t^2} = \frac{1}{s_{11}^E} \frac{\partial^2 u}{\partial x^2} \tag{4.8}$$

4.2.2.2 The Longitudinal Vibration Mode Based on the Longitudinal Piezoelectric Effect

Let us consider next the longitudinal vibration mode based on the longitudinal piezoelectric effect d_{33}. When the resonator is long in the z direction and the electrodes are deposited on each end of the rod as depicted for case (b) in Table 4.1, the following conditions are satisfied:

$$X_1 = X_2 = X_4 = X_5 = X_6 = 0 \text{ and } X_3 \neq 0. \tag{4.9}$$

So,

$$X_3 = (x_3 - d_{33}\, E_z)/s_{33}^E \tag{4.10}$$

for this configuration. Assuming a local displacement v in the z direction, the dynamic equation is given by:

$$\rho \frac{\partial^2 v}{\partial t^2} = \frac{1}{s_{33}^E} \left[\frac{\partial^2 v}{\partial z^2} - d_{33} \frac{\partial E_z}{\partial z} \right] \tag{4.11}$$

The electrical condition for the longitudinal vibration is not $\partial E_z/\partial z = 0$, but rather $\partial D_z/\partial z = 0$, such that:

$$\varepsilon_0 K_3^x \times \frac{\partial E_z}{\partial z} + \frac{d_{33}}{s_{33}^E} \frac{\partial^2 v}{\partial z^2} = 0 \tag{4.12}$$

Substituting equation (4.12) into equation (4.11), we obtain:

$$\rho \frac{\partial^2 v}{\partial t^2} = \frac{1}{s_{33}^D} \frac{\partial^2 v}{\partial z^2} \tag{4.13}$$

where

$$s_{33}^D = \frac{s_{33}^E}{1 + (d_{33}^2 / s_{33}^E\, \varepsilon_0 K_3^X)} = \frac{s_{33}^E}{1 + k_{33}^2} \tag{4.14}$$

The dynamic equations for other vibration modes can be obtained in a similar fashion using the elastic boundary conditions in Table 4.1.

4.2.2.3 Consideration of the Loss

When we consider the mechanical loss of the piezoelectric material, which is viscoelastic and proportional to the strain, the dynamic equations (4.8) and (4.13) must be modified.[2]

The transverse vibration mode is now described by:

$$\rho \frac{\partial^2 u}{\partial t^2} = \frac{1}{s_{11}^E}\left[\frac{\partial^2 (u + \delta(\partial u / \partial t)}{\partial x^2}\right] \tag{4.15}$$

and the longitudinal vibration mode by:

$$\rho \frac{\partial^2 v}{\partial t^2} = \frac{1}{s_{33}^D}\left[\frac{\partial^2 (v + \delta(\partial v / \partial t)}{\partial z^2}\right] \tag{4.16}$$

Because equations (4.8) and (4.13) (and, therefore, equations (4.15) and (4.16)) have very similar forms, we will consider just the longitudinal vibration k_{31} mode (transverse effect) here as an example of how both cases might be treated.

4.2.2.4 Solution for Longitudinal Vibration k_{31} Mode

Let us solve equation (4.8) using the Laplace transform. Denoting the Laplace transforms of $u(t,x)$ and $E_z(t)$ as $U(s,x)$ and $\tilde{E}\,(s)$, respectively, equation (4.8) is transformed to:

$$\rho s_{11}^E s^2 U(s,x) = \frac{\partial^2 U(s,x)}{\partial x^2} \tag{4.17}$$

We will assume the following *initial conditions*:

$$u(0,x) = 0 \quad and \quad \frac{\partial [u(0,x)]}{\partial t} = 0 \tag{4.18}$$

We may also make use of the fact that:

$$\rho\, s_{11}{}^{E} = 1/v^2, \tag{4.19}$$

where v is the speed of sound in the piezoelectric ceramic, to obtain a general solution:

$$U(s,x) = A\, e^{(sx/v)} + B\, e^{-(sx/v)} \tag{4.20}$$

The constants A and B can be determined by applying the *boundary conditions* at $x = 0$ and l:

$$X_1 = \frac{(x_1 - d_{31} E_z)}{s_{11}^E} = 0 \tag{4.21}$$

We may also make use of the fact that

$$L[x_1] = (\partial U/\partial x) = A(s/v)\, e^{(sx/v)} - B\,(s/v)\, e^{-(sx/v)} \tag{4.22}$$

Consequently, equations (4.20) and (4.22) become:

$$U(s,x) = \frac{d_{31}\tilde{E}\,(v/s)[e^{-s(l-x)/v} + e^{-s(l+x)/v} - e^{-sx/v} - e^{-s(2l-x)/v}]}{(1 - e^{-2sl/v})} \tag{4.23}$$

$$L[x_1] = \frac{d_{31}\tilde{E}\,[e^{-s(l-x)/v} + e^{-s(l+x)/v} - e^{-sx/v} - e^{-s(2l-x)/v}]}{(1 - e^{-2sl/v})} \tag{4.24}$$

The inverse Laplace transforms of equations (4.23) and (4.24) now provide the displacement $u(t,x)$ and strain $x_1(t,x)$. Making use of the expansion series

$$1/(1 - e^{-2sl/v}) = 1 + e^{-2sl/v} + e^{-4sl/v} + e^{-6sl/v} \ldots, \tag{4.25}$$

the strain, $x_1(t,x)$, can now be obtained by shifting the $d_{31}E_z(t)$ curves with respect to t according to Laplace Transform Theorem. We may also consider that because $u(t, l/2) = 0$ (from $U(s, l/2) = 0$) and $u(t,0) = -u(t, l)$ (from $U(s, 0) = -U(s, l)$), the total displacement of the plate device Δl becomes equal to $2u(t, l)$. We finally arrive at the following:

$$U(s,l) = \frac{d_{31}\tilde{E}(v/s)(1-e^{-sl/v})}{(1+e^{-sl/v})} = d_{31}\tilde{E}(v/s)[\tanh(sl/2v)] \tag{4.26}$$

(i) Response to a rectangular pulse voltage

We will now consider the response to a rectangular pulse voltage such as the one pictured in the lower left-hand corner of Figure 4.4. We begin by substituting

$$\tilde{E}(s) = (E_o/s)(1 - e^{-(nls)/v}) \tag{4.27}$$

into equation (4.26), which will allow us to obtain the displacement Δl for $n = 1$, 2, and 3. The quantity **n** is a time scale based on half the resonance period of the piezoelectric plate.

For **n** = 1,

$$U(s,l) = \frac{d_{31}E_o(v/s^2)(1-e^{-sl/v})^2}{(1+e^{-sl/v})}$$

$$= d_{31}E_o(v/s^2)[1 - 3e^{-sl/v} + 4e^{-2sl/v} - 4e^{-3sl/v} + \cdots] \tag{4.28}$$

Notice that the base function of $U(s,l) \propto 1/s^2$ gives the base function of $u(t, l)$ in terms of t. The inverse Laplace transform of equation (4.28) yields:

$$u(t, l) = d_{31}E_o v t \qquad 0 < t < l/v$$

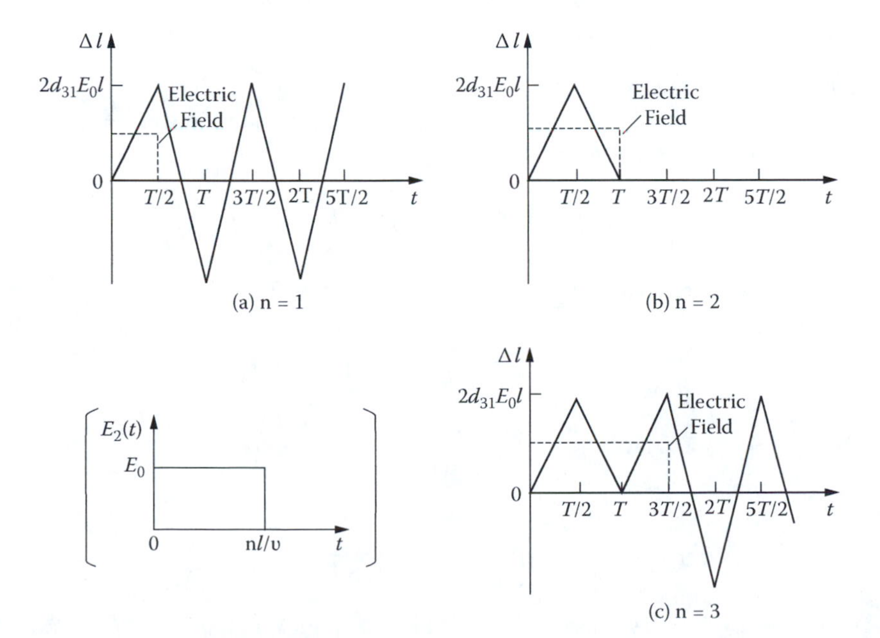

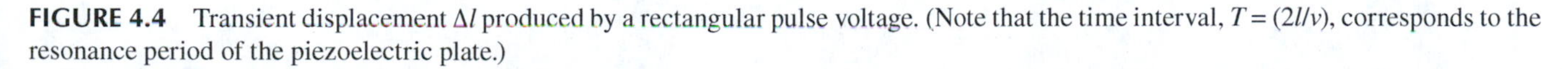

FIGURE 4.4 Transient displacement Δl produced by a rectangular pulse voltage. (Note that the time interval, $T = (2l/v)$, corresponds to the resonance period of the piezoelectric plate.)

$$u(t, l) = d_{31} E_o v\,[t - 3\,(t - l/v)] \qquad l/v < t < 2\,l/v$$

$$u(t, l) = d_{31} E_o v\,[t - 3\,(t - l/v) + 4\,(t - 2\,l/v)] \qquad 2\,l/v < t < 3\,l/v \tag{4.29}$$

The transient displacement, Δl, produced by the rectangular pulse voltage is pictured in Figure 4.4a for **n** = 1. The resonance period of this piezoelectric plate corresponds to $(2l/v)$. Notice how under these conditions continuous ringing occurs.

For **n** = 2,

$$U(s, l) = d_{31}E_o\,(v/s^2)\,(1 - 2e^{-sl/v} + e^{-2sl/v}) \tag{4.30}$$

Thus,

$$u(t, l) = d_{31}E_o v\,t \qquad 0 < t < l/v$$

$$u(t, l) = d_{31} E_o v\,[t - 2\,(t - l/v)] \qquad l/v < t < 2\,l/v$$

$$u(t, l) = d_{31} E_o v\,[t - 2\,(t - l/v) + (t - 2\,l/v)] = 0 \qquad 2\,l/v < t \tag{4.31}$$

In this case, the displacement, Δl, occurs in a single pulse and does not exhibit ringing as depicted in Figure 4.4b.

For **n** = 3, $U(s, l)$ is again expanded as an infinite series:

$$U(s,l) = \frac{d_{31} E_o (v/s^2)(1 - e^{-3sl/v})(1 - e^{-sl/v})}{(1 + e^{-sl/v})}$$

$$= d_{31}E_o(v/s^2)[1 - 2e^{-sl/v} + 2e^{-2sl/v} - 3e^{-3sl/v} + 4e^{-4sl/v} - 4e^{-5sl/v} + \cdots] \tag{4.32}$$

The displacement response for this case is pictured in Figure 4.4c.

So we see from this example that ***when a piezoelectric actuator is driven by a rectangular pulse, the mechanical ringing is completely suppressed when the pulse width is adjusted exactly to the resonance period of the sample*** (that is, when $T = 2\,l/v$, or integral multiples of the resonance period).

(ii) Response to a pseudostep voltage

We can take our discussion one step further by considering the response to a pseudostep voltage such as is pictured in the upper left-hand corner of Figure 4.5. We begin this time by substituting

$$\tilde{E}(s) = (E_o v/nls^2)\,(1 - e^{-nls/v}) \tag{4.33}$$

into equation (4.26). Let us again obtain the displacement Δl for **n** = 1, 2, and 3.

For **n** = 1,

$$U(s,l) = \frac{(d_{31} E_o / l)(v^2/s^3)(1 - e^{-sl/v})^2}{(1 + e^{-sl/v})}$$

$$= (d_{31} E_o / l)(v^2/s^3)[1 - 3e^{-sl/v} + 4e^{-2sl/v} - 4e^{-3sl/v} + \cdots] \tag{4.34}$$

Note that the base function of $U(s,l) \propto 1/s^3$ will lead to a base function of $u(t, l)$ in the form $t^2/2$ (parabolic curve), such that:

$$u(t, l) = (d_{31}E_o v^2/2\,l)\,t^2 \qquad 0 < t < l/v$$

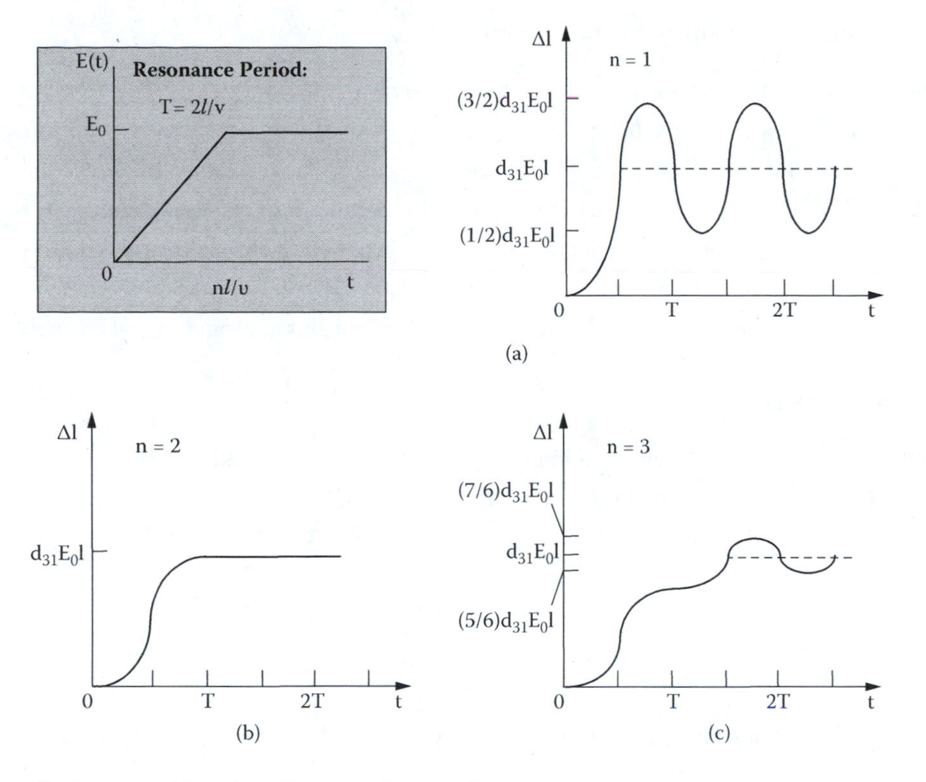

FIGURE 4.5 Transient displacement Δl produced by a pseudostep voltage.

$$u(t, l) = (d_{31}E_o v^2/2\ l)\ [t^2 - 3(t - l/v)^2] \qquad l/v < t < 2l/v$$

$$u(t, l) = (d_{31}E_o v^2/2L)\ [t^2 - 3(t - l/v)^2 + 4(t - 2l/v)^2] \qquad 2\ l/v < t < 3\ l/v \tag{4.35}$$

The transient displacement for an actuator driven by the pseudostep voltage pictured in Figure 4.5a is seen to exhibit continuous ringing. Note that this curve is actually a sequence of parabolic curves. It is not sinusoidal.

For **n** = 2,

$$U(s, l) = (d_{31}E_o/2\ l)\ (v^2/s^3)\ [1 - 2e^{-sl/v} + e^{-2sl/v}] \tag{4.36}$$

Thus,

$$u(t, l) = (d_{31}E_o v^2/4\ l)\ t^2 \qquad 0 < t < l/v$$

$$u(t, l) = (d_{31}E_o v^2/4\ l)\ [t^2 - 2\ (t - l/v)^2] \qquad l/v < t < 2l/v$$

$$u(t, l) = (d_{31}E_o v^2/4\ l)\ [t^2 - 2\ (t - l/v)^2 + (t - 2\ l/v)^2]$$

$$= (d_{31}E_o\ l\ /2) \qquad 2l/v < t \tag{4.37}$$

No ringing is apparent in the response for this case represented in Figure 4.5b. ***When the applied field*** $\tilde{E}$ ***includes the term*** $(1 + e^{-sl/v})$***, the expansion series terminates in finite terms, leading to a complete suppression of mechanical ringing.***

For **n** = 3, $U(s, l)$ is again expanded as an infinite series:

$$U(s, l) = (d_{31}E_o/3l)(v^2/s^3)[1 - 2e^{-sl/v} + 2e^{-2sl/v} - 3e^{-3sl/v} + 4e^{-4sl/v} - 4e^{-5sl/v} \ldots] \tag{4.38}$$

The displacement response for this condition is represented by the curve appearing in Figure 4.5c.

4.2.2.5 Experimental Observations of Transient Vibrations

Experimental results are presented here for the transient displacement response of a piezoelectric actuator. A PZT-based bimorph actuator with a mechanical quality factor (Q_m) of 1,000 (low loss) was used for the experiments.[3] The bimorph tip displacement was monitored with an eddy-current-type noncontact sensor. The resonance frequency of this bimorph is about 100 Hz. The response of the bimorph to a rectangular pulse and a pseudostep voltage are shown in Figures 4.6a and 4.6b, respectively. Note how ringing is completely eliminated in both cases when the pulse width or the rise time of the pseudostep is adjusted exactly to the resonance period of the bimorph. It should be noted as well how the displacement response is essentially a sequence of triangular or parabolic curves, for the rectangular pulse or the pseudostep inputs, respectively.

4.2.2.6 Consideration of the Loss

Recall that to properly account for loss effects equation (4.15) rather than equation (4.8) must be used for the transverse vibration. The displacement $U(s,l)$ can then be obtained by making the substitution:

$$s \rightarrow \frac{s}{\sqrt{1+\delta s}} \tag{4.39}$$

in equation (4.26). This solution has not yet been obtained in an explicit form. Approximate solutions for the piezoelectric resonance state have been determined by Ogawa.[2]

Experimental results appear in Figure 4.7.[4] The displacement Δl produced by a rectangular pulse voltage observed in this investigation are similar to the results shown in Figure 4.6, except for the vibrational damping. Once again we see that when the pulse width of this rectangular voltage is adjusted to the piezoelectric resonance period, T, or integral multiples of it, the vibrational ringing is eliminated.

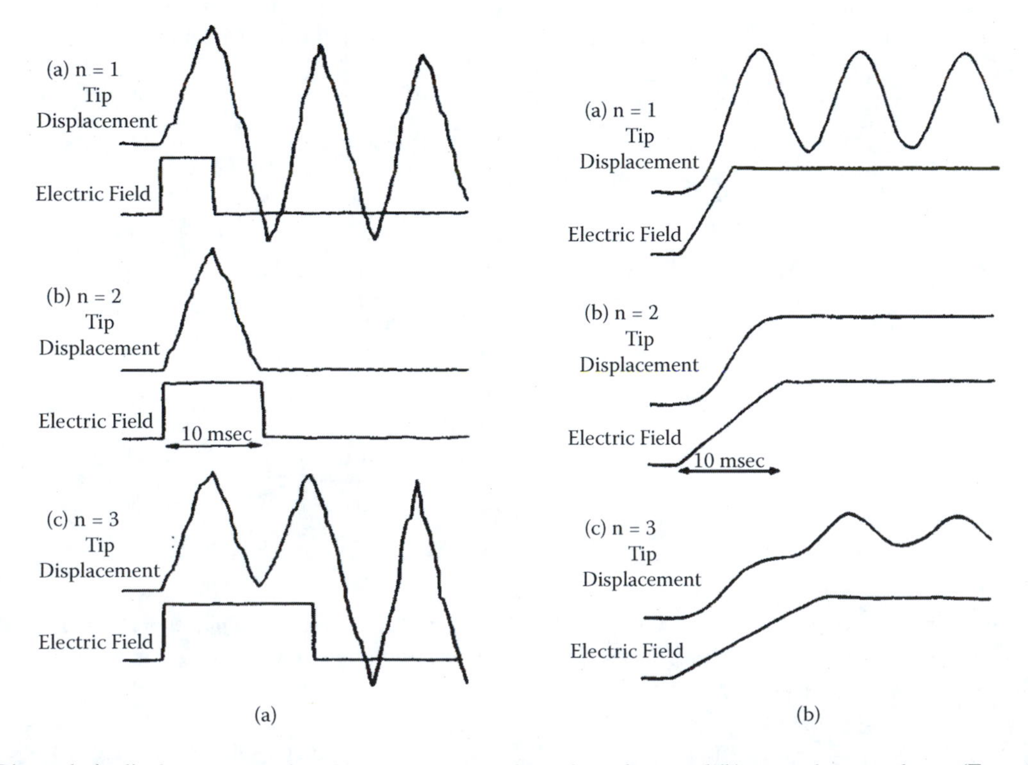

FIGURE 4.6 Bimorph tip displacement produced by: (a) a rectangular pulse voltage and (b) a pseudostep voltage. (From S. Sugiyama and K. Uchino: *Proc. 6th IEEE Int'l Symp. Appl. Ferroelectrics*, p. 637 [1986]. With permission.)

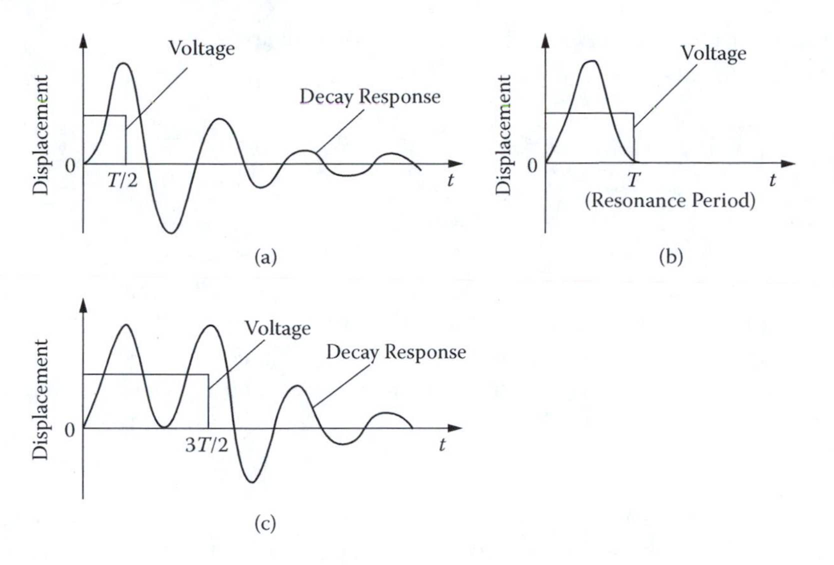

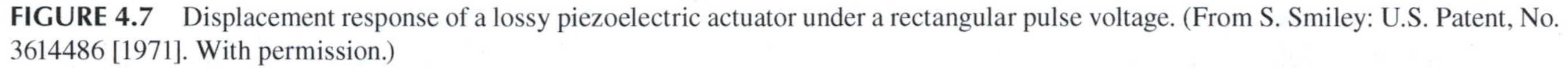

FIGURE 4.7 Displacement response of a lossy piezoelectric actuator under a rectangular pulse voltage. (From S. Smiley: U.S. Patent, No. 3614486 [1971]. With permission.)

4.3 RESONANCE DRIVE

When an alternating electric field is applied to a piezoelectric ceramic, mechanical vibration is excited, and if the drive frequency is adjusted to the mechanical resonance frequency of the device, a large resonant strain is generated. This phenomenon is called ***piezoelectric resonance*** and is very useful for applications such as energy trap devices and actuators. We will consider in this section the steady-state response of such a device to a sinusoidally varying electric field. Note that the steady state is not realized in the initial couple of sinusoidal waves, but after a minimum of 7–10 waves of the sinusoidal voltage. The amplification of the induced displacement by a factor of the mechanical quality factor Q_m under the piezoelectric resonance can be considered an amplification mechanism in terms of time.

4.3.1 The Electromechanical Coupling Factor

Let us consider first the ***electromechanical coupling factor,*** *k*, which represents the energy transduction rate in an electromechanical transducer (see section 2.1.) When an external field is applied on a piezoelectric device to generate a mechanical deformation, a formal definition of *k* may be written in terms of the stored and input energies as:

$$k^2 = \frac{Mechanical\ Stored\ Energy}{Electrical\ Input\ Energy} \tag{4.40a}$$

Alternatively, it can also be defined by:

$$k^2 = \frac{Electrical\ Stored\ Energy}{Mechanical\ Input\ Energy} \tag{4.40b}$$

The internal energy of a piezoelectric vibrator is given by the summation of mechanical energy U_M ($=\int x dX$) and electrical energy U_E ($=\int D dE$). The total energy, U, is calculated as follows, when the linear piezoelectric equations (equations (4.1) and (4.2)) apply:

$$U = U_M + U_E = [(1/2)s_{ij}{}^{E}X_jX_i + (1/2)d_{mi}E_mX_i] + [(1/2)d_{mi}X_iE_m + (1/2)\varepsilon_0\, K_{mk}{}^{X}\, E_k\, E_m] \tag{4.41}$$

Here, the terms including the elastic compliance, *s*, and the dielectric constant, K, represent the purely mechanical and electrical energies (U_{MM} and U_{EE}), respectively. The terms including the piezoelectric strain coefficient, *d*, represent the energy transduction from electrical to mechanical energy and vice versa through the piezoelectric effect. Thus, the electromechanical coupling factor, k, can also be defined by:

$$k = \frac{U_{ME}}{\sqrt{U_{MM}\,U_{EE}}} \tag{4.42}$$

The value of k will depend on the vibration mode (that is, several values could be obtained from a single ceramic sample corresponding to the different modes of vibration), and can be positive or negative. (The absolute value $|k|$ can be used in the discussion of energy.) Some typical vibration modes appear in Table 4.1 with their associated coupling factors.[5]

It is important to note that the electromechanical coupling factor can also be expressed in the general form:

$$k^2 = d^2 / (\varepsilon_0\, K^X\, s^E) \tag{4.43}$$

which better reflects how the value of k will depend on the mode of vibration and the anisotropic dielectric and elastic properties of the material. It should not be surprising then to find that different values of the electromechanical coupling factor can be obtained for a single specimen.

As an example, let us examine the following data measured for a typical PZT-based piezoelectric ceramic sample. If we first compare the magnitudes of k_{33} and k_{31}, it is generally observed that k_{33} (69%) > k_{31} (39%) because the transverse piezoelectric strain is about 1/3 the longitudinal strain for a given applied electric field strength. Comparing k_{31} and k_p (the planar coupling factor), we generally find that k_{31} (39%) < k_p (57%) because k_p is a measure of the material's response that includes the contribution of the (two-dimensional) planar mode, which effectively leads to a higher energy conversion. When we compare k_{33} and k_t (the thickness coupling factor), it is found in most cases that k_{33} (69%) > k_t (48%). This is because the thickness mode is two-dimensionally clamped, which leads to a lower energy conversion. When we characterize the action of a bimorph, the effective electromechanical coupling factor k_{eff} can be defined for the bending vibration mode. We find in this case that k_{eff} is much smaller (10%) than k_{31} (39%), even though the bending action occurs through the d_{31} piezoelectric strain coefficient. It should be noted as well that k_{15} (61%) tends to be very large in perovskite-type piezoelectric ceramics, because of their relatively large shear strain coefficient, d_{15}.

The electromechanical coupling factor is not related to the efficiency of the actuator, which is defined by the ratio of the output mechanical energy to the consumed electrical energy. When the ceramic is deformed by an external electric field, the input electrical energy is, of course, larger than the output mechanical energy. However, the ineffective electrical energy is stored as electrostatic energy in the actuator (the ceramic actuator is also a capacitor), and reverts to the power supply in the final process of an operating cycle. The efficiency is determined only by the losses manifested as hysteresis in the polarization versus electric field, strain versus stress, and strain versus electric field curves (see section 2.3). Thus, the efficiency of a piezoelectric device is usually more than 97%.

4.3.2 Piezoelectric Resonance

Let us consider once again a piezoelectric rectangular plate in which the transverse piezoelectric effect excites the length extensional mode, as depicted in Figure 4.3. We will apply a sinusoidal electric field ($E_z = E_o e^{j\omega t}$) to the plate and consider the ***standing longitudinal vibration mode based on the transverse piezoelectric effect d_{31}*** for t >> 0. The following expression for the strain coefficient, $x_1(t, x)$, can be obtained by means of equation (4.24):

$$x_1(t,x) = d_{31}\, E_o\, e^{j\omega t}\left[\frac{(e^{-j\omega(l-x)/v} - e^{-j\omega(l+x)/v} + e^{-j\omega x/v} - e^{-j\omega(2l-x)/v})}{(1 - e^{-2j\omega l/v})}\right]$$

$$= d_{31}\, E_o\, e^{j\omega t}\left[\frac{\sin[\omega(l-x)/v] + \sin[\omega x/v]}{\sin[\omega l/v]}\right] \tag{4.44}$$

where $s = j\omega$ (this corresponds to Fourier transform).

Here, the speed of sound, v, is given by:

$$v = \frac{1}{\sqrt{\rho s_{11}^E}} \tag{4.45}$$

4.3.2.1 Electrical Impedance

The piezoelectric actuator is an electronic component from the viewpoint of the driving power supply; thus, its ***electrical impedance*** [= v(applied)/i(induced)] plays an important role. The induced current is just the rate change of the electric displacement D_z, which can be calculated by making use of the definition for D given by equation (4.2):

$$i = w\left[\frac{\partial}{\partial t}\left(\int_0^l D_z dx\right)\right] = j\omega w \int_0^l \left[\left(\varepsilon_0 K_{33}^X - \left(\frac{d_{31}^2}{s_{11}^E}\right)\right) E_z + \left(\frac{d_{31}\, x_1}{s_{11}^E}\right)\right] dx \tag{4.46}$$

Substituting the expression given by equation (4.44) for x_1 in this last equation and this expression for current into the defining equation for the admittance, Y, yields:

$$Y = \frac{1}{Z} = \frac{i}{v} = \frac{i}{E_z b}$$

$$= \left(\frac{j\omega w l}{b}\right)\varepsilon_0 K_{33}^{LC}\left[1 + \left(\frac{d_{31}^2}{\varepsilon_0 K_{33}^{LC}\, s_{11}^E}\right)\left(\frac{\tan(\omega l/2v)}{(\omega l/2v)}\right)\right] \tag{4.47}$$

This describes the admittance of a mechanically free (unclamped) sample. Here, the width of the plate is w, its length l, and its thickness, b. The applied voltage is v, and the induced current is i. The quantity K_{33}^{LC} is called the ***longitudinally clamped dielectric constant***, which is given by:

$$\varepsilon_0\, K_{33}^{LC} = \varepsilon_0\, K_{33}^{X} - (d_{31}^2/s_{11}^E) \tag{4.48}$$

The resonance state is defined when the admittance becomes infinite (or the impedance becomes zero). It is described by equation (4.47) when $\tan(\omega l/2v) = \infty$ and $\omega l/2v = \pi/2$. The resonance frequency, f_R, can thus be defined as:

$$f_R = \frac{v}{2l} = \frac{1}{2l\sqrt{\rho s_{11}^E}} \tag{4.49}$$

In contrast, the antiresonance state is realized when the admittance becomes zero (and the impedance infinite). Under these conditions:

$$\left(\frac{\omega_A l}{2v}\right)\cot\left(\frac{\omega_A l}{2v}\right) = \frac{-d_{31}^2}{\varepsilon_0 K_{33}^{LC}\, s_{11}^E} = \frac{-k_{31}^2}{(1-k_{31}^2)} \tag{4.50}$$

where ω_A is the angular antiresonance frequency and, according to Table 4.1, the electromechanical coupling factor, k_{31}, is given by:

$$k_{31} = \frac{d_{31}}{\sqrt{s_{11}^E \varepsilon_0 K_{33}^X}} \tag{4.51}$$

4.3.2.2 Strain Distribution in the Sample

The positional dependence of x_1 may be described by the following expression derived from equation (4.44):

$$x_1(x) = d_{31} E_o \left[\frac{\cos[\omega(l-2x)/2v]}{\cos(\omega l/2v)} \right] \tag{4.52}$$

The resonance and antiresonance states may be interpreted in terms of the following conceptual model. In a material with high electromechanical coupling (k close to 1), the resonance or antiresonance states appear for $\tan(\omega l/2v) = \infty$ or 0 (that is, when $[\omega l/2v = (\mathbf{m}-1/2)\pi]$ or $[\mathbf{m}\pi]$, where **m** is an integer), respectively. The strain distribution, $x_1(x)$, as given by equation (4.52) for each state, is illustrated in Figure 4.8. In the resonance state, large strains and large changes in the "motional" capacitance are induced, and current can flow easily. This is manifested in a large ***motional admittance***. On the other hand, at antiresonance, the strains induced in the device exactly cancel, leading to no change in the capacitance and a high impedance. This condition is manifested by a motional admittance of zero.

When we consider a typical case of $k_{31} = 0.3$, we find that the antiresonance frequency is closer to the resonance frequency. A material with low coupling coefficients exhibits an antiresonance mode in which the capacitance change due to the size change is compensated completely by the charging current. In other words, the total admittance is the sum of the motional admittance and the damped admittance.

The general procedure for determining the electromechanical parameters (k_{31}, d_{31}, s_{11}^E, and K_{33}^X) is as follows:

1. The speed of sound in the specimen, v, is obtained from the resonance frequency f_R (see Figure 4.9), using equation (4.49).
2. The elastic compliance, s_{11}^E, can be calculated using this speed and the density of the specimen, ρ, in equation (4.45).
3. The electromechanical coupling factor k_{31} can be calculated using the calculated v and the measured antiresonance frequency, f_A, in equation (4.50). When characterizing low-coupling piezoelectric materials, the following approximate equation is also applicable:

$$\frac{k_{31}^2}{(1-k_{31}^2)} = \left(\frac{\pi^2}{4}\right)\frac{\Delta f}{f_R} \tag{4.53}$$

   where: $\Delta f = f_A - f_R$.
4. The strain piezoelectric coefficient, d_{31}, is calculated using equation (4.51) and the measured dielectric constant, K_{33}^X.

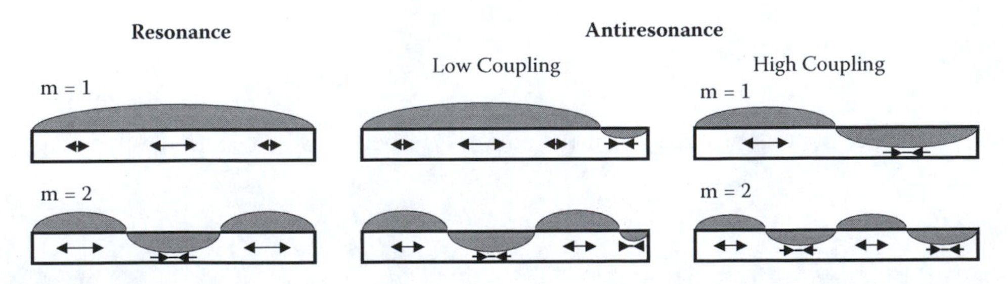

FIGURE 4.8 The strain distribution, $x_1(x)$, for the resonance and antiresonance states of a piezoelectric rectangular plate actuator, when driven based on d_{31}.

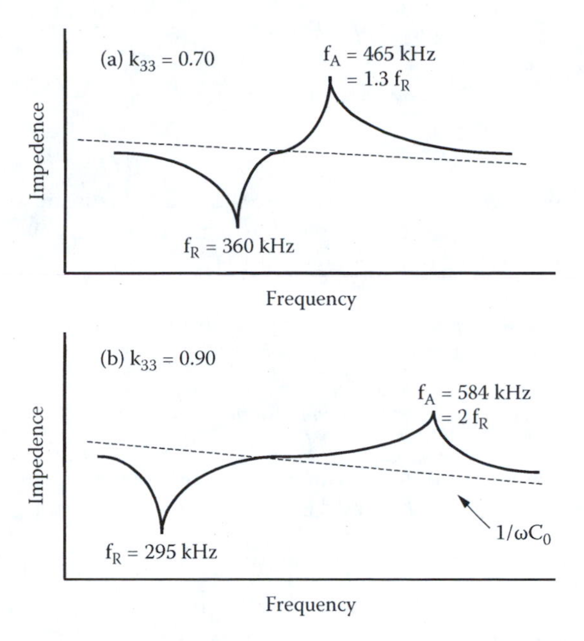

FIGURE 4.9 Impedance curves for: (a) a PZT 5H ceramic with $k_{33} = 0.70$ and (b) a PZN-PT single crystal with $k_{33} = 0.90$. (From J. Kuwata, K. Uchino, and S. Nomura: *Ferroelectrics* **37**, 579 (1981). With permission.)

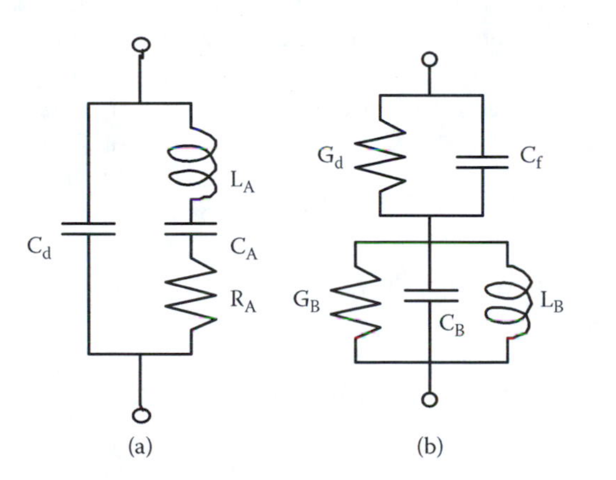

FIGURE 4.10 Equivalent circuits for a piezoelectric device at (a) resonance and (b) antiresonance.

Measured impedance curves for two common, relatively high-k materials appear in Figure 4.9.[6] Note the large separation between the resonance and antiresonance peaks for the higher-k material (Figure 4.9b). In this case: $f_A = 2f_R$.

4.3.2.3 Equivalent Circuits for Piezoelectric Vibrators

The equivalent circuit for a piezoelectric actuator is typically represented by a network comprising inductive (L), capacitive (C), and resistive (R) components. The equivalent circuit for the resonance state, which has very high admittance (or low impedance), is pictured in Figure 4.10a. The electrostatic capacitance (***damped capacitance***) is labeled C_d in the diagram, and the components L_A and C_A in this series resonance circuit represent the piezoelectric response of the actuator. In the case of the transverse piezoelectric vibration of a rectangular plate, for example, these quantities are defined by the following equations:

$$L_A = \left[\frac{\rho}{8}\right]\left[\frac{l\,b}{w}\right]\left[\frac{(s_{11}^E)^2}{d_{31}^2}\right] \tag{4.54}$$

$$C_A = \left[\frac{8}{\pi^2}\right]\left[\frac{lw}{b}\right]\left[\frac{d_{31}^2}{s_{11}^E}\right] \quad (4.55)$$

The resistance, R_A, is associated with mechanical loss (in particular, the intensive elastic loss tan ϕ' described in Section 2.3.4). In contrast, the equivalent circuit for the antiresonance state of the same actuator is shown in Figure 4.10b, which has high impedance as compared to that of the network pictured in Figure 4.10a.

In the ATILA software, the analyses based on static, resonance, and pulse drives can be made by choosing the "Static," "Harmonic or Modal," and "Transient" in the Problem Data. Refer to Figure 4.11.

The load condition (resistance, capacitance, and inductance) can also be changed in ATILA.

Figure 4.12 shows an example for a Rosen-type transformer. Figure 4.12b shows a "condition" setting (1 MΩ) on the right-hand-side electrode in Figure 4.12a. Figures 4.13a and 4.13b show the resonance curve difference for 1 GΩ and 1 MΩ, respectively.

4.4 PIEZOELECTRIC DAMPER

The mechanical damper is another important component in a mechatronic system. In addition to electro- and magnetorheological fluids, piezoelectric materials can also be used in mechanical damping devices. Consider a piezoelectric material attached to an object whose vibration is to be damped. When the vibration is transmitted to the piezoelectric material, the mechanical energy of the vibration is converted into electrical energy through the piezoelectric effect, and an AC voltage is generated. If the piezoelectric material is in an open- or short-circuit condition, the generated electrical energy is converted back into vibration energy without loss. This cycle is repeated continuously, producing a sustained vibration. If a proper resistance is connected, however, the electrical energy is consumed through Joule heating, leaving less energy to be converted into mechanical energy, and the vibration is rapidly damped. The damping takes place most rapidly when the series combination of the resistance, R, and capacitance, C, of the piezoelectric material is selected such that ***electrical impedance matching*** occurs. The optimum choice will satisfy the condition: $R = 1/(2\pi fC)$, where f is the vibration frequency.[7] A collaborative effort between ACX Company and K2, has led to the production of the new "smart ski," a part of which is pictured in Figure 4.14, that makes use of this principle. The blade design incorporates PZT patches that effectively suppress excess vibration in the ski as it slides over the snow.[8]

The electric energy, U_E, generated in the first part of the cycle can be expressed in terms of the electromechanical coupling factor, k, and the mechanical energy, U_M, as:

$$U_E = U_M (k^2) \quad (4.56)$$

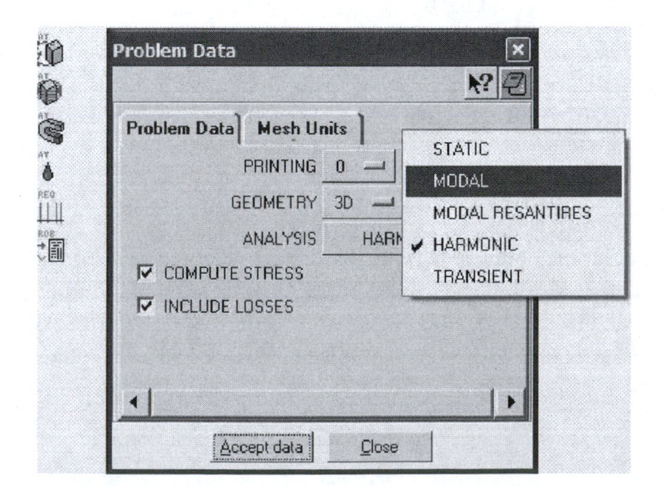

FIGURE 4.11 Selection of Problem Data: "Static," "Modal or Harmonic," and "Transient" correspond to DC, AC resonance, and pulse drive conditions, respectively.

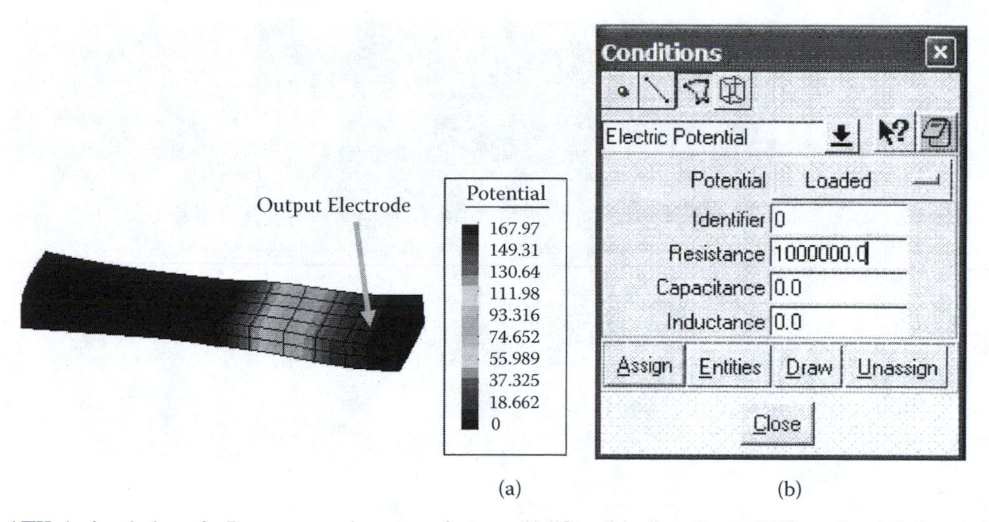

FIGURE 4.12 (a) ATILA simulation of a Rosen-type piezo-transformer; (b) "Condition" setting (1 MΩ) on the right-hand side electrode in (a).

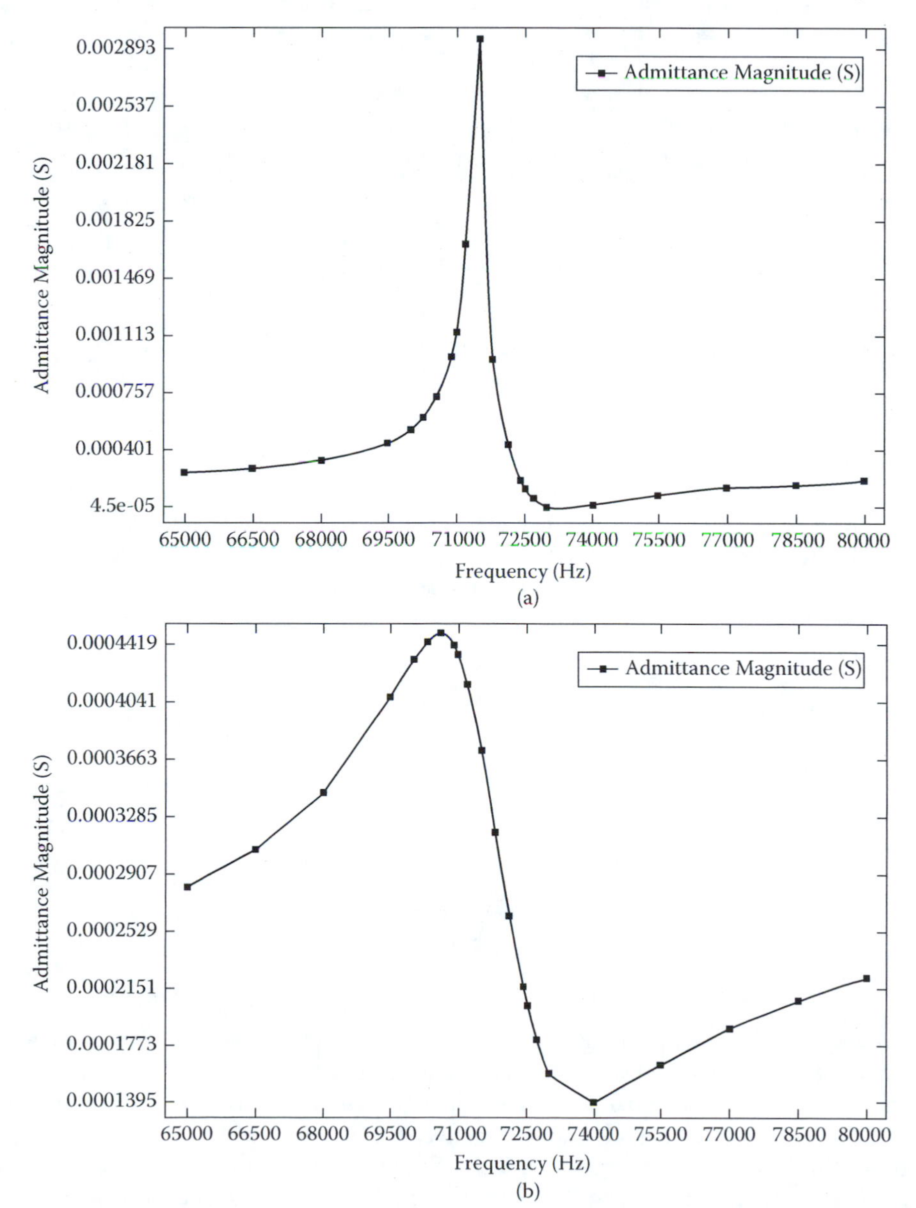

FIGURE 4.13 The resonance curves for (a) 1 GΩ load and (b) 1 MΩ _load.

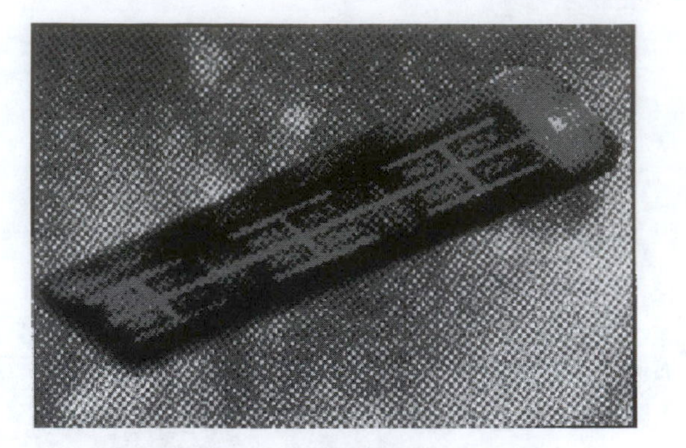

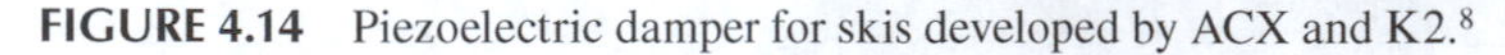

FIGURE 4.14 Piezoelectric damper for skis developed by ACX and K2.[8]

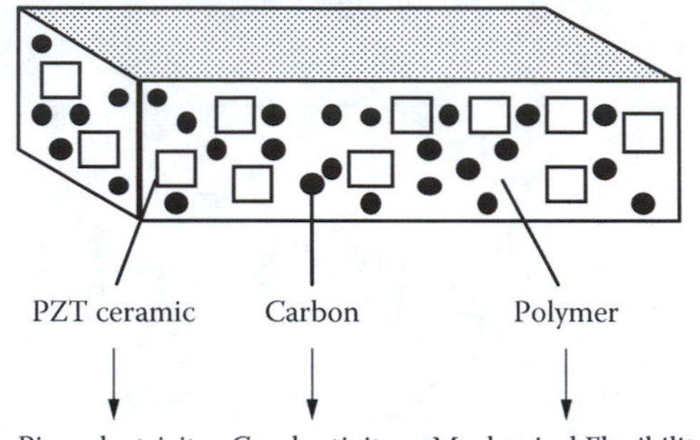

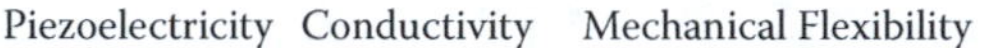

FIGURE 4.15 A piezoceramic:carbon black:polymer composite for vibration damping. (From Y. Suzuki, K. Uchino, H. Gouda, M. Sumita, R. E. Newnham, and A. R. Ramachandran: *J. Ceram. Soc. Jpn.*, Int'l Edition **99**, 1096 [1991]. With permission.)

Making use of a suitable series resistance to transform the electrical energy into heat, transforming efficiency as high as 50% can be achieved with the piezoelectric damper. Accordingly, the vibration energy is decreased at a rate of $(1 - k^2/2)$ times per vibration cycle, because an energy equal to $(k^2/2)$ multiplied by the mechanical vibration energy is dissipated as heat energy. As the square of the amplitude is equivalent to the amount of vibration energy, the amplitude will decrease at a rate of $(1 - k^2/2)^{1/2}$ times per vibration cycle. Assuming a resonance period of T_o, the number of vibrations occurring over a time interval t will just be $2t/T_o$ and the amplitude of the vibration after t s will be:

$$\left[1-\frac{k^2}{2}\right]^{t/T_o} = e^{-t/\tau} \tag{4.57}$$

or

$$\tau = -T_o \ln\left[1-\frac{k^2}{2}\right] \tag{4.58}$$

where τ is the time constant of the exponential decay in the vibration amplitude. So, we see from this analysis that the vibration will be more quickly damped when a material with a higher k value is used.

Where ceramics tend to be rather brittle and hard, they can be difficult to incorporate directly into a mechanical system. Hence, a flexible piezoelectric composite is a practical alternative. A composite material comprising a piezoceramic powder and carbon black suspended in a suitable polymer, as pictured schematically in Figure 4.15, can be fabricated with an electrical conductivity that is highly sensitive to small changes in the concentration of carbon black.[9] The conductivity of the composite changes by more than 10 orders of magnitude around a certain fraction of carbon black called the ***percolation threshold***, where links between the carbon particles begin to form. The conducting pathways that form become essentially internal resistances

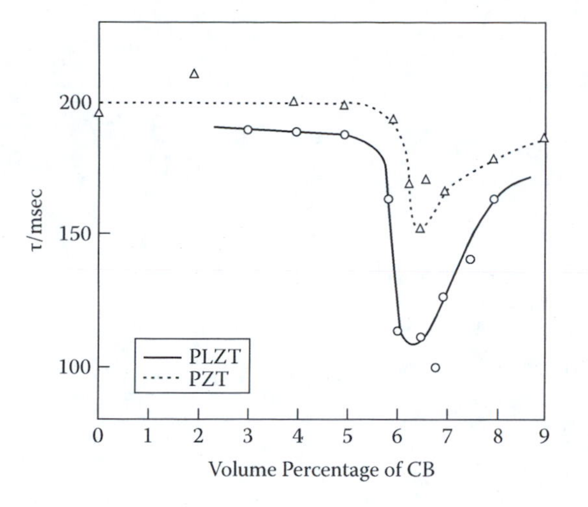

FIGURE 4.16 Damping time constant, τ, as a function of volume percentage of carbon black in PZT:PVDF and PLZT:PVDF composites. (Note that the minimum time constant [quickest damping] occurs at the percolation threshold for both composites, and the higher-*k* material [PLZT:PVDF] is the more effective damper.)

within the material. A concentration of carbon black can thus be established that effectively leads to the formation of a series resistance favorable for significant dissipation of the vibrational energy.

The dependence of the damping time constant, τ, on the volume percentage of carbon black in PLZT:PVDF and PZT:PVDF composites is shown in Figure 4.16. The minimum time constant and, therefore, the most rapid vibration damping is seen to occur with a volume percentage of about 7% carbon black for both composites. Note that the PLZT:PVDF material, with a higher electromechanical coupling, k, exhibits the larger dip in τ and, thus, more effective damping.

REFERENCES

1. T. Ota, T. Uchikawa, and T. Mizutani: *Jpn. J. Appl. Phys.* **24**, Suppl. 24-3, 193 (1985).
2. T. Ogawa: *Crystal Physical Engineering*, Shoka-bo Pub., Tokyo (1976).
3. Sugiyama and K. Uchino: *Proc. 6th IEEE Int'l Symp. Appl. Ferroelectrics*, p. 637 (1986).
4. S. Smiley: U.S. Patent, No.3614486 (1971).
5. T. Ikeda: *Fundamentals of Piezoelectric Materials*, Ohm Publication, Tokyo (1984).
6. J. Kuwata, K. Uchino, and S. Nomura: *Ferroelectrics* **37**, 579 (1981).
7. K. Uchino and T. Ishii: *J. Ceram. Soc. Jpn.* **96**, 963 (1988).
8. ACX Company catalogue: Passive Damping Ski, (2000).
9. Y. Suzuki, K. Uchino, H. Gouda, M. Sumita, R. E. Newnham, and A. R. Ramachandran: *J. Ceram. Soc. Jpn.*, Int'l Edition **99**, 1096 (1991).

5 Finite Element Analysis for Smart Transducers

The finite element method and its application to smart transducer systems are introduced in this chapter.

5.1 FUNDAMENTALS OF FINITE ELEMENT ANALYSIS

Consider the piezoelectric domain Ω pictured in Figure 5.1, within which the displacement field, **u**, and electric potential field, ϕ, are to be determined. The **u** and ϕ fields satisfy a set of differential equations that represent the physics of the continuum problem considered. Boundary conditions are usually imposed on the domain's boundary, Γ, to complete the definition of the problem.

The finite element method is an approximation technique for finding solution functions.[1] The method consists of subdividing the domain Ω into subdomains, or ***finite elements***, as illustrated in Figure 5.2. These finite elements are interconnected at a finite number of points, or ***nodes***, along their peripheries. The ensemble of finite elements defines the ***problem mesh***. Note that because the subdivision of Ω into finite elements is arbitrary, there is no unique mesh for a given problem.

Within each finite element, the displacement and electric potential fields are uniquely defined by the values they assume at the element nodes. This is achieved by a process of ***interpolation*** or ***weighing*** in which ***shape functions*** are associated with the element. By combining, or ***assembling***, these local definitions throughout the whole mesh, we obtain a trial function for Ω that depends only on the nodal values of **u** and ϕ and that is "piecewise" defined over all the interconnected elementary domains. Unlike the domain Ω, these elementary domains may have a simple geometric shape and homogeneous composition.

We will show in the following sections how this trial function is evaluated in terms of the variation principle to produce a system of linear equations whose unknowns are the nodal values of **u** and ϕ.[2]

5.2 DEFINING THE EQUATIONS FOR THE PROBLEM

5.2.1 The Constitutive and Equilibrium Equations

The constitutive relations for piezoelectric media may be derived in terms of their associated thermodynamic potentials.[3,4] Assuming the strain, x, and electric field, E, are independent variables, the basic equations of state for the converse and direct piezoelectric effects are written:

$$\begin{cases} X_{ij} = c_{ijkl}^{E} x_{kl} - e_{kij} E_k \\ D_i = e_{ikl} x_{kl} + \chi_{ij}^{x} E_j \end{cases} \tag{5.1}$$

The quantities c^E (elastic stiffness at constant electric field), e (piezoelectric stress coefficients), and χ^x (dielectric susceptibility at constant strain) are assumed to be constant, which is reasonable for piezoelectric materials subjected to small deformations and moderate electric fields. Furthermore, no distinction is made between isothermal and adiabatic constants.

On the domain, Ω, and its boundary, Γ, (where the normal is directed outward from the domain), the fundamental dynamic relation must be verified:

$$\rho \frac{\partial^2 u_i}{\partial t^2} = \frac{\partial x_{ij}}{\partial r_j} \tag{5.2}$$

where **u** is the displacement vector, ρ the mass density of the material, t the time, **X** is the stress tensor, and $\mathbf{r} = <r_1\ r_2\ r_3>$ is a unit vector in the Cartesian coordinate system.

When no macroscopic charges are present in the medium, Gauss' theorem imposes for the electric displacement vector, **D**:

$$\frac{\partial D_i}{\partial r_i} = 0 \tag{5.3}$$

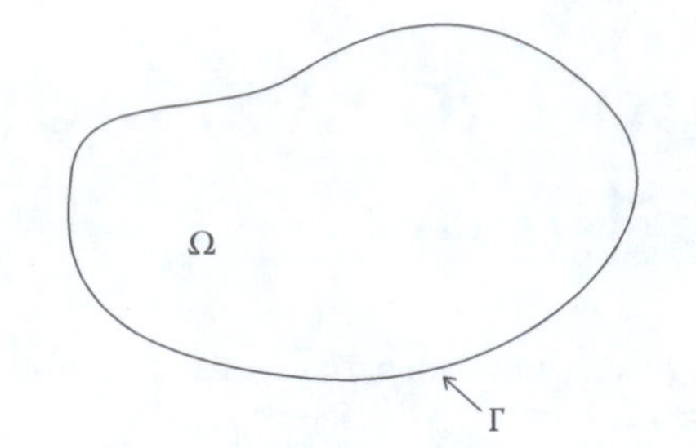

FIGURE 5.1 Schematic representation of the problem domain Ω with boundary Γ.

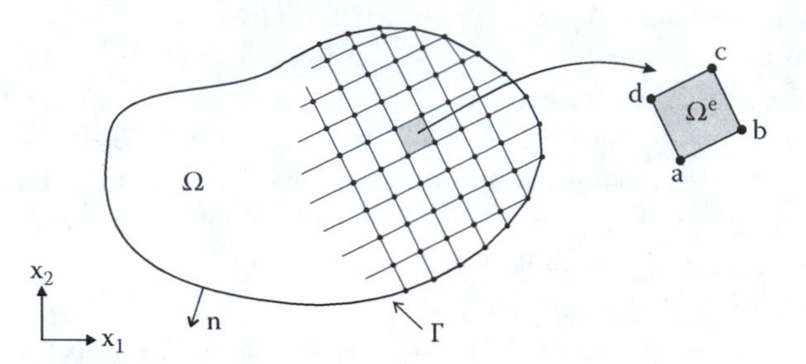

FIGURE 5.2 Discretization of the domain Ω.

Considering small deformations, the strain tensor, **x,** is written as:

$$x_{kl} = \frac{1}{2}\left(\frac{\partial u_k}{\partial r_l} + \frac{\partial u_l}{\partial r_k}\right) \tag{5.4}$$

Assuming electrostatic conditions, the electrostatic potential, ϕ, is related to the electric field **E** by

$$\mathbf{E} = -\text{grad } \phi \tag{5.5}$$

or, equivalently

$$E_i = -\frac{\partial \phi}{\partial r_i} \tag{5.6}$$

Using equations (5.2), (5.3), and (5.6) in combination with (5.1) yields:

$$\begin{cases} -\rho\omega^2 u_i = \dfrac{\partial}{\partial r_j}\left(c^E_{ijkl} x_{kl} - e_{kij} E_k\right) \\ \dfrac{\partial}{\partial r_j}\left(e_{ikl} x_{kl} + \chi^x_{ij} E_j\right) = 0 \end{cases} \tag{5.7}$$

5.2.2 Boundary Conditions

Mechanical and electrical boundary conditions complete the definition of the problem.

The mechanical conditions are as follows:

- The Dirichlet condition on the displacement field, **u**, is given by

$$u_i = u_i^o \tag{5.8}$$

where $\mathbf{u}^o$ is a known vector. For convenience, we name the ensemble of surface elements subjected to this condition S_u.
- The Neumann condition on the stress field, **X**, is given by

$$X_{ij} \cdot n_j = f_i^o \tag{5.9}$$

where **n** is the vector normal to Γ, directed outward, and $\mathbf{f}^o$ is a known vector. For convenience, we name the ensemble of surface elements subjected to this condition S_X.

The electrical conditions are as follows:

- The conditions for the excitation of the electric field between those surfaces of the piezoelectric material that are not covered with an electrode and are, therefore, free of surface charges is given by:

$$D_i \cdot n_i = 0 \tag{5.10}$$

where **n** is the vector normal to the surface. For convenience, we name the ensemble of surface elements subjected to this condition S_σ. Note that with condition (5.9), we assume that the electric field outside Ω is negligible, which is easily verified for piezoelectric ceramics.
- When considering the conditions for the potential and excitation of the electric field between those surfaces of the piezoelectric material that are covered with electrodes, we assume that there are p electrodes in the system. The potential on the whole surface of the pth electrode is

$$\phi = \phi_p \tag{5.11}$$

The charge on that electrode is

$$-\iint_{S_p} D_i n_i dS_p = Q_p \tag{5.12}$$

In some cases, the potential is used, and in others it is the charge. In the former case, ϕ_p is known, and equation (5.11) is used to determine Q_p. In the latter case, Q_p is known, and equation (5.10) is used to determine ϕ. Finally, to define the origin of the potentials, it is necessary to impose the condition that the potential at one of the electrodes be zero ($\phi_o = 0$).

5.2.3 The Variational Principle

The ***variational principle*** identifies a scalar quantity Π, typically named the ***functional***, which is defined by an integral expression involving the unknown function, **w**, and its derivatives over the domain Ω and its boundary Γ. The solution to the continuum problem is a function **w** such that

$$\delta\Pi = 0 \tag{5.13}$$

Π is said to be stationary with respect to small changes in **w**, δ**w**. When the variational principle is applied, the solution can be approximated in an integral form that is suitable for finite element analysis (FEM). In general, the matrices derived from the variational principle are always symmetric.

Equation (5.7) and the boundary conditions expressed by equations (5.8) to (5.12) allow us to define the so-called Euler equations to which the variational principle is applied such that a functional of the following form is defined that is stationary with respect to small variations in **w**.

$$\Pi = \iiint_{\Omega} \frac{1}{2}(x_{ij}\, c^{E}_{ijkl}\, x_{kl} - \rho\,\omega^2\, u_i^2)\, d\Omega - \iint_{S_u} (u_i - u_i^o) n_j (c^{E}_{ijkl} x_{kl} - e_{kij} E_k) dS_u$$

$$- \iiint_{\Omega} \frac{1}{2}(2 x_{kl} e_{ikl} E_i + E_i \chi^{x}_{ij} E_j) d\Omega - \iint_{S_X} f_i u_i dS_X$$

$$- \sum_{p=0}^{M} \iint_{S_p} (\phi - \phi_p) n_i (e_{ikl} x_{kl} + \chi^{x}_{ij} E_j) dS_p + \sum_{p=0}^{P} \phi_p Q_p \tag{5.14}$$

Note that the first term of this expression for Π represents the Lagrangian of the mechanical state. Satisfying the stationary condition for Π implies that all the conditions described by equations (5.7) through (5.12) are satisfied.

5.3 APPLICATION OF THE FINITE ELEMENT METHOD

5.3.1 Discretization of the Domain

The domain Ω is divided into subdomains Ω^e, or finite elements (Figure 5.2), such that:

$$\Omega = \sum_{e} \Omega^e \tag{5.15}$$

Common finite elements available for discretizing the domain are shown in Figure 5.3. As a result of the discretization, the functional Π can then be written as:

$$\Pi = \sum_{e} \Pi^e \tag{5.16}$$

Note that the term Π^e contains only volume integral terms if the element Ω^e is inside the domain Ω. Elements having a boundary coincident with Γ will have a term Π^e that contains both volume and surface integrals.

5.3.2 Shape Functions

The finite element method, being an approximation process, will result in the determination of an approximate solution of the form

$$\mathbf{w} \approx \hat{\mathbf{w}} = \sum \mathbf{N_i a_i} \tag{5.17}$$

where the N_i is a shape function prescribed in terms of independent variables (such as coordinates) and the a_i is a nodal parameter, known or unknown. The shape functions must guarantee the continuity of the geometry between elements. Moreover, to ensure convergence, it is necessary that the shape functions be at least C^m continuous if derivatives of the mth degree exist in the integral form. This condition is automatically met if the shape functions are polynomials complete to the mth order.

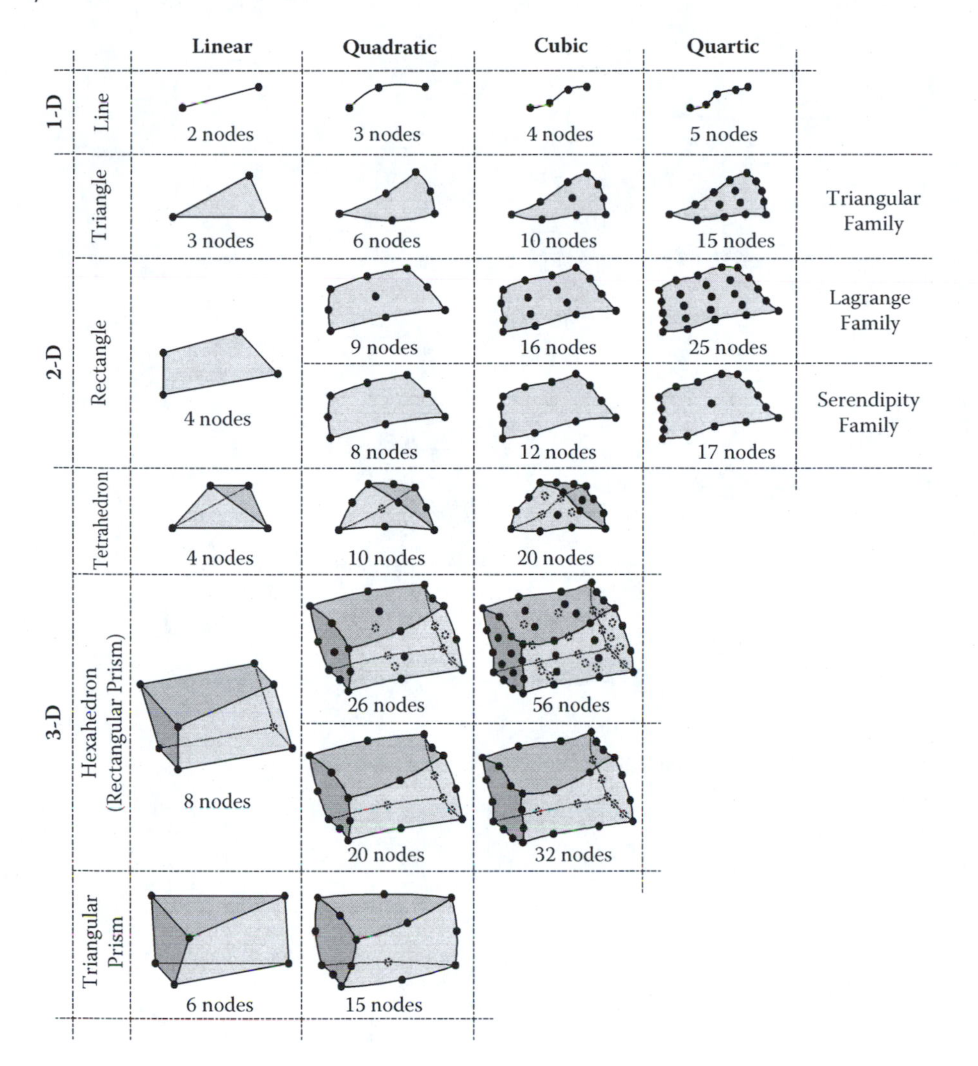

FIGURE 5.3 Common finite elements.

The construction of shape functions for an element Ω^e, defined by n nodes, usually requires that if N_i is the shape function for node i, then $N_i = 1$ at node i, and $N_i = 0$ at the other nodes. Also, for any point p in Ω^e we must have

$$\sum_{1}^{n} N_i(\mathbf{p}) = 1 \tag{5.18}$$

Polynomials are commonly used to construct shape functions. For instance, a Lagrange polynomial

$$N_i(\xi) = \prod_{\substack{k=1 \\ k \neq i}}^{n} \frac{\xi_k - \xi}{\xi_k - \xi_i} \tag{5.19}$$

can be used at node i of a one-dimensional element containing n nodes (see Figure 5.4). It verifies the following conditions:

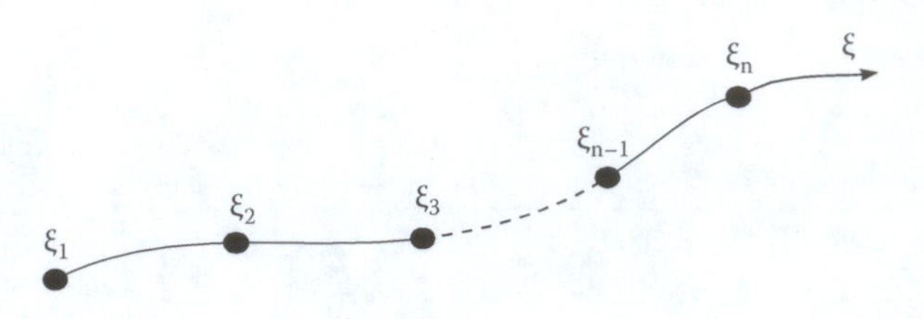

FIGURE 5.4 A generalized n-node linear element.

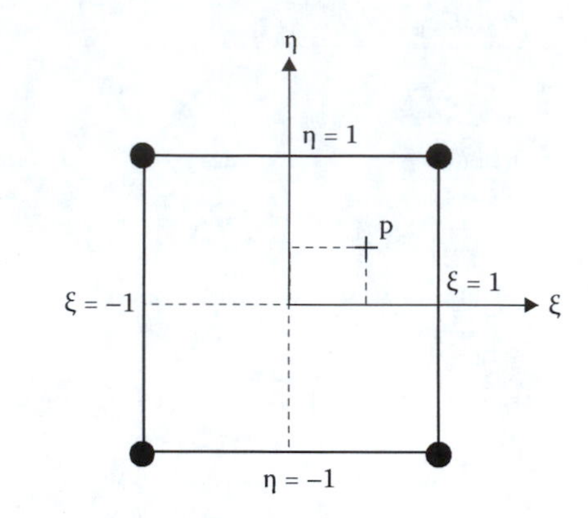

FIGURE 5.5 An example of a four-node quadrilateral element.

$$\begin{cases} N_i(\xi_i) = 1 \\ N_i(\xi_{k\neq 1}) = 0 \\ \sum_1^n N_i(\xi) = 1 \end{cases} \tag{5.20}$$

For the four-node rectangular element of Figure 5.5, we can write the four shape functions as follows:

$$N(\xi,\eta) = N_i(\xi)\,N_j(\eta) \tag{5.21}$$

where i and j indicate the row and column of the node in the element and $i,j = 1,2$. The conditions described by equation (5.20) can be verified for these functions. Finally, the position of any point p with coordinates (ξ, η) in the element is given by:

$$\begin{bmatrix} \xi \\ \eta \end{bmatrix} = \left\langle \left(\frac{1}{4}\right)[(-1)+\xi][(-1)+\eta]\left(-\frac{1}{4}\right)[(1)+\xi][(-1)+\eta] \right.$$

$$\left(-\frac{1}{4}\right)[(-1)+\xi][(1)+\eta]\left(\frac{1}{4}\right)[(1)+\xi][(1)+\eta]\right\rangle\begin{bmatrix}-1 & -1\\ 1 & -1\\ -1 & 1\\ 1 & 1\end{bmatrix} \tag{5.22}$$

In most cases, the same shape functions are used to describe the element geometry and to represent the solution $\hat{\mathbf{w}}$.

5.3.3 Parent Elements

In order to better represent the actual geometry, it is generally useful to use curvilinear finite elements for the discretization of Ω. These elements are then mapped into parent finite elements (Figure 5.6) to facilitate the computation of Π^e.

An isoparametric representation is commonly used to perform the mapping of the actual elements into the reference elements. Consider a volume element Ω^e of the domain Ω defined by n nodes. The position vector $\mathbf{R}$ of a point p of Ω^e can be written as a function of parameters ξ, η, and ξ:

$$\xi \rightarrow R = R(\xi) \tag{5.23}$$

which is the same as:

$$R = \begin{Bmatrix} x \\ y \\ z \end{Bmatrix} = \begin{Bmatrix} x(\xi,\eta,\zeta) \\ y(\xi,\eta,\zeta) \\ z(\xi,\eta,\zeta) \end{Bmatrix} \tag{5.24}$$

The finite element representation can be written in the following form:

$$\begin{aligned} x &= \mathbf{N}(\xi,\eta,\zeta)\ \mathbf{x_n} \\ y &= \mathbf{N}(\xi,\eta,\zeta)\ \mathbf{y_n} \\ z &= \mathbf{N}(\xi,\eta,\zeta)\ \mathbf{z_n} \end{aligned} \tag{5.25}$$

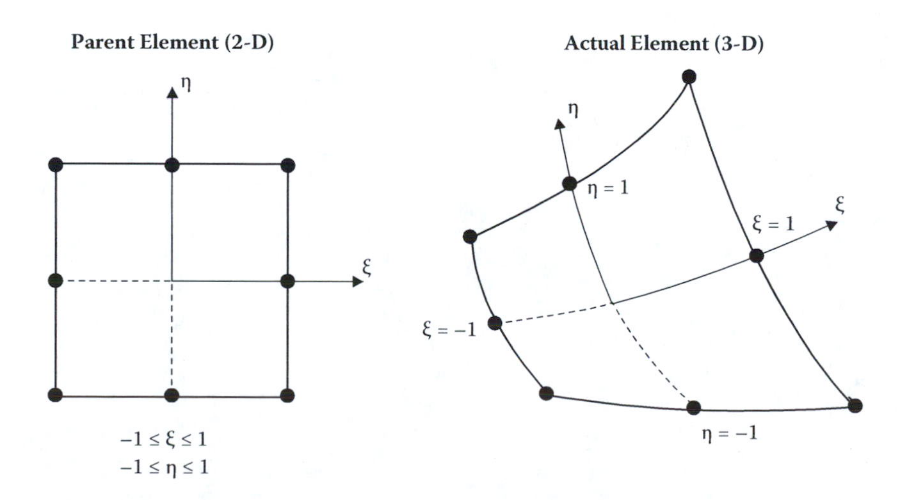

FIGURE 5.6 An example of a parent element for an eight-node quadrilateral element.

where:

$$\begin{aligned} \mathbf{x_n} &= \langle x_1 \quad x_2 \quad \dots \quad x_n \rangle \\ \mathbf{y_n} &= \langle y_1 \quad y_2 \quad \dots \quad y_n \rangle \\ \mathbf{z_n} &= \langle z_1 \quad z_2 \quad \dots \quad z_n \rangle \\ \mathbf{N} &= \langle N_1 \quad N_2 \quad \dots \quad N_n \rangle \end{aligned} \tag{5.26}$$

The scalars x_i, y_i, and z_i represent the Cartesian coordinates at node i, and N_i is the shape function at node i. A differential element at point p is thus defined by:

$$d\mathbf{R} = \mathbf{F}_\xi \, d_\xi \tag{5.27}$$

and

$$\frac{\partial}{\partial_\xi} = \mathbf{F}_\xi^{\,X} \frac{\partial}{\partial \mathbf{R}} = \mathbf{J} \frac{\partial}{\partial \mathbf{R}} \tag{5.28}$$

where:

$$\mathbf{F}_\xi = \begin{bmatrix} x_{,\xi} & x_{,\eta} & x_{,\zeta} \\ y_{,\xi} & y_{,\eta} & y_{,\zeta} \\ z_{,\xi} & z_{,\eta} & z_{,\zeta} \end{bmatrix} = [\mathbf{a}_1 \quad \mathbf{a}_2 \quad \mathbf{a}_3] \tag{5.29}$$

and

$$d\mathbf{R} = \mathbf{a}_1 d\xi + \mathbf{a}_2 d\eta + \mathbf{a}_3 d\zeta \tag{5.30}$$

Vectors $\mathbf{a}_1$, $\mathbf{a}_2$, and $\mathbf{a}_3$ are the base vectors associated with the parametric space, $\mathbf{J}$ is the Jacobian matrix of the transformation, and $\mathbf{F}_\xi$ is the transformation matrix from the parametric space to the Cartesian space.

From equations (5.27) and (5.28), we obtain:

$$d_\xi = \mathbf{F}_\xi^{-1} d\mathbf{R} \tag{5.31}$$

and

$$\frac{\partial}{\partial \mathbf{R}} = \mathbf{J}^{-1} \frac{\partial}{\partial_\xi} = \mathbf{j} \frac{\partial}{\partial_\xi} \tag{5.32}$$

A volume element is defined by

$$dV = \left| (d\mathbf{x} \wedge d\mathbf{y}) \cdot d\mathbf{z} \right| \tag{5.33}$$

which, in Cartesian space, is $dV = dx\, dy\, dz$, and in parametric space:

$$dV = \left|\left(\mathbf{a}_1 d\xi \wedge \mathbf{a}_2 d\eta\right) \cdot \mathbf{a}_3 d\zeta\right| \tag{5.34}$$

which is the same as

$$dV = J d\xi d\eta d\zeta \qquad \text{where: } J = \left|\det \mathbf{J}\right| \tag{5.35}$$

Therefore, the quantity Π^e can be expressed in parametric space as

$$\int_{\Omega^e} (\ldots) dx\, dy\, dz = \int_{\Omega^{parent}} (\ldots) J\, d\xi\, d\eta\, d\zeta \tag{5.36}$$

which allows the computation of each element to be performed on the parent element rather than on the real element.

5.3.4 Discretization of the Variational Form

We now write the solution functions for the piezoelectric problem in terms of the shape functions. For each element Ω^e defined by n nodes, the electric field is obtained from equation (7.6) and can be written in the form:

$$\mathbf{E} = -\mathbf{B}_\phi^e \mathbf{\Phi} \tag{5.37}$$

where $\mathbf{\Phi}$ is the vector associated with the nodal values of the electrostatic potential:

$$\mathbf{B}_\phi^e = \left[\mathbf{B}_{\phi 1}^e \quad \mathbf{B}_{\phi 2}^e \quad \cdots \quad \mathbf{B}_{\phi n}^e\right] \tag{5.38}$$

and

$$\mathbf{B}_{\phi i}^e = \begin{bmatrix} \dfrac{\partial N_i^e}{\partial x} \\ \dfrac{\partial N_i^e}{\partial y} \\ \dfrac{\partial N_i^e}{\partial z} \end{bmatrix} \tag{5.39}$$

The terms $\mathbf{B}^e_{\Phi i}$ are the first spatial derivatives of the shape functions. Similarly, for each Ω^e the strain tensor defined by equation (5.4) becomes

$$\mathbf{x} = -\mathbf{B}_u^e \mathbf{U} \tag{5.40}$$

where $\mathbf{U}$ is the vector of the nodal values of the displacement:

$$\mathbf{B}_u^e = \left[\mathbf{B}_{u1}^e \quad \mathbf{B}_{u2}^e \quad \cdots \quad \mathbf{B}_{un}^e\right] \tag{5.41}$$

and

$$\mathbf{B}^{\mathbf{e}}_{\mathbf{ui}} = \begin{bmatrix} \frac{\partial N_i^e}{\partial x} & 0 & 0 \\ 0 & \frac{\partial N_i^e}{\partial y} & 0 \\ 0 & 0 & \frac{\partial N_i^e}{\partial z} \\ 0 & \frac{\partial N_i^e}{\partial z} & \frac{\partial N_i^e}{\partial y} \\ \frac{\partial N_i^e}{\partial z} & 0 & \frac{\partial N_i^e}{\partial x} \\ \frac{\partial N_i^e}{\partial y} & \frac{\partial N_i^e}{\partial x} & 0 \end{bmatrix} \tag{5.42}$$

Consequently, equation (5.1) becomes:

$$\begin{cases} \mathbf{X} = \mathbf{c}^{\mathbf{E}}\mathbf{B}^{\mathbf{e}}_{\mathbf{u}}\mathbf{U} + \mathbf{e}^{X}\ \mathbf{B}^{\mathbf{e}}_{\phi}\ \mathbf{\Phi} \\ \mathbf{D} = \mathbf{e}\ \mathbf{B}^{\mathbf{e}}_{\mathbf{u}}\mathbf{U} + \chi^{x}\mathbf{B}^{\mathbf{e}}_{\phi}\ \mathbf{\Phi} \end{cases} \tag{5.43}$$

Finally, we can rewrite the functional Π^e on the element, as

$$\Pi^e = \frac{1}{2}\iiint\limits_{\Omega^e} U^{e^X}(B_u^{e^X} c^E B_u^e - \rho\,\omega^2 N^{e^X} N^e)U^e d\Omega^e$$

$$+\iiint\limits_{\Omega^e} U^{e^X} B_u^{e^X} e B_\phi^e \Phi^e d\Omega^e - \frac{1}{2}\iiint\limits_{\Omega^e} \Phi^{e^X} B_\phi^{e^X} \chi^x B_\phi^e \Phi^e d\Omega^e$$

$$-\iint\limits_{S_X^e} U^{e^X} N^{e^X} f\, dS_X^e + \sum_{p=0}^{p} \phi_p Q_p \tag{5.44}$$

After integrating the shape function matrices and their derivatives, we can write:

$$\Pi^e = \frac{1}{2}U^{e^X}(K_{uu}^e - \omega^2 M^e)U^e + U^{e^X} K_{u\phi}^e \Phi^e + \frac{1}{2}\phi^{e^X} K_{\phi\phi}^e \Phi^e - U^{eX} F^e + \sum_{p=0}^{p} \phi_p Q_p \tag{5.45}$$

where

$$\mathbf{K}^{\mathbf{e}}_{\mathbf{uu}} = \iiint\limits_{\Omega^e} \mathbf{B}^{\mathbf{e}\,x}_{\mathbf{u}} \mathbf{c}^{\mathbf{E}} \mathbf{B}^{\mathbf{e}}_{\mathbf{u}} d\Omega^e$$

$$\mathbf{M}^{\mathbf{e}} = \iiint\limits_{\Omega^e} \rho\omega^2 \mathbf{N}^{\mathbf{e}\,x} \mathbf{N}^{\mathbf{e}} d\Omega^e$$

$$\mathbf{K}^{\mathbf{e}}_{\boldsymbol{\phi}\mathbf{u}} = \iiint\limits_{\Omega^e} \mathbf{B}^{\mathbf{e}\,x}_{\mathbf{u}} \mathbf{e}\, \mathbf{B}^{\mathbf{e}}_{\boldsymbol{\phi}} d\Omega^e \tag{5.46}$$

$$\mathbf{K}^{\mathbf{e}}_{\boldsymbol{\phi\phi}} = -\iiint\limits_{\Omega^e} \mathbf{B}^{\mathbf{e}\,x}_{\boldsymbol{\phi}} \chi^x \mathbf{B}^{\mathbf{e}}_{\boldsymbol{\phi}} d\Omega^e$$

$$\mathbf{F}^{\mathbf{e}} = \iint\limits_{S^e_X} \mathbf{N}^{\mathbf{e}\,x} \mathbf{f}\, dS^e_X$$

and K^{e}_{uu}, $K^{e}_{\Phi u}$, and $K^{e}_{\Phi\Phi}$ are the elastic, piezoelectric, and dielectric susceptibility matrices, and M^e is the consistent mass matrix.

5.3.5 Assembly

The matrices in equation (5.47) must be rearranged for the whole domain Ω by a process called *assembly*. From this process, we obtain the following matrices:

$$\mathbf{K}_{\mathbf{uu}} = \sum_e \mathbf{K}^{\mathbf{e}}_{\mathbf{uu}}$$

$$\mathbf{M} = \sum_e \mathbf{M}^{\mathbf{e}}$$

$$\mathbf{K}_{\boldsymbol{\phi}\mathbf{u}} = \sum_e \mathbf{K}^{\mathbf{e}}_{\boldsymbol{\phi}\mathbf{u}} \tag{5.47}$$

$$\mathbf{K}_{\boldsymbol{\phi\phi}} = \sum_e \mathbf{K}^{\mathbf{e}}_{\boldsymbol{\phi\phi}}$$

$$\mathbf{F} = \sum_e \mathbf{F}^{\mathbf{e}}$$

The application of the variational principle implies the minimization of the functional Π with respect to variations of the nodal values **U** and Φ. Therefore:

$$\frac{\partial \Pi}{\partial u_i} = 0 \quad \forall\ i \tag{5.48}$$

and

$$\frac{\partial \Pi}{\partial \phi_j} = 0 \quad \forall\ j \tag{5.49}$$

Making use of equations (5.46) and (5.16), and applying the stationary condition to Π, we obtain:

$$\begin{bmatrix} \mathbf{K_{uu}} - \omega^2\mathbf{M} & \mathbf{K_{u\phi}} \\ \mathbf{K_{u\phi}}^X & \mathbf{K_{\phi\phi}} \end{bmatrix} \begin{bmatrix} \mathbf{U} \\ \mathbf{\Phi} \end{bmatrix} = \begin{bmatrix} \mathbf{F} \\ -\mathbf{Q} \end{bmatrix} \tag{5.50}$$

The vector for the nodal charges, **Q**, is such that for all nodes i that belong to an electrode p with potential ϕ_p, the sum of the charges Q_i is equal to ϕ_p. For all other nodes j that do not belong to an electrode, $Q_j = 0$.

5.3.6 Computation

Specific integration, diagonalization, and elimination techniques are employed to solve the system (equation 5.51) on a computer. A full description of these techniques is a topic that extends well beyond the scope of this text and thus will not be presented here. The matrix equation (equation 5.51) may be adapted for a variety of different analyses, such as the static, modal, harmonic, and transient types:

- Static analysis

$$\begin{bmatrix} \mathbf{K_{uu}} & \mathbf{K_{u\phi}} \\ \mathbf{K_{u\phi}}^X & \mathbf{K_{\phi\phi}} \end{bmatrix} \begin{bmatrix} \mathbf{U} \\ \mathbf{\Phi} \end{bmatrix} = \begin{bmatrix} \mathbf{F} \\ -\mathbf{Q} \end{bmatrix} \tag{5.51}$$

- Modal analysis

$$\begin{bmatrix} \mathbf{K_{uu}} - \omega^2\mathbf{M} & \mathbf{K_{u\phi}} \\ \mathbf{K_{u\phi}}^X & \mathbf{K_{\phi\phi}} \end{bmatrix} \begin{bmatrix} \mathbf{U} \\ \mathbf{\Phi} \end{bmatrix} = \begin{bmatrix} 0 \\ -\mathbf{Q} \end{bmatrix} \tag{5.52}$$

- Harmonic analysis

$$\begin{bmatrix} \mathbf{K_{uu}} - \omega^2\mathbf{M} & \mathbf{K_{u\phi}} \\ \mathbf{K_{u\phi}}^X & \mathbf{K_{\phi\phi}} \end{bmatrix} \begin{bmatrix} \mathbf{U} \\ \mathbf{\Phi} \end{bmatrix} = \begin{bmatrix} \mathbf{F} \\ -\mathbf{Q} \end{bmatrix} \tag{5.53}$$

- Transient analysis

$$\begin{bmatrix} \mathbf{M} & \mathbf{0} \\ \mathbf{0} & \mathbf{0} \end{bmatrix} \begin{bmatrix} \ddot{\mathbf{U}} \\ \ddot{\mathbf{\Phi}} \end{bmatrix} + \frac{1}{\omega_o} \begin{bmatrix} \mathbf{K}'_{\mathbf{uu}} & \mathbf{K}'_{\mathbf{u\phi}} \\ \mathbf{K}'^{\,X}_{\mathbf{u\phi}} & \mathbf{K}'_{\mathbf{\phi\phi}} \end{bmatrix} \begin{bmatrix} \dot{\mathbf{U}} \\ \dot{\mathbf{\Phi}} \end{bmatrix} + \begin{bmatrix} \mathbf{K_{uu}} & \mathbf{K_{u\phi}} \\ \mathbf{K_{u\phi}}^X & \mathbf{K_{\phi\phi}} \end{bmatrix} \begin{bmatrix} \mathbf{U} \\ \mathbf{\Phi} \end{bmatrix} = \begin{bmatrix} \mathbf{F} \\ -\mathbf{Q} \end{bmatrix} \tag{5.54}$$

Each of these cases requires specific conditioning and computation techniques. Some of the techniques used in the commercial program ATILA are demonstrated in the CD supplement to the text.

5.4 FEM SIMULATION EXAMPLES

Application examples of the finite element method ATILA to piezoelectric actuators and transducers are introduced in this chapter, including multilayer actuators, Π-type linear, "windmill," metal tube ultrasonic motors, piezoelectric transformers and "cymbal" underwater transducers.

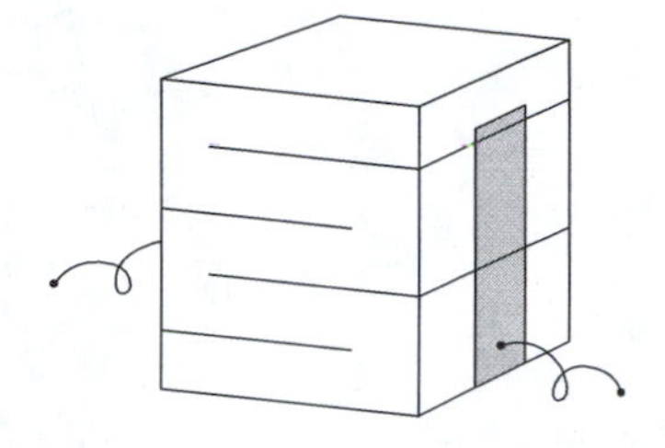

FIGURE 5.7 Multilayer piezoelectric actuator with an interdigital electrode pattern.

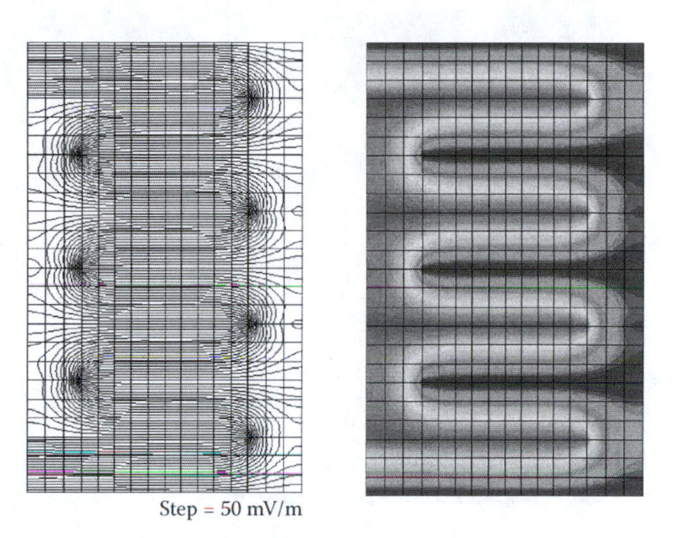

FIGURE 5.8 Potential distribution around the internal electrode edges in an eight-layered multilayer actuator (2-D simulation results).

5.4.1 Multilayer Actuator

As shown in Figure 5.7, a multilayer piezoelectric actuator is composed of active and inactive areas, corresponding to both the electrode-overlapped and electrode-unoverlapped portions. Accordingly, concentration of electric field and stress is expected around the internal electrode edges.

Figures 5.8 and 5.9 are the 2-D calculation results with ATILA under pseudo-DC drive in an eight-layered multilayer actuator in terms of the potential distribution and stress concentration around the internal electrode edges. The maximum tensile stress should be lower than the fracture strength of this PZT ceramic.

5.4.2 Π-Type Linear Ultrasonic Motor

Using a Longitudinal 1st-Bending 4th coupled mode linear motor as illustrated in Figure 5.10, we can learn how to optimize the motor dimensions.[5] Figure 5.11 shows the dependence of the Longitudinal 1 mode and Bending 4 mode resonance frequencies on the thickness of the elastic bar around 2.8 mm. The L1 mode resonance frequency is rather insensitive to the bar thickness, whereas the B4 mode frequency varies significantly with the thickness. When the thickness is 2.5 mm, driving at a point (a) provides a symmetrical large leg-wagging motion, whereas driving at (b) provides a leg up-down action. Neither can provide the motor movement. When the thickness is adjusted to 2.8 mm, driving at a point (c), which superposes L1 and B4 modes simultaneously, we observe a horse-trotting mode with a 90° phase lag between two leg-wagging motions, which must be the optimized condition.

Figure 5.12 compares the ATILA calculation and experimental results for a brass motor. The optimum bar thicknesses are 2.75 and 2.80 mm, respectively, which differs only by less than 2%.

5.4.3 Windmill Ultrasonic Motor

Figure 5.13a illustrates the windmill motor structure, and Figure 5.13b shows the corresponding metal ring/finger coupled vibration mode (radial-bending coupling), calculated by ATILA (3-D calculation).[6] Note that when the radial and bonding vibration modes are synchronized, the motion seems to be a grasping-twisting action of four fingers.

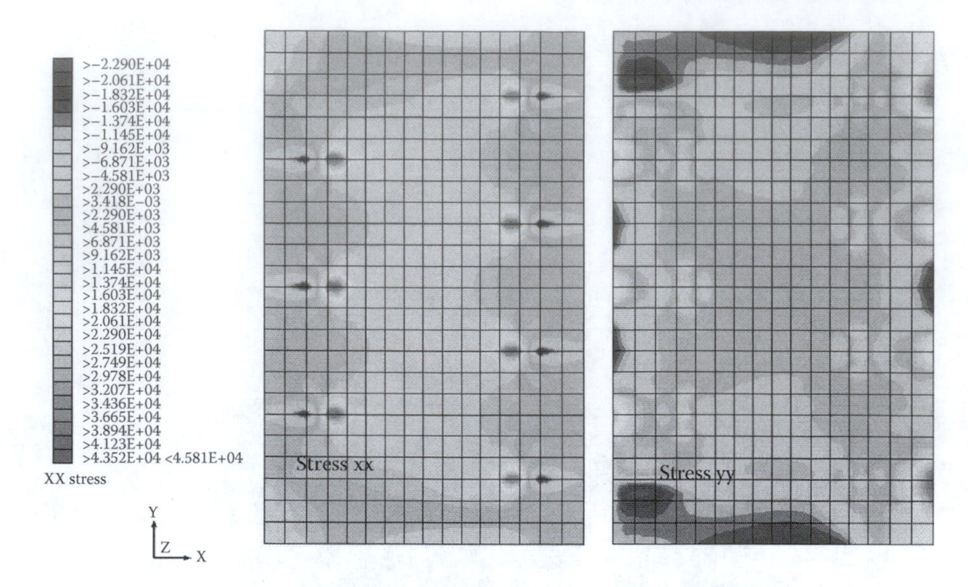

FIGURE 5.9 Stress concentration (XX and YY) around the internal electrode edges (2-D simulation results).

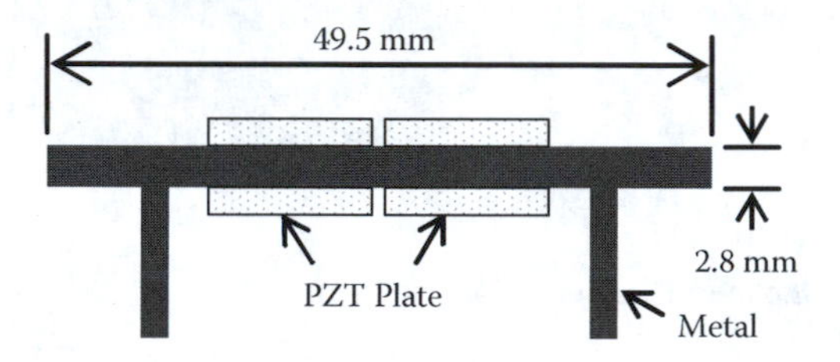

FIGURE 5.10 Dimensions of a π-shaped linear motor optimally determined by the finite element analysis (FEM): the bimorph type.

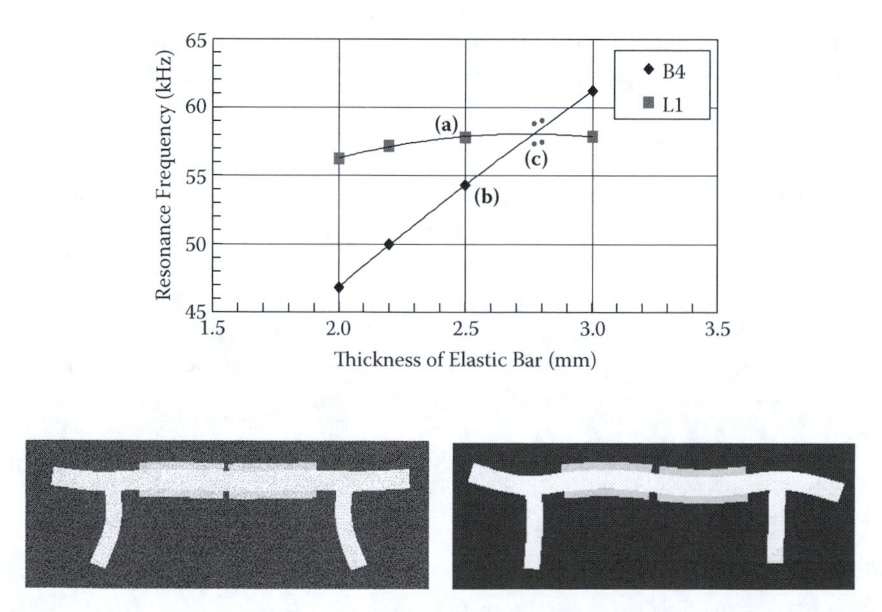

FIGURE 5.11 Dependence of the L1 and B4 resonance frequencies on the thickness of the elastic bar around 2.8 mm. Driving at point (a) or at point (b) corresponds to a symmetrical wagging or an up-down action. Driving at (c) results in a horse-trotting motion.

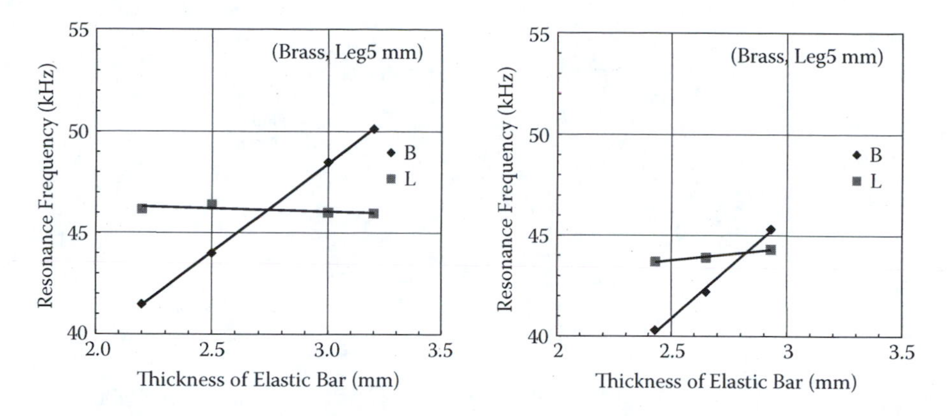

FIGURE 5.12 (a) ATILA calculation results and (b) experimental results.

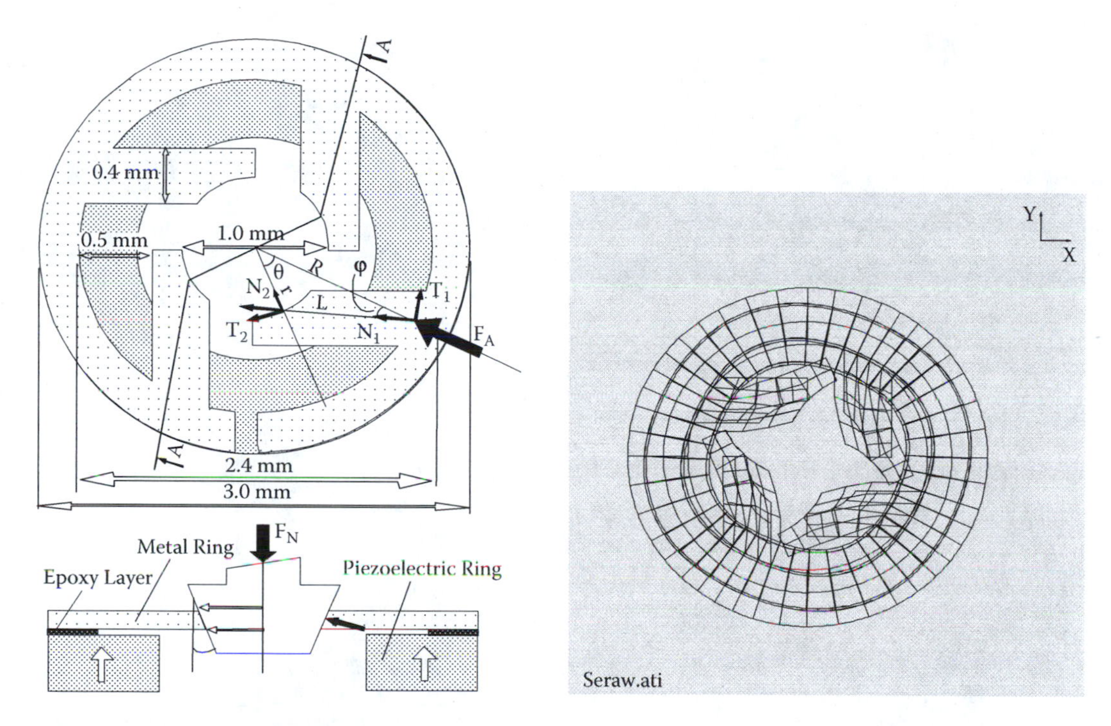

FIGURE 5.13 (a) Windmill motor illustration, and (b) a metal ring/finger coupled vibration mode.

5.4.4 Metal Tube Ultrasonic Motor

A metal tube motor is composed of two PZT rectangular plates bonded on a metal tube, as depicted in Figure 5.14.[7] Because of this asymmetric configuration of the two PZT plates, the resonance frequencies of the two orthogonal bending modes along x' and y' deviate slightly. Thus, driving the motor at the intermediate frequency exhibits a superposed vibration, similar to a hula-hoop mode, because of 90° phase lag between the preceding two split modes. Exciting either X or Y plate generates counterclockwise or clockwise wobbling motion of the metal tube, respectively. (See the ATILA calculation in Figure 5.15. Time sequence is from top to bottom.) Refer to chapter 6 for the design optimization of the metal tube motor.

The ATILA software can calculate the stator vibration under mechanically free or forced condition, but without coupling another software that treats the surface friction model, the motor characteristics (speed, torque, etc.) cannot be calculated.

5.4.5 Piezoelectric Transformer

With reference to equation (2.56), disk-type transformers have advantages over the rectangular plate Rosen type because of the usage of k_p, instead of k_{31} ($k_p > k_{31}$). One of the disk types shown in Figure 5.16 exhibits an enhanced step-up voltage ratio

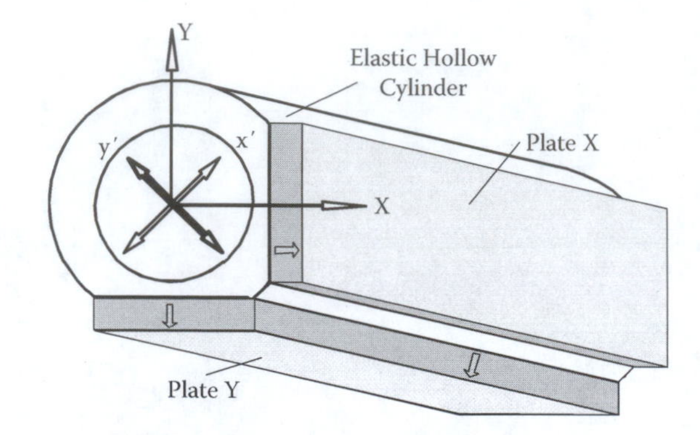

FIGURE 5.14 Structure of a metal tube motor. Two PZT plates are bonded asymmetrically on a metal tube.

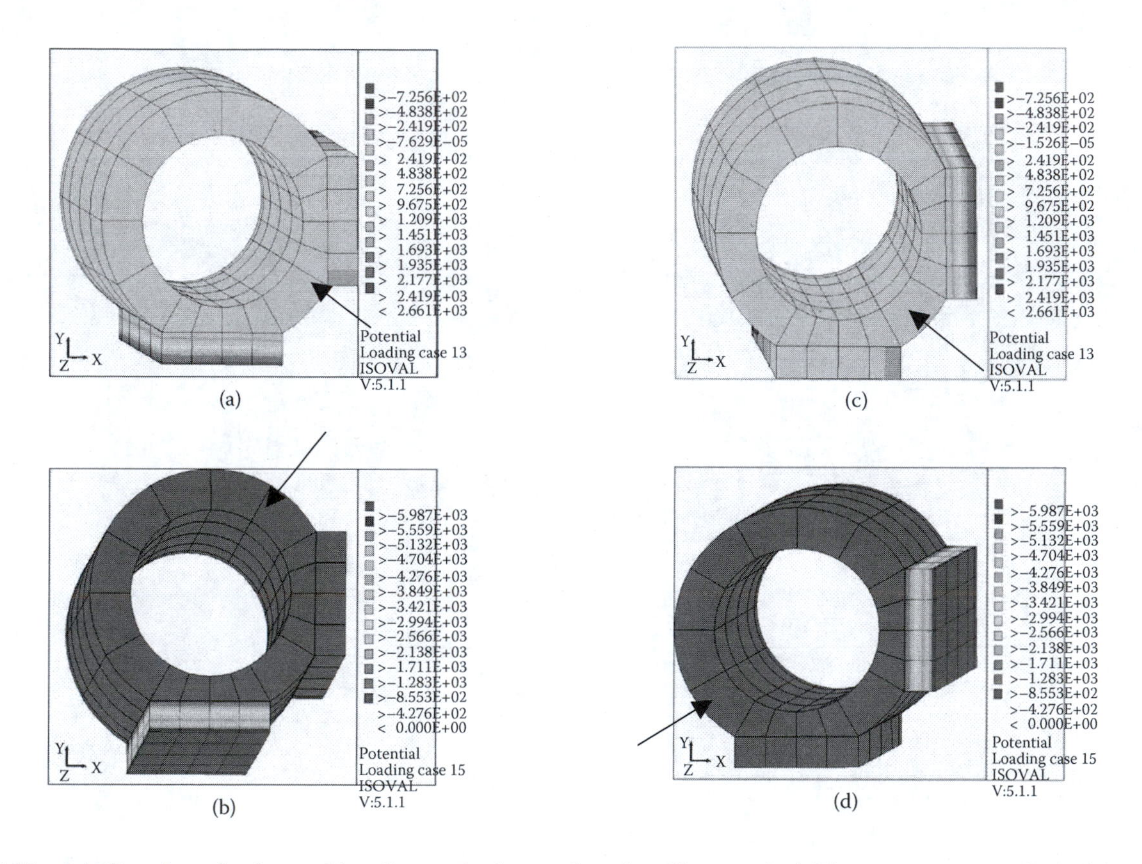

FIGURE 5.15 (a,b) Top view of orthogonal bending mode shapes when plate X was excited (Time sequence: a → b); (c,d) bending mode shapes when plate Y was excited (Time sequence: c → d). X or Y plate excitation generates counterclockwise or clockwise rotation.

of more than 300 under zero load, because the curved electrode configuration excites the k_{15} mode superposing on k_p mode (Note again that $k_{15} > k_p$).[8] Furthermore, curved electrodes reduce the stress concentration, which can be calculated by ATILA.[1] Figure 5.17 demonstrates higher-order vibration modes and the calculated potential distribution.

The most recent version of ATILA can simulate step-up/-down voltage ratio even under a certain electrical load condition.

We demonstrated lighting up the 4 W CCFL directly by 500 V using a single disk sample (25 mm in diameter and 1 mm in thickness) without using an additional booster coil.

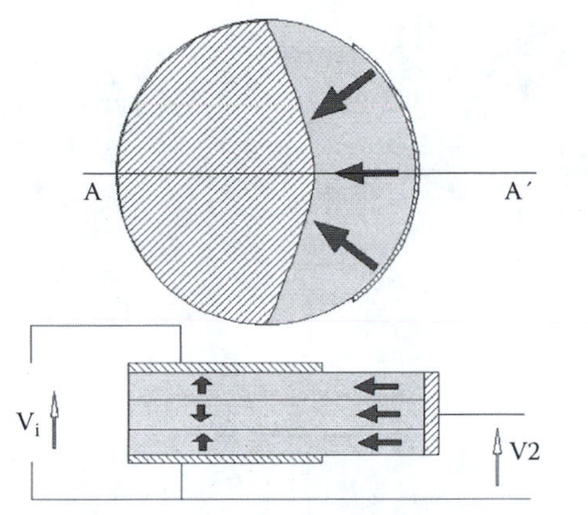

FIGURE 5.16 Disk-shaped piezoelectric transformer with crescent curved electrodes (Penn State). Different from the conventional Rosen-type, the circular shape (k_p) enhances the energy conversion rate and voltage step-up ratio. Also, the curved electrode contour excites local shear mode (k_{15}), which further enhances the voltage step-up ratio to more than 300. A PZT disk with 25 mm diameter and 1 mm thickness can generate 10 W or higher, so that this three-layer transformer can be applied to a small laptop computer AC/DC adaptor.

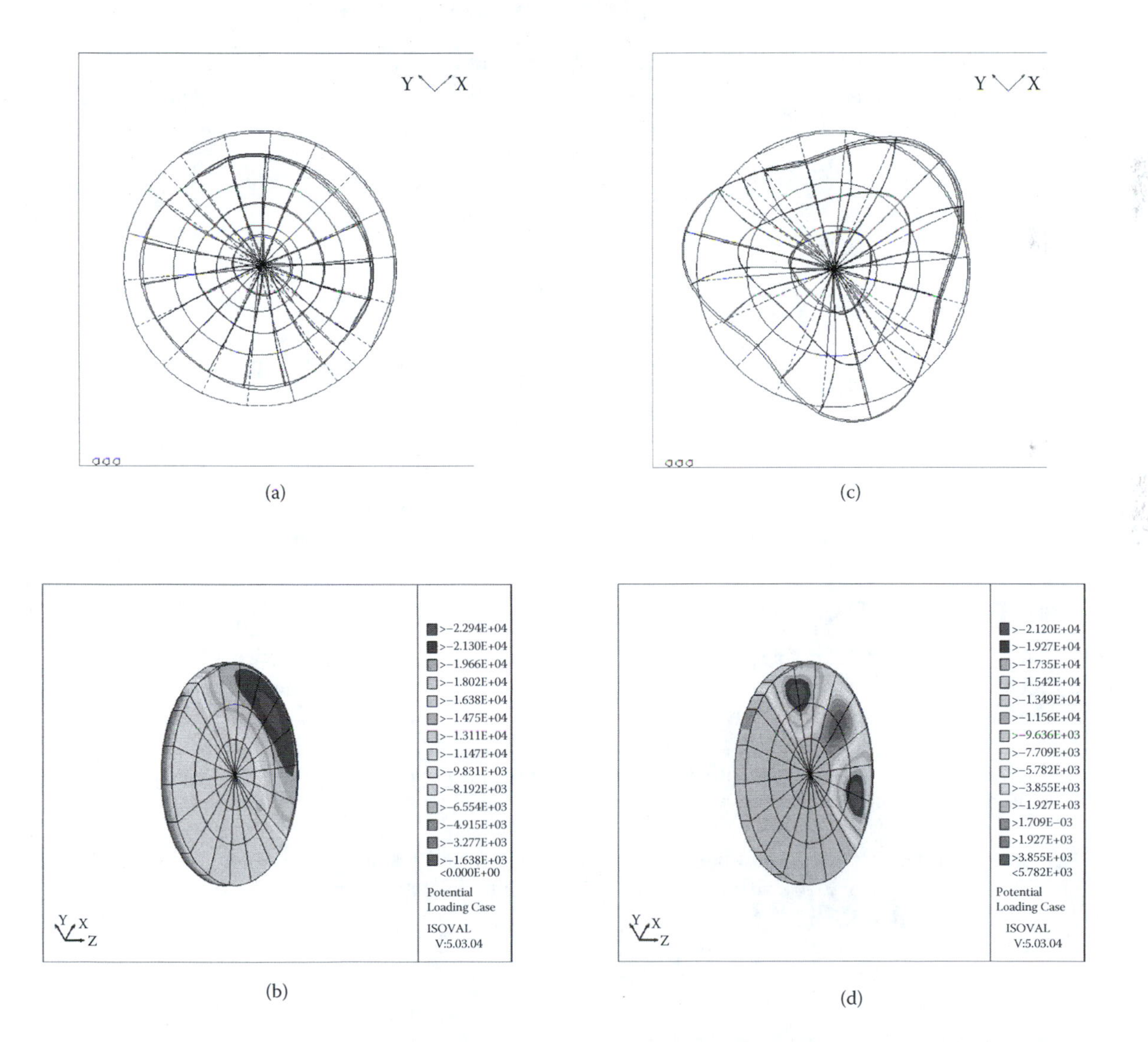

FIGURE 5.17 The first and third vibration modes [(a) and (c)] of a crescent electrode–type disk transformer, and their corresponding voltage distribution calculation by ATILA [(b) and (d)], without load (open circuit).

FIGURE 5.18 (a) Cymbal appearance and (b) a half-cross-section model for the ATILA calculation.

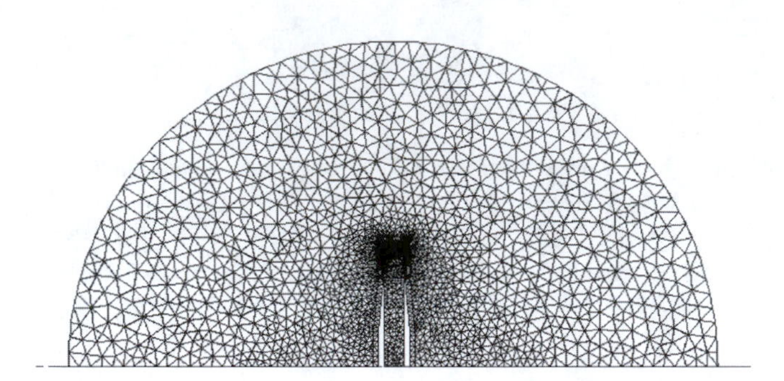

FIGURE 5.19 Automatically generated mesh for water surrounding the Cymbal transducer.

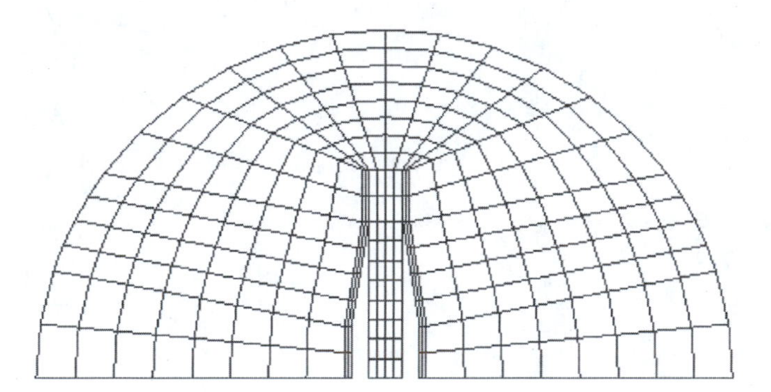

FIGURE 5.20 Structured mesh for water surrounding the Cymbal transducer, taking the symmetry into account.

5.4.6 Cymbal Underwater Transducer

A cymbal (Figure 5.18a) is a type of displacement amplifier, and also fits well with water in view of its relatively low acoustic impedance. Figures 5.18b and 5.19 exhibit mesh structures of the cymbal and surrounding water, respectively. Figure 5.20 shows an alternative mesh structure, structured to minimize the node numbers, leading to reduced calculation time.

In this textbook, using the provided ATILA-Light education version, the reader is recommended to learn the most effective structured mesh to simulate the device with the minimum node number.

Figure 5.21 compares the resulting ATILA simulation and the experimental result in terms of transmitting voltage response as a function of operating frequency. A sharp dip of the *TVR* (transmitted voltage response) signal around 65 kHz is also successfully traced (refer to section 3.7.3).[9]

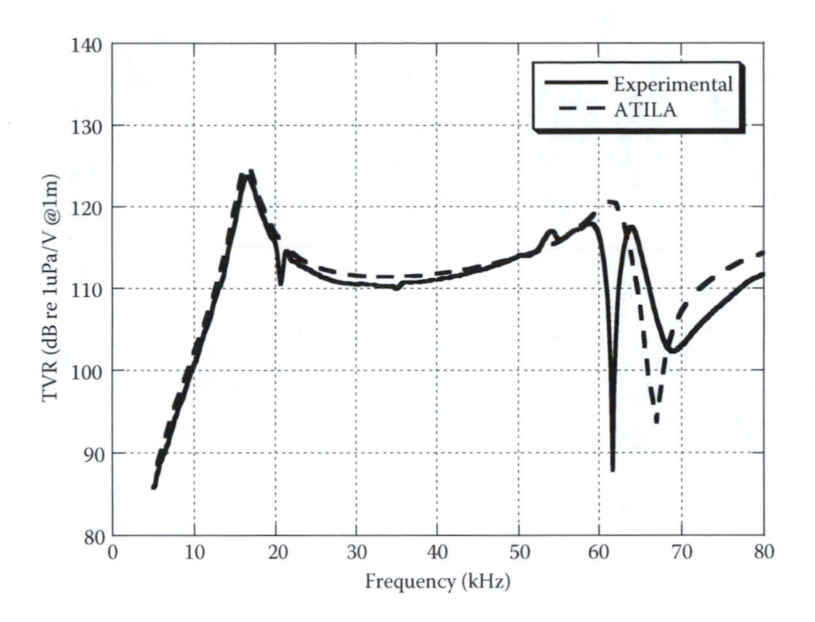

FIGURE 5.21 Comparison of the ATILA simulation and the experimental result in terms of transmitted voltage response (TVR) as a function of frequency.

REFERENCES

1. O. C. Zienkiewicz: The Finite Element Method—3rd expanded and revised ed., ISBN 0-07-084072-5, McGraw-Hill Book Company (U.K.) Limited (1977).
2. J.-N. Decarpigny: Application de la Méthode des Eléments Finis à l'Etude de Transducteurs Piézoélectriques, Doctoral thesis, ISEN, Lille, France (1984).
3. T. Ikeda: *Fundamentals of Piezoelectricity*, ISBN: 0-19-856339-6, Oxford University Press (1990).
4. B. Dubus, J.-C. Debus, and J. Coutte: Modélisation de Matériaux Piézoélectriques et Electrostrictifs par la Méthode des Eléments Finis, Revue Européenne des Eléments Finis, Vol. 8, No. 5–6, pp. 581–606 (1999).
5. T.-Y. Kim, B.-J. Kim, T.-G. Park, M.-H. Kim, and K. Uchino: Design and Driving Characteristics of Ultrasonic Linear Motor, *Ferroelectrics*, **263**, 113–118 (2001).
6. B. Koc., P. Bouchilloux, and K. Uchino: Piezoelectric Micromotor Using A Metal-Ceramic Composite Structure, *IEEE Transactions Ultrasonic, Ferroelectrics, and Frequency Control* **47** (4), 836–843 (2000).
7. B. Koc, S. Cagatay, and K. Uchino: A Piezoelectric Motor Using Two Orthogonal Bending Modes of a Hollow Cylinder, *IEEE Ultrasonic, Ferroelectric, Frequency Control Transactions*, **49**(4), 495–500 (2002).
8. B. Koc and K. Uchino: A Disk Type Piezoelectric Transformer with Crescent Shape Input Electrodes, *Proceedings of Piezoelectric Materials: Advances in Science, Technology and Applications*, pp. 375–382 (2000).
9. J. Zhang, W. J. Hughes, P. Bouchilloux, R. J. Meyer Jr., K. Uchino, and R. E. Newnham: A Class V Flextensional Transducer: The Cymbal, *Ultrasonics*, 37, 387–393 (1999).

6 Design Optimization with FEM

To successfully design piezoelectric transducers such as an ultrasonic motor, it is often necessary to simultaneously determine a large number of parameters. This is particularly true with mixed-mode ultrasonic motors because they involve at least two structural modes of vibration of different natures. Typically, these modes must display similar eigen frequencies, so that the resulting vibrations can be effectively combined in an elliptical field. Moreover, these modes are excited by means of piezoelectric elements. Maximizing the electromechanical coupling for each mode is of great importance, because it is proportional to the force factor of the ultrasonic device. Finally, the amplitude of the vibration in each direction must be evaluated to determine the viability of the system.

This list of parameters is clearly not exhaustive. The complete analysis of the ultrasonic motor requires further analysis of the tribology of the contact mechanism at the stator–rotor interface, for instance. However, at this stage of our work, we will focus our attention on the stator (or vibrating element) of the ultrasonic motor alone. As will be shown later, the method presented here allows for including higher levels of complexity in the system, and it will eventually be extended to include complete ultrasonic motor analysis and optimization capabilities.

We will consider two types of optimization processes: one is a conventional parameter scan type, and the other is a genetic optimization process. When we scan variable parameters, sometimes the optimized design is not globally best, but only locally better. To obtain the globally best design, a genetic algorithm can be coupled in the simulation process.

6.1 OPTIMIZATION OF THE METAL TUBE MOTOR[1]

Figure 6.1 shows geometrical and material parameters for a metal tube motor optimization. The main purpose of the optimization study is to investigate the characteristics of the stator when the piezoelectric plates are placed at an optimum angle so that the two orthogonal bending mode frequencies are equivalent to each other. There are two main groups under consideration: geometrical and material parameters. The geometrical parameters are divided into two subgroups in terms of the elastic material: with circular cross section and with square cross section. In the circular cross-section group, the rod and tube are compared considering the optimum angle between the ceramic plates. In the square cross-section group, instead of the angle, the behavior of the stator with either two or four ceramic plates was inspected. In terms of material parameters, aluminum and brass are evaluated for each circular cross section.

6.1.1 Circular Cross Section

The stator of the ultrasonic motor designed had two flattened surfaces on which piezoelectric ceramics were bonded. Because of the partially circular and partially square cross sections, the stator had two slightly split bending modes originating from the degenerated modes. The difference in the resonance frequencies of these modes was in the 3–5 kHz range.

Originally, the angle between these two surfaces was 90°. However, the ceramics can be oriented at angles other than 90°. To equate slightly the split resonance frequencies, the ceramic plates must be positioned at an optimum angle. Theoretical calculations by ATILA revealed that this angle was equal to 83° when the densities of the ceramic and elastic material were assumed to be equal (Figure 6.2). On the basis of these results, the behavior of the stator with the following dimensions was examined: outer diameter = 1.6 mm, inner diameter = 0.8 mm, length = 14 mm, and thickness of the ceramic plates = 0.3 mm. Elastic material was chosen as brass, and the ceramic plates used were soft PZT.

Figure 6.3 shows an optimization process of the angle between the two PZT plates for a brass metal tube. The key is to adjust two split resonance peaks corresponding to two orthogonal bending modes. Taking an angle at 83°, two peaks almost overlap, providing the maximum wobbling vibration amplitude in a brass tube. The optimum angle differs depending on the metal species.

Brass rod and tube. The tendency of the admittance curves when the ceramic plates were placed at two different angles is shown in Figure 6.4. A single peak was observed when the angle was at 80°. When the angle diverged from 80°, the two bending mode frequencies were slightly split. The optimum angle was independent of the geometry. For both the *brass rod* and the *tube*, the angle remained the same. However, the resonance frequency shifted to a higher value in the case of the tube. For the rod, the resonance frequency was slightly less than 35 kHz, whereas for the tube, it was around 39 kHz. This variation was due to the difference in the densities of the rod and the tube.

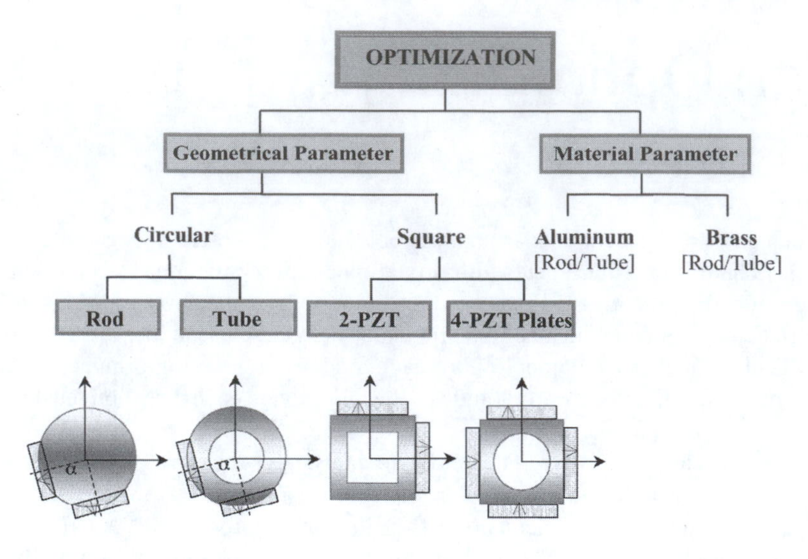

FIGURE 6.1 Geometrical and material parameters for a metal tube motor optimization.

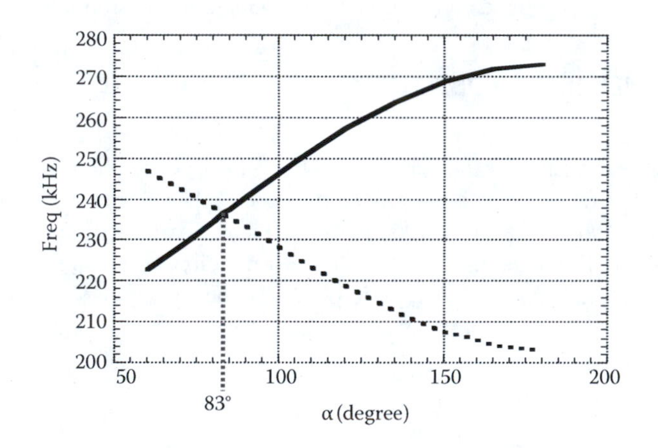

FIGURE 6.2 Change in bending mode frequency with respect to the orientation angle of ceramic plates.

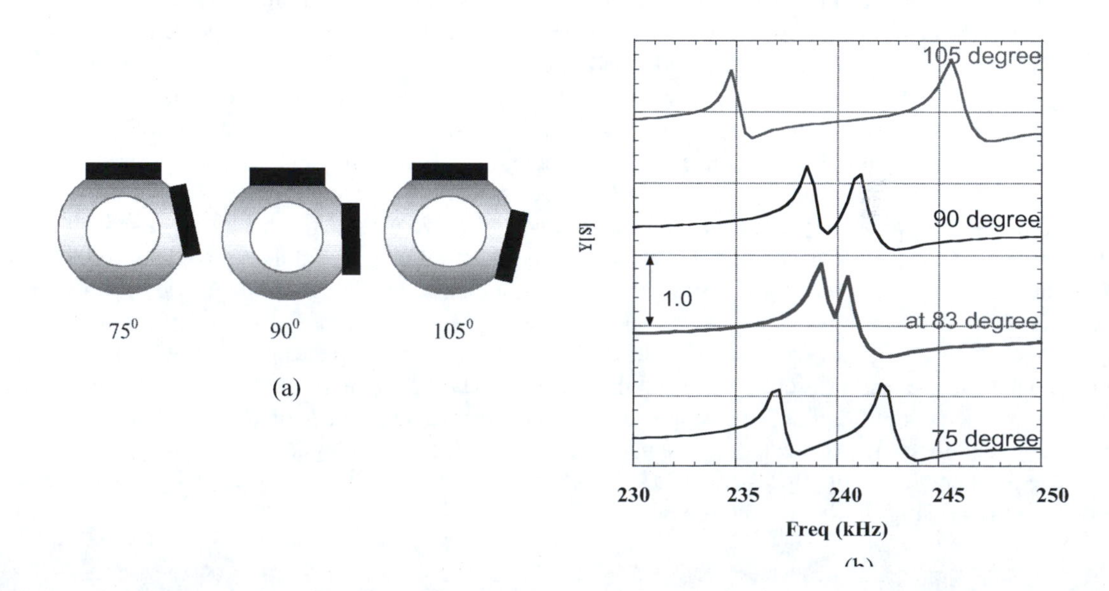

FIGURE 6.3 Angle optimization of the two PZT plates for enhancement of the wobbling vibration amplitude.

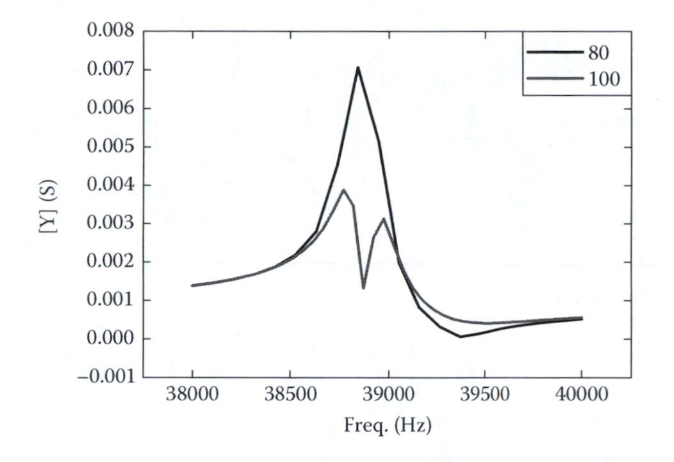

FIGURE 6.4 Admittance curves with respect to the angle for brass rod and tube.

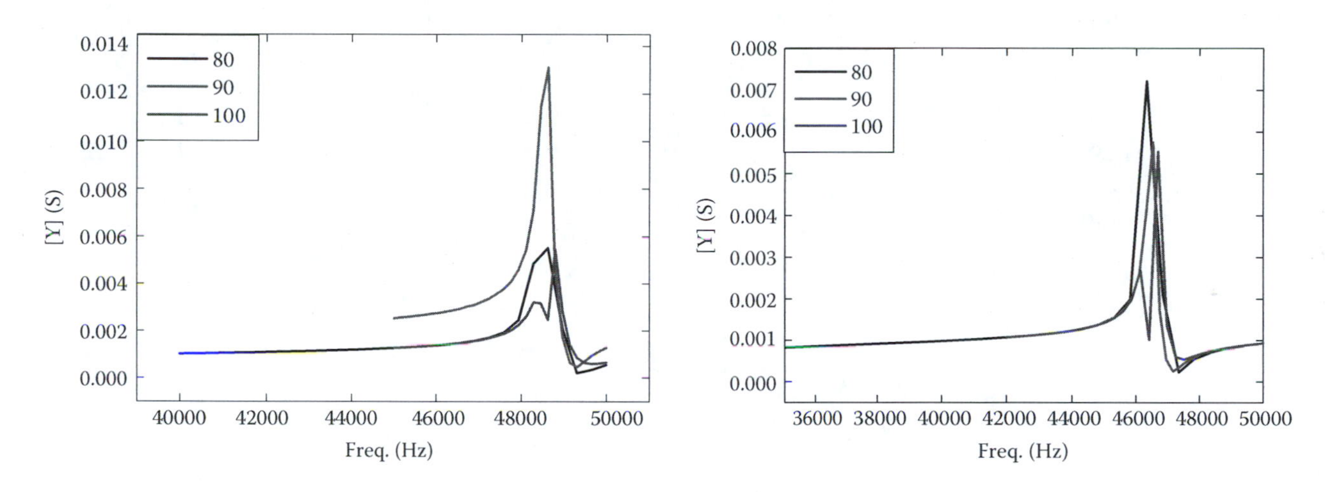

FIGURE 6.5 Admittance curves for (a) aluminum tube and (b) rod motors.

Aluminum rod and tube. Figure 6.5a illustrates the dependence of bending mode resonance frequencies on various angles if the elastic material was an *aluminum tube*. From the graph, it can be observed that the optimum angle for this structure was 90°. When the angle was equal to 100°, split of the resonance frequencies occurred. Although a single peak was observed at 80°, the peak observed at 90° was sharper, and the resonance frequencies matched one another. Despite the coupled appearance in the admittance curves of the two bending modes at 90°, one of the modes was not excited. The stator composed of aluminum tube can only excite one bending mode in one direction. Hence, elliptical motion cannot be observed. This result was due to the softness of the elastic material. When the elastic material was stiffer, such as brass, the ceramic plate, which was connected to the ground, acted as a mass, resulting in two orthogonal bending modes. On the other hand, in the case of an *aluminum rod* at either 80° or 90°, the two modes were excited at the same resonance frequency, as shown in Figure 6.5b. At 100°, split of the resonance frequencies can be realized. Because of the stiffness of the rod, elliptical motion was observed.

6.1.2 Square Cross Section

A stator with a square cross section and 4-PZT plates excited with two-phase was considered. The structure of the stator is shown in Figure 6.6. The length was 13 mm. The admittance graph in Figure 6.7 concludes that the two bending modes were coupled. Simulation results also proved that elliptical motion was generated for this kind of stator. As the width of both elastic material and PZT plates increased, the stator became more rigid and the displacement values dropped. However, elliptical motion was still observable.

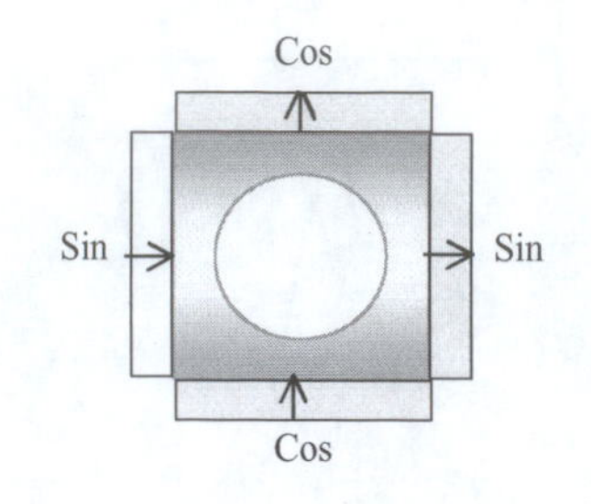

FIGURE 6.6 Stator with a square cross section.

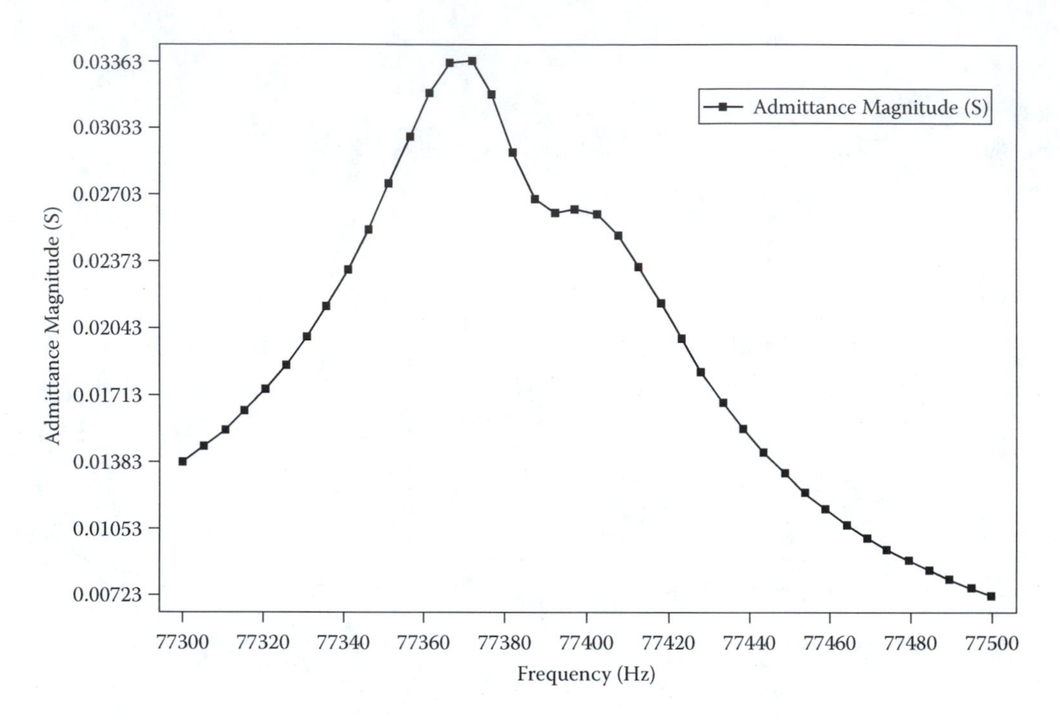

FIGURE 6.7 Admittance curves of the stator with a square cross section.

6.2 GENETIC OPTIMIZATION

The method we introduce here combines the genetic algorithm to finite element analysis. The finite element analysis method suggests itself naturally for modeling a large variety of structures. Commercial FE software, such as ATILA,[2] specialize in the analysis of smart structures and are efficient at solving a variety of piezoelectric structures.[3,4] For this reason, ATILA was chosen as the numerical analysis tool to evaluate the quality of ultrasonic stator designs. The genetic algorithm (GA) was chosen among other optimization procedures for its multidimensionality and ease of implementation. The GA was also attractive for its capability of avoiding local optima and rapidly finding the global optimum in the searched domain.

6.2.1 GENETIC ALGORITHM

The basic principles of genetic algorithms were first proposed by J. H. Holland.[5] These algorithms are inspired by the mechanism of natural selection, in which stronger individuals are the likely winners in a competitive environment.

Genetic algorithms presume that the potential solution of any problem is an individual and can be represented by a set of parameters. These parameters are regarded as the "genes" of a "chromosome," which is a representation of the individual (Figure 6.8). The chromosome can be structured, and it is often represented by a string of values in binary form. A positive value, generally known as a "fitness" value, is used to reflect the degree of "fitness" of the chromosome for the problem. The fitness value is clearly closely related to the objective value.

Throughout a genetic evolution, the fitter chromosome has a tendency to yield good quality "offspring," which means a better solution to the problem. A "population" pool of chromosomes has to be installed, and these can be randomly set initially. In each cycle of the genetic operation a subsequent "generation" is created from the chromosomes in the current population.

Genetic Algorthim Lingo	Example	Description and Comments
Genes	P1, P2,..., PM	Design Parameters
Alleles	P1 0, 1, 2, ..., 7 P2 0, 1, 2, ..., 31 ⋮ ⋮ PM 0, 1, 2, ...	Values Taken by the Parameters (Length, Thickness, Radius, ...)
	Binary Representation 0 1 0 = P1 = 2 0 0 1 1 0 = P2 = 6 ⋮ ⋮ ⋮ 0 1 ··· 0 1 = PM	
Chromosome or Individual	A Tonpilz Transducer	A Solution Described by the Design Parameters
Genotype	P1 P2 PM 0 1 0 | 0 0 1 1 0 | ··· | 0 1 ··· 0 1	A Binary String
Phenotype		A Possible Design

FIGURE 6.8 Brief glossary of genetic algorithms terminology.

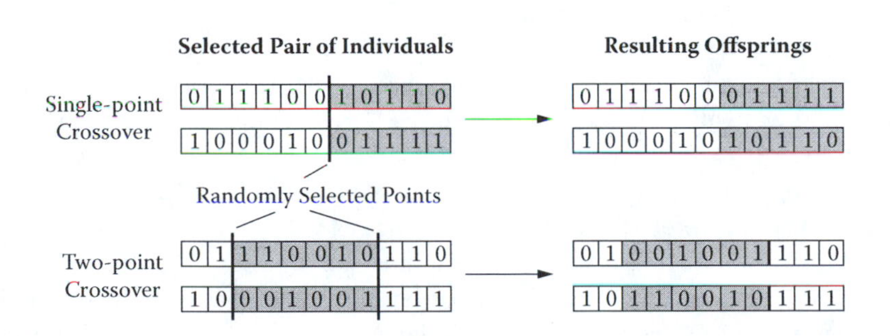

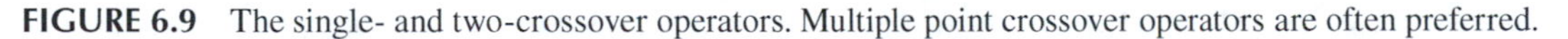

FIGURE 6.9 The single- and two-crossover operators. Multiple point crossover operators are often preferred.

Typically, this step involves three operators, "selection," "crossover," and "mutation" to direct the population toward convergence at the global optimum. The selection operator attempts to apply pressure upon the population in a manner similar to that of natural selection found in biological systems. Poorer-performing individuals are weeded out and better-performing, or fitter, individuals, have a greater than average chance of transmitting the information they contain to the next generation.

The crossover operator allows solutions to exchange information in a way similar to that used by a natural organism (Figure 6.9). The method consists of choosing pairs of individuals promoted by the selection operator, randomly selecting one or more points within the binary strings, and swapping the information (bits) from one point to another between the two individuals. The mutation operator (Figure 6.10) is used to randomly change the value of single bits within individual strings. Excessive use of this operator may disrupt the evolution process.

The application of genetic algorithms to ultrasonic devices presents some difficulties. One of them concerns the coding of the genes, which may not always be regularly spaced integer numbers. Rather, most problems require that design parameters be real numbers, and sometimes, discrete values. Discrete values can easily be mapped, and a lookup table set to code and decode the corresponding gene. A simple and efficient solution for real values consists of applying a tolerance on the parameter value.

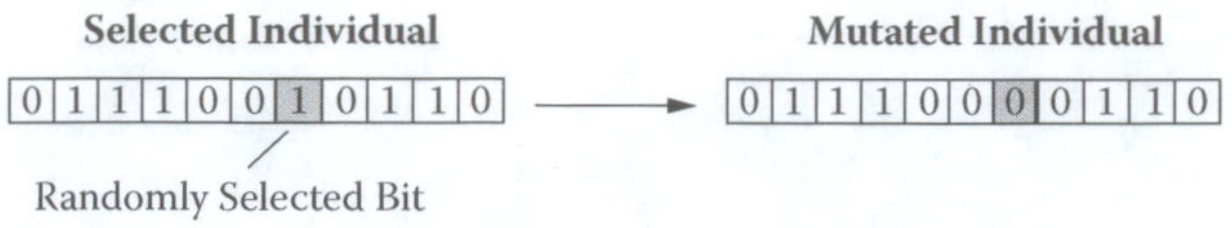

Selected Individual

0 1 1 1 0 0 1 0 1 1 0 → Mutated Individual 0 1 1 1 0 0 0 0 1 1 0

Randomly Selected Bit

FIGURE 6.10 The mutation operator causes an individual to change.

For instance, if one parameter is used to represent the geometrical length of a section of the structure, and assuming that this parameter can take values between 10 and 20 mm, then it is usually sufficient to consider the real value with a tolerance of 1%. Therefore, scaling and shifting the alleles to map them with the integer values 0 to 10,000 provides a satisfactory solution. Further, it is important to note that regular binary numbers may not be best suited to the genetic process. Indeed, observe that the integers 7 and 8 are different by just 1 but that their respective binary representations, 0111 and 1,000, are different by 4 bits. To correct for this, and allow a difference of only 1 bit between two successive integers, Gray codes[6] are used instead of standard binary representations.

A genetic algorithm code following the general rules described earlier has been programmed in object-oriented Pascal on a PC platform. The complete algorithm is shown in Figure 6.11. The code was tested against more or less ill-behaved mathematical functions, as is classically done.[7] The algorithm combining the genetic procedure and the finite element method is shown in Figure 6.12. The general flow diagram is similar to that of the genetic algorithm itself, except for the evaluation function, which is now more complex and includes the finite element analysis.

It is important to emphasize at this point that the evaluation function may include more than one single finite element run. Indeed, to evaluate the performance of the stator of an ultrasonic motor, it is required to determine, in general, the resonance frequency of each participating mode, its electromechanical coupling factor, and the resulting motion produced at the contact interface. Usually, one or two modal analyses are required to determine the characteristics of the modes, and a harmonic analysis is necessary to determine the shape of the ellipse.

There are, therefore, more than one call to the finite element program (ATILA) for the evaluation of each individual. This increases the order of the algorithm rather significantly, and may result in long computation times. To alleviate this problem, great care should be taken in minimizing the size and complexity of each finite element mesh. As a result, each finite element case may use a different mesh that enhances the computation time for that run.

Finally, the genetic algorithm provides some advantages with respect to parallel computation. It is clear that each individual can be evaluated, for instance, independently of the other individuals. Therefore, in a multiprocessor environment, it is possible to evaluate two or more individuals in parallel (Figure 6.13). Such a technique was implemented in the program that we developed.

The program that was developed to implement the aforementioned technique is rather general. The genetic algorithm itself does not require modifications from one problem to another. Only the general settings (population size, gene size, probabilities of crossover, and mutation) can be modified, depending on the problem at hand.

The finite element mesh (or meshes) is, of course, problem dependent. However, using meta-functions (that we also developed) for the creation of the mesh makes it relatively easy to generate a mesh for any given problem, whether two- or three-dimensional.

Finally, the most sensitive part that must be modified for every problem is the function that evaluates the performance of each individual based on the finite element results obtained. Even for one given problem, this function may also depend on the type of result that is sought. Usually, this evaluation function will contain terms associated with the operating frequency of the stator, the electromechanical coupling associated with each participating mode, and the elliptical factor associated with the vibration. Other terms, such as stresses, mass, volume, etc., can also be added.

6.2.2 Application Example: Mixed-Mode Ultrasonic Motor (MMUM)

The example that we have chosen is that of a commercial mixed-mode ultrasonic motor manufactured by Cedrat Technologies, in France.[8] The schematic of the stator is shown in Figure 6.14. It includes two ceramic multilayer actuators, one center mass, and one elliptical shell. One of the greatest advantages of such motors is that the two modes—within certain limits—can be controlled independently. Therefore, fine-tuning of the elliptical vibration field in the contact area can be performed, thus enabling various operating modes of the motor to be defined.

The operating mode of the structure is shown in Figure 6.15. The modes of interest here are a flexure mode of the shell, which provides the normal component of the vibration at the contact interface, and a translation mode of the center mass, which provides the tangential component of the vibration (Figure 6.16).

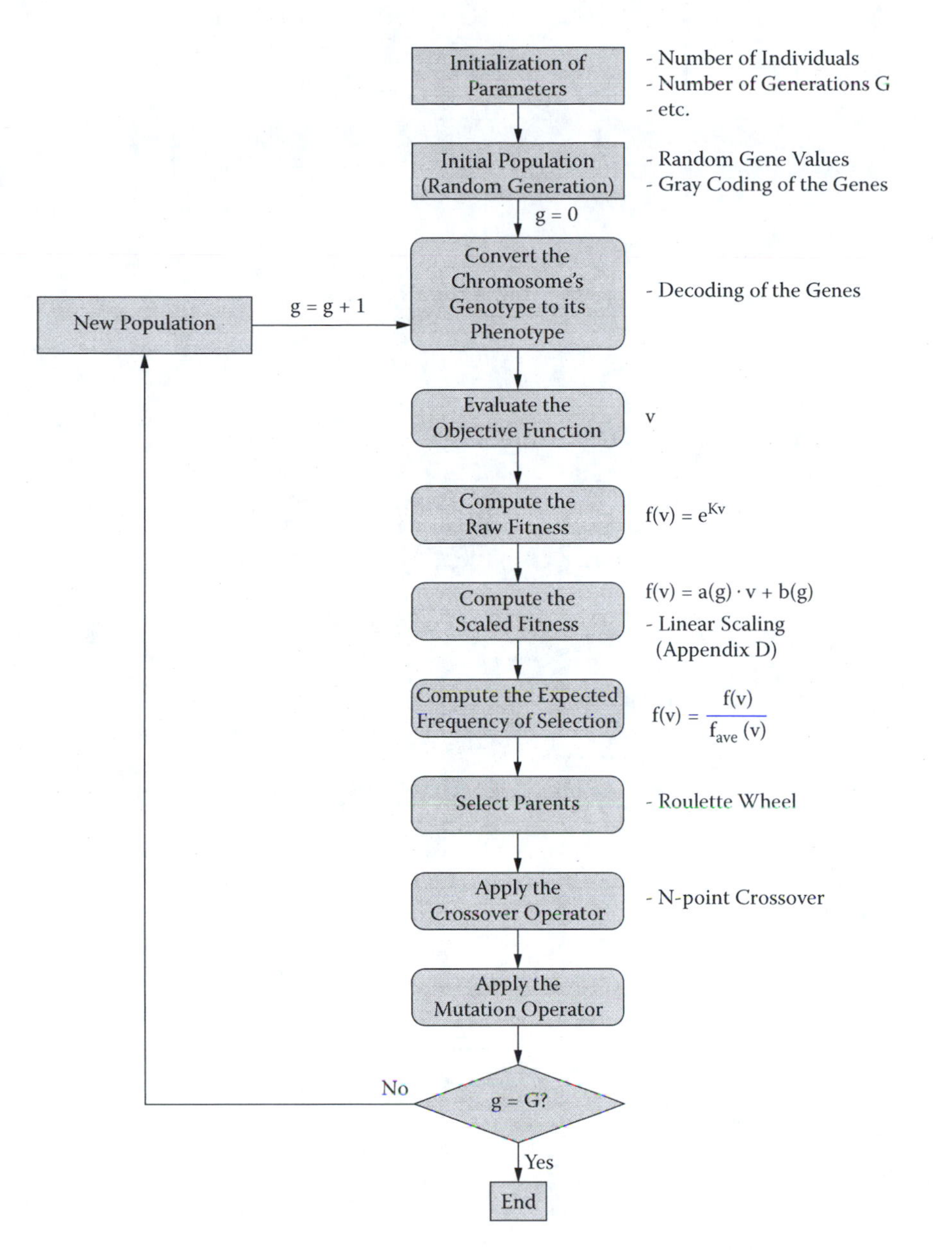

FIGURE 6.11 General algorithm of the evolutionary process.

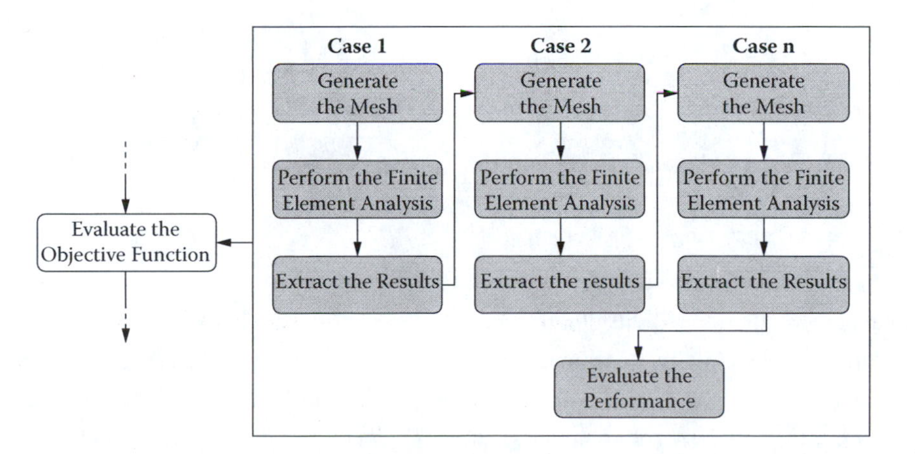

FIGURE 6.12 The combination of the FEA and GA methods takes place at the level of the evaluation function.

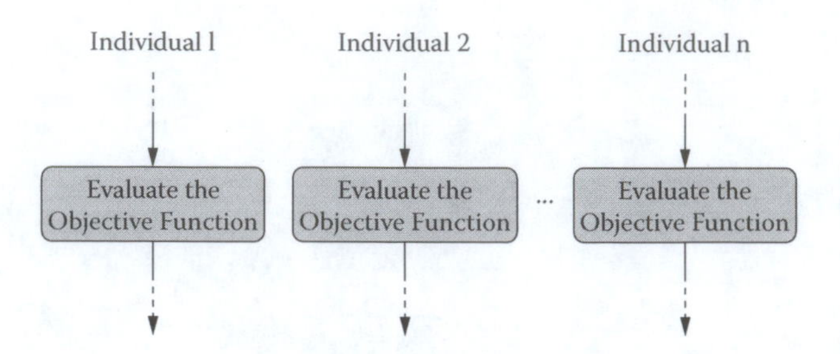

FIGURE 6.13 Parallel evaluation of individuals allows for speeding up the evolutionary process on a multiprocessor environment.

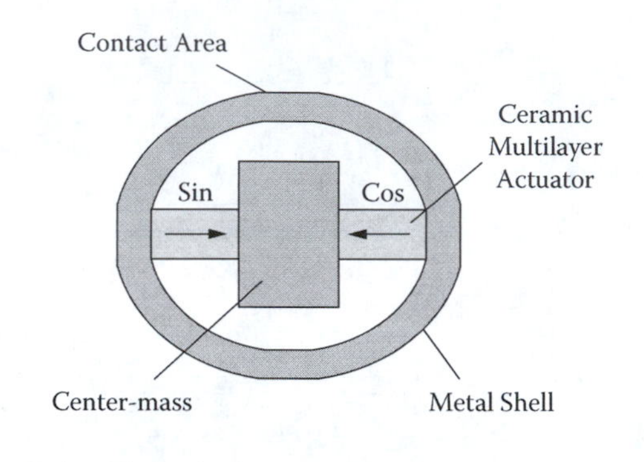

FIGURE 6.14 Schematic of MMUM by Cedrat Technologies.

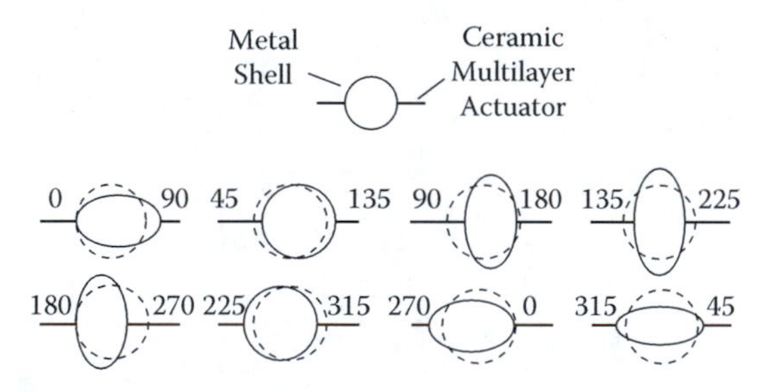

FIGURE 6.15 Operating principle of the MMUM.

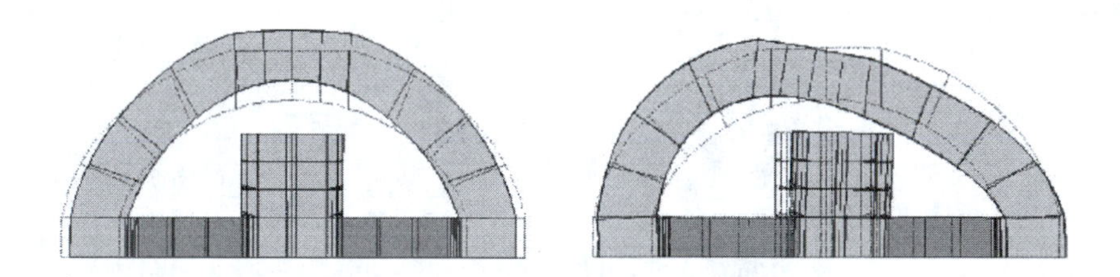

FIGURE 6.16 (Left) Flexure and (right) tangential vibration modes.

The finite element analysis is therefore divided into three steps. Two modal analyses are required to evaluate the frequency of the flexure and translation modes. Finally, a harmonic analysis combining the two effects is used to determine the amplitude of the vibration as well as the ellipse factor. This procedure is summarized in Figure 6.17. The finite element meshes used for the modal analyses (Figure 6.18) minimize the computation time. They are performed in two dimensions, and make use of the planes of symmetry and antisymmetry of the structure. Note that the flexure mode is pure when both ceramic multilayer actuators are excited in phase, and that the translation mode is pure when both are excited in opposition to the phase.

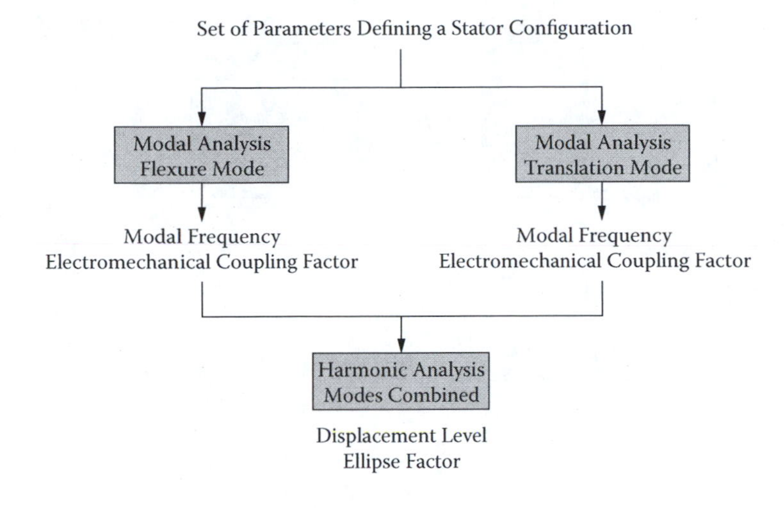

FIGURE 6.17 The evaluation of each individual requires three FE runs.

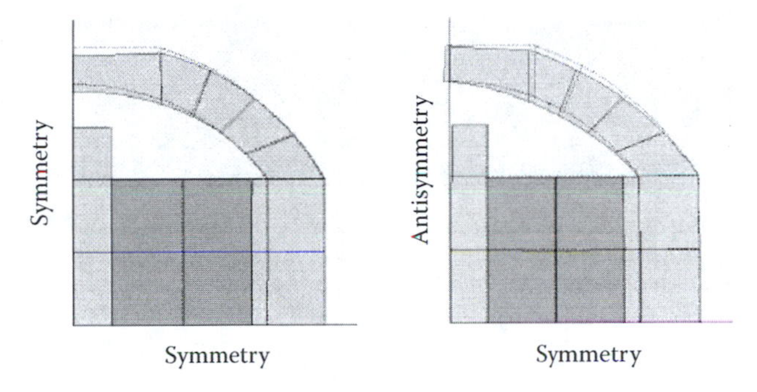

FIGURE 6.18 Finite element meshes used for the stator analysis.

A total of six parameters have been identified from the general structure of the stator (Figure 6.19). These parameters consist of the thickness, and inner and outer radii of the metal shell, the length of the contact area, as well as the dimensions of the center mass. All parts are made of stainless steel. Because two-dimensional models are used and the out-of-plane dimensions are not the same for all the parts, equivalent material properties are necessary, in particular for the center mass. The width of the metal shell is fixed to the width of the multilayer actuators, whose dimensions are considered to be fixed as well (they are 6.3 mm high and have a square cross section with sides of 12.7 mm). These actuators are composed of six plates of piezoelectric material poled in the thickness direction. Because all the layers are not represented individually in the finite element models (to reduce the model size), equivalent material properties must be used for the piezoelectric material as well.

The supports represented in Figure 6.19 play an important role in this structure. Indeed, as can be easily seen from the mode shapes, there is no nodal location that can be used to fix the stator to a mechanical reference. Therefore, these supports must be able to hold the structure and allow for forces to be transmitted to a rotor, but they should not dissipate the vibration energy. As can be readily noted, these supports are in fact simple metal plates. They are placed such that the contact interface is in a direction parallel to the plane of those plates, thereby allowing for large forces to be transmitted in that direction. On the other hand, these plates are relatively thin, so that they decouple the stator from the mechanical reference and thus avoid dispersing the vibration energy.

The evaluation function for the performance of the individuals (defined by the values of the six parameters) is such that it favors:

- A total length (in the direction of the multilayer actuators) less than 1 in. (25.4 mm).
- An operating frequency as low as possible.
- An electromechanical coupling coefficient as high as possible for each mode.
- Vibration amplitudes as large as possible.
- An ellipse factor close to unity.

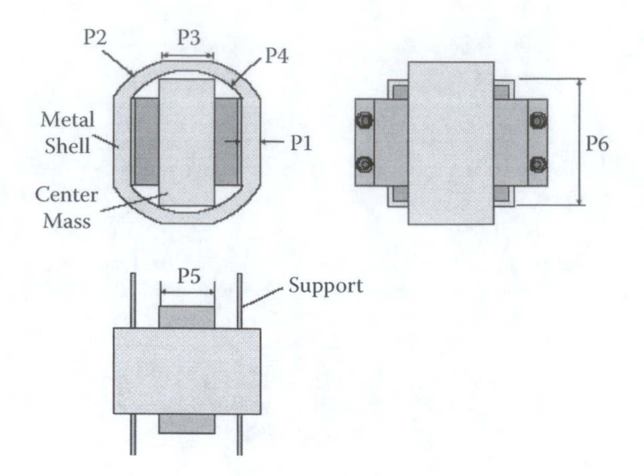

FIGURE 6.19 The six design parameters.

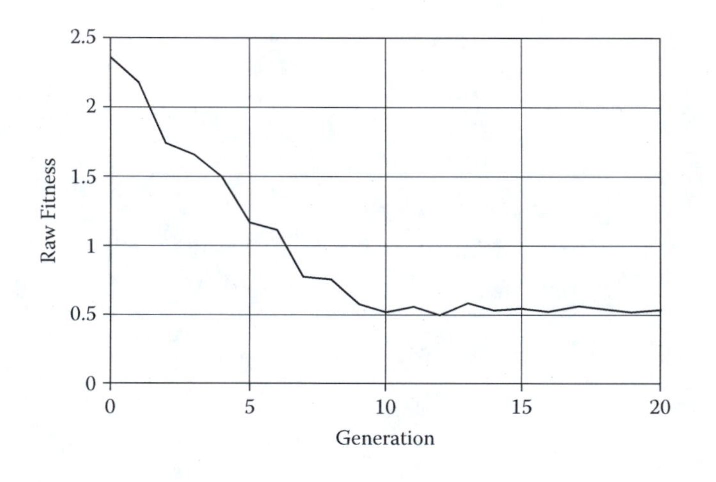

FIGURE 6.20 Progression of the algorithm.

The final performance, which combines all the criteria described in the previous section, should typically be a number between 0 and 1, 0 being the optimum value. To perform this mapping, it could be useful to consider upper and lower bounds for each criterion. For instance, frequencies between 20 and 100 kHz could be mapped to the interval [0,1]. Individuals exhibiting frequencies out of this range could be considered degenerate, and penalized. This same penalization idea could be used for those individuals having a total length greater than 25.4 mm. It is, however, obvious that the length (as small as possible) and frequency (as low as possible) criteria are contradictory. Therefore, it can be expected that a natural equilibrium will be reached between these parameters. The genetic algorithm is reputed for converging rapidly toward an optimum of the domain space, but to experience difficulties in "fine-tuning" the values of the parameters (Figure 6.20).

The solution generated by the combined FEA+GA method is summarized in Table 6.1. Values of the computed resonance and antiresonance frequencies of the flexure and tangential modes are given in Table 6.2.

6.2.3 Comparison with Experiments

The parts of the stator are fabricated using the electrodischarge machining technique (Figure 6.21a), except for the metal supports, which are made from gauged steel plates 0.5 mm thick. It was decided to assemble the multilayer actuators at the lab (Figure 6.21b). Each actuator is composed of six piezoelectric plates having a thickness of 1 mm (the material used is the APC841 from American Piezo Ceramics Inc.). To provide electrical connections, a brass shim of thickness 25 μm is used in between each plate and at the ends of the stack. All these elements are bonded together using the Stycast 45LV epoxy from Emerson & Cuming. The stator was then assembled (Figure 6.21c). Aligning the different parts (center mass, ceramic stacks, supports, and shell) was not a trivial task and required several attempts. Finally, the stator was attached to a rigid support by means of #1 screws, and electrical wires were connected.

TABLE 6.1
Solution obtained

Parameter	Dimension (mm)
P1	2.49
P2	12.52
P3	7.39
P4	10.44
P5	10.70

TABLE 6.2
Computed modal response for the final individual

	Resonance Frequency (kHz)	Antiresonance Frequency (kHz)	Electromechanical coupling coefficient (%)
Flexure mode	42,551.7	43,951.8	25.04%
Translation mode	42,556.3	43,922.8	24.75%

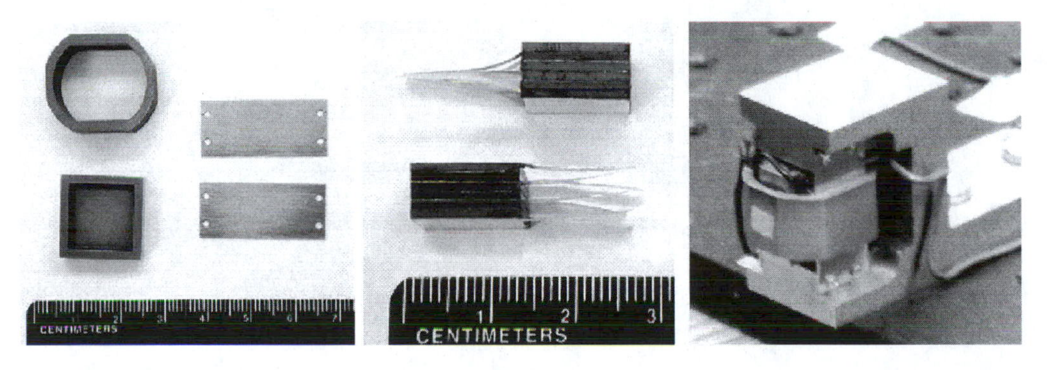

FIGURE 6.21 (a) Metal parts for the stator prototype; (b) piezoelectric stacks assembled; (c) assembled prototype motor.

Tests performed with a laser interferometer revealed the presence of the desired modes in the neighborhood of 40 kHz (Figure 6.22). The tangential mode was obtained at 43 kHz and the flexure mode at 39.7 kHz. Given that this was the first iteration of this prototype, the results were rather good. Although the separation between the two modes did not allow using this prototype as an efficient motor, by selecting a frequency between 39.7 and 43 kHz, it was possible to measure the elliptical trajectory of a point on the contact interface. Figure 6.23 shows some of these ellipses obtained at 39 kHz and for a varying phase angle between the signals exciting the two ceramic stacks. It is clearly seen from these results that the design is successful and that good control of the vibration ellipse can be achieved. When both stacks are excited in phase, for instance, the ellipse is nearly reduced to its normal component. By increasing the phase between the signals, a rounder ellipse forms, until it becomes almost reduced to its tangential component when both signals are in opposition to the phase. In this last case, there remains a normal component to the vibration (the ellipse is not quite horizontal), most likely because of spurious flexure at the contact interface.

These last results clearly show that the design was successful in that it allowed defining a structure with the desired modes of vibration, and that these modes operated as expected; i.e., they made it possible to control the shape of the vibration ellipse. Although this structure is commercially available, its performance will be further improved by taking into account the optimization process introduced in this chapter.

Finally, a natural extension of the method would be to implement the analysis of the contact mechanism within the evaluation of an individual. To achieve this, methods based on modal decomposition and equivalent circuit representations of the

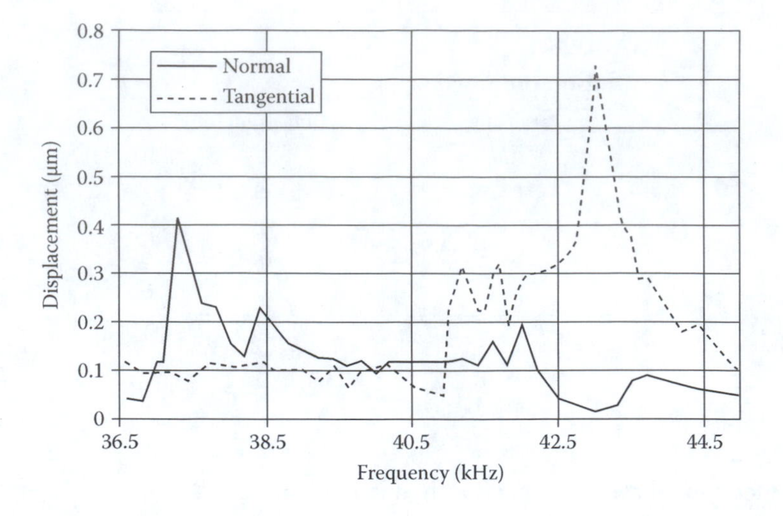

FIGURE 6.22 Vibration measurement.

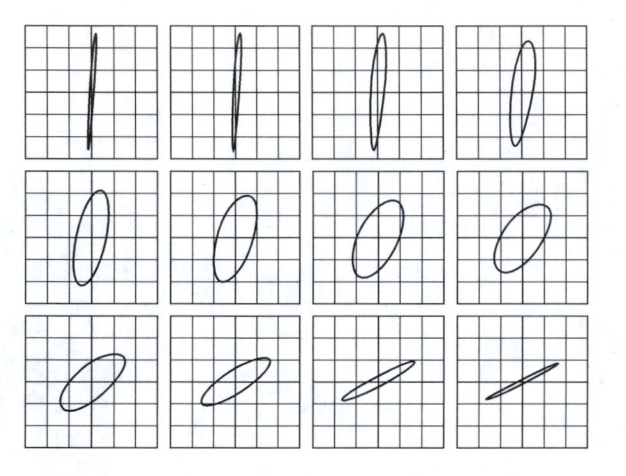

FIGURE 6.23 Elliptical vibration trajectories recorded at the contact interface for varying phase angle between the excitation signals. In the top left, the phase angle is 0°, and in the bottom right, the phase angle is 165° with an increment of 15° for each locus.

contact mechanism could be added to the general program. After evaluating the performance of an individual, the electromechanical characteristics of the system (computed by ATILA) would be used to determine the values of the equivalent circuit (which may be problem dependent). An analysis of the contact problem would reveal the transmitted force and velocity to the rotor or linear guide. Therefore, the evaluation of an individual could be performed directly on the performance it yields at the rotor level.

6.3 CYMBAL ARRAY

A 3 × 3 cymbal array shown in Figure 6.24a was developed for underwater sonar application in The Penn State University, and analyzed by ATILA simulation and experimentally measured. The resonance frequency of the unit cymbal was 17 kHz.

When all the cymbals are identical and driven at 17 kHz under water, the displacements of the center, edge, and corner cymbals are calculated as in Figure 6.25. The definitions of the center, edge, and corner transducers are illustrated in Figure 6.24b. Interestingly, the center cymbal excites a significant magnitude, whereas the corner cymbal does not generate a displacement. Thus, experimentally, the collapse of the center cymbal during the operation was observed occasionally. This nonuniform distribution of the displacement is due to the mutual interaction of the cymbal operation through water pressure. This problem can be solved by slightly modifying the cymbal performance; that is, when the resonance frequencies of the center, edge, and corner cymbals are adjusted to 17, 16, and 18 kHz, respectively, as shown in Figure 6.26, uniform displacements are obtained for these cymbal transducers. In conclusion, the significant interaction is due to the totally identical cymbal transducer choice and—although this may not be observed in practice—due to a slight deviation of the transducer performance of each cymbal.

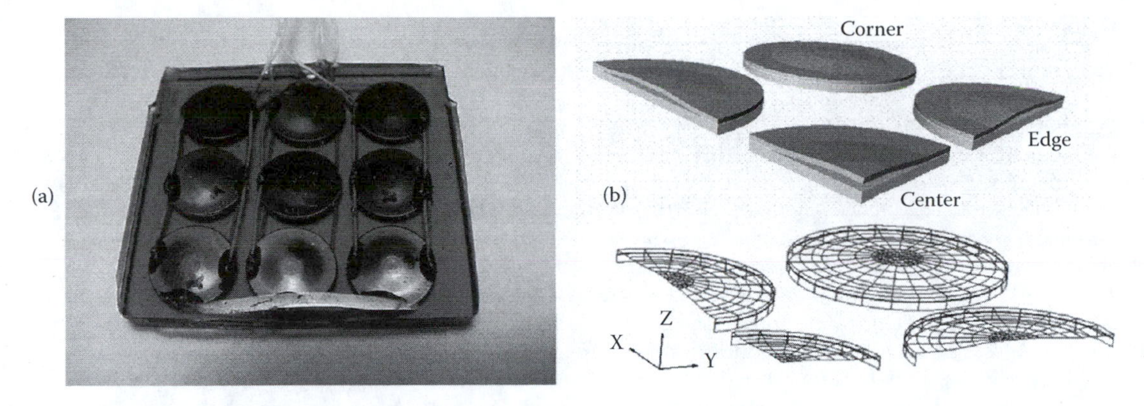

FIGURE 6.24 (a) A 3 × 3 cymbal array, and (b) ATILA simulation model of the cymbal array.

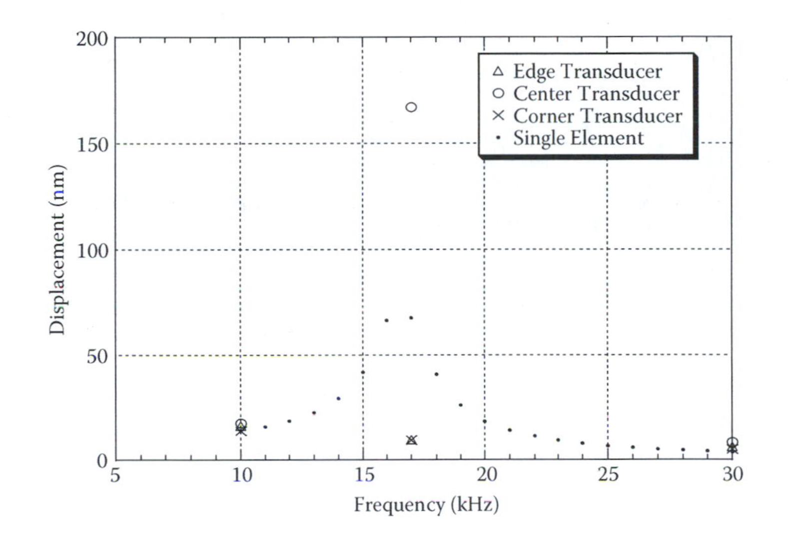

FIGURE 6.25 The displacements of the center, edge, and corner identical cymbals in a 3 × 3 array driven at 17 kHz under water. Note that the center cymbal generates a significant magnitude.

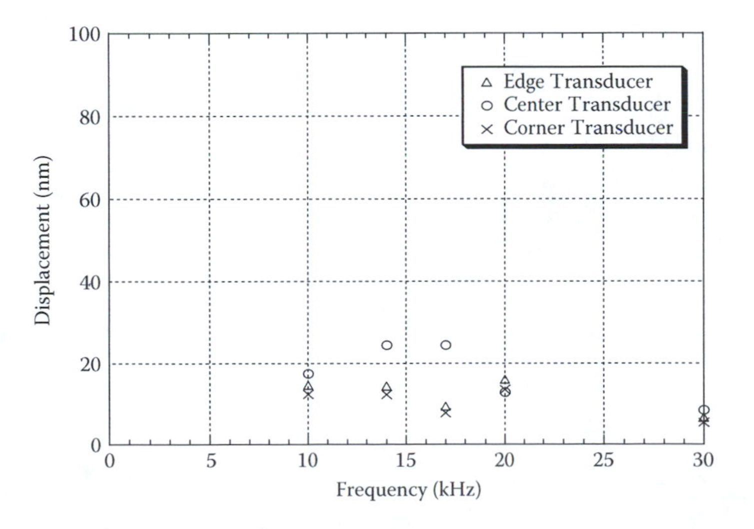

FIGURE 6.26 The displacements of the center, edge, and corner identical cymbals in a 3 × 3 array driven at 17 kHz under water. Three slightly deviated cymbals are arrayed in this structure (center 17 kHz, edge 16 kHz, and corner 18 kHz).

REFERENCES

1. S. Cagatay, B. Koc, and K. Uchino: Piezoelectric Ultrasonic Micromotors for Mechatronic Applications, *Proceedings of the 9th International Conference on New Actuators*, A4.2, pp. 132–134, Bremen, Germany, June 14–16 (2004).
2. ATILA: Finite Element Analysis Program for Piezoelectric Structures. Distributed by Micromechatronics, Inc., State College, PA.
3. P. Bouchilloux, R. Le Letty, N. Lhermet, and F. Claeyssen: Computer Aided Modeling of Induced Strain Actuators and Other Piezoelectric Structures, *Proceedings of the ASME: Dynamic Systems and Control Division*, DSC-Vol. **61**, pp. 307–12 (1997).
4. B. Koc, P. Bouchilloux, and K. Uchino: Piezoelectric Micromotor Using a Metal-Ceramic Composite Structure, IEEE Transactions on Ultrasonics, *Ferroelectrics and Frequency Control*, Vol. **47**, No. 4, pp. 836–43, 2000.
5. J. H. Holland: *Adaptation in Natural and Artificial Systems*, University of Michigan Press, Ann Arbor, 1975.
6. E. T. Lee and M. E. Lee: Algorithms for generating generalized Gray codes, *Kybernetes*, Vol. **28**, No. 6–7, pp. 837–44, 1999.
7. J. Andre, P. Siarry, and T. Dognon: An improvement of the standard genetic algorithm fighting premature convergence in continuous optimization, *Advances in Engineering Software*, Vol. **32**, No. 1, pp. 49–60, 2001.
8. R. Le Letty, F. Claeyssen, F. Barillot, M. F. Six, and P. Bouchilloux: New Linear Piezomotors for High Force/Precise Positioning Applications, 1998 IEEE Industry Applications Conference, Thirty-Third IAS Annual Meeting, Vol. **1**, pp. 213–217, October 1998.

7 Future of FEM in Smart Structures

Finite element method (FEM) can be adapted to smart structures with piezoelectric or magnetostrictive materials rather successfully when neither the applying electric field nor the generating AC strain is very large and when the linear relation can be assumed in the strain versus electric field or the strain versus stress. However, further improvement in the FEM algorithm is required for high-field or high-power drive of the piezoelectric system, where nonlinear and hysteretic characteristics should be taken into account, as well as heat generation. In this chapter, we will discuss the high-power issues.

7.1 NONLINEAR/HYSTERESIS CHARACTERISTICS

Ferroelectric materials tend to exhibit ***nonlinear*** and ***hysteretic*** characteristics in the relationships of polarization versus electric field, strain versus electric field, and strain versus stress. Nonlinearity is inevitable for ferroelectrics, because ferroelectricity originates from the nonlinear characteristics of the atomic lattice elastic properties. On the contrary, the hysteretic behavior primarily originates from domain reorientation. The situation is illustrated in Figure 7.1.

Hysteresis is simulated in the FEM as an intensive loss factor: dielectric loss tan δ', elastic loss tan ϕ', and piezoelectric loss tan θ'. However, it is notable that the hysteresis curve due to this kind of loss factor is elliptic, which is different from the actual hysteresis. Also, note that the value of the loss factor should be less than 10% (or 0.1) in theory. The treatment for large hysteresis, such as a butterfly-shape strain curve, has not been established yet.

Also, note that the loss factors should have tensor properties similar to the real parameters. For example, there are 10 independent parameters in a PZT ceramic; elastic compliances, s_{11}^E, s_{12}^E, s_{13}^E, s_{33}^E, s_{44}^E; piezoelectric constants, d_{33}, d_{31}, d_{15}; and dielectric constants $\varepsilon_{11}^X/\varepsilon_0$, $\varepsilon_{33}^X/\varepsilon_0$. Accordingly, there must be 10 loss tensor components. However, at present only three loss factors—dielectric loss tan δ', elastic loss tan ϕ', and piezoelectric loss tan θ'—have been considered in the calculation in the ATILA software.

7.1.1 Elastic Nonlinearity

Elastic nonlinearity provides theoretically asymmetric (or skewed) impedance or strain spectrum under frequency sweep.[1] With increasing further electric field level, ***jump*** and ***hysteresis*** during rising and falling frequency are observed, as exemplified in Figures 7.2 and 7.3. The PZN-PT single crystal is known as a very high electromechanical coupling material; however, its significant nonlinearity is also known. On increasing the drive electric field level, the material becomes stiffer; the material also becomes stiffer on increasing the stress. This nonlinearity introduces the hysteresis phenomenon with drive frequency in the impedance and the strain spectrum.

ATILA cannot provide this kind of asymmetric impedance spectrum at present even under a large field drive, because no information is put on the elastic nonlinearity of PZTs.

7.1.2 Loss Anisotropy

The importance of the introduction to loss anisotropy is exemplified in the development of the Rosen-type piezoelectric transformer. In the present FEM simulation, the mechanical quality factor Q_m is for the data obtained from the k_{31} mode. However, as shown in Table 7.1, the loss tangent tan ϕ' for s^E_{33} may be larger than that for s^E_{11}. Therefore, when we calculate the admittance curve for a Rosen-type transformer with two polarization configurations (shown in Figure 7.4), we should use two mechanical quality factors for s^E_{11} and s^E_{11}, for the left and right portions, respectively. Figures 7.5a and 7.5b depict the simulated admittance curves for the isotropic loss case (only one Q_m) and anisotropic loss case (two Q_ms), respectively. Note that maximum admittance value is roughly double for case (a).

7.2 HEAT GENERATION

With increasing drive voltage, the hysteretic property is accompanied inevitably by heat generation from the piezoelectric material. Heat generation in various types of PZT-based actuators has been studied under a relatively large electric field applied (1 kV/mm or more) at an off-resonance frequency, and a simple analytical method was established to evaluate the temperature rise, which is very useful for the design of piezoelectric high-power actuators.

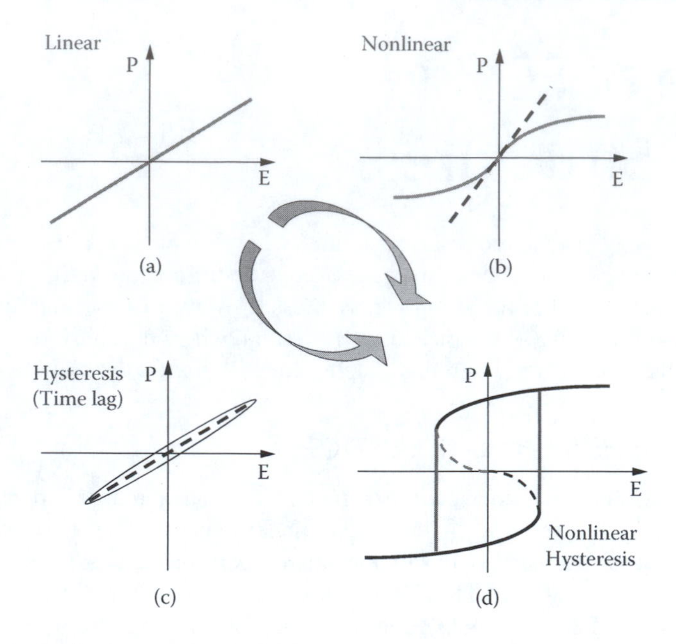

FIGURE 7.1 The relationship of polarization versus electric field in (a) a linear material, (b) a nonlinear material, (c) a hysteretic material, or (d) an actual material.

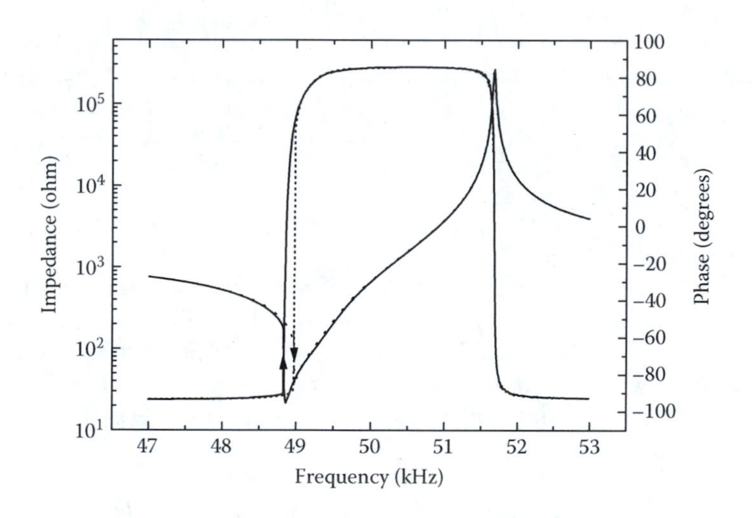

FIGURE 7.2 Impedance jump and hysteresis during rising and falling frequency around the electromechanical resonance point (observed under a relatively large electric field in a doped $Pb(Zn_{1/3}Nb_{2/3})O_3$-$PbTiO_3$ single crystal).

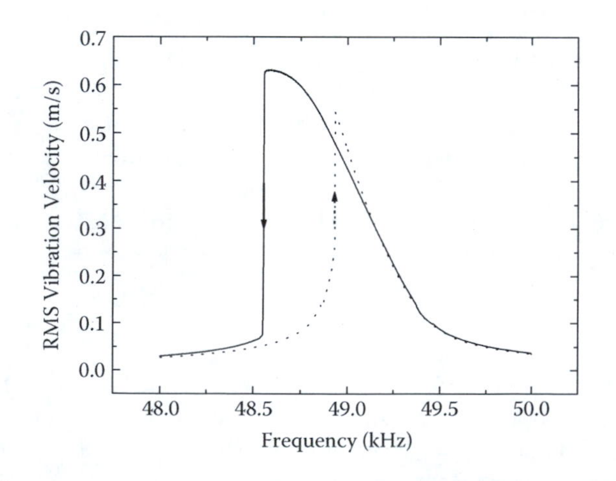

FIGURE 7.3 Vibration velocity (strain) jump and hysteresis during rising and falling frequency around the electromechanical resonance point (observed under a relatively large electric field in a doped $Pb(Zn_{1/3}Nb_{2/3})O_3$-$PbTiO_3$ single crystal).

TABLE 7.1
Loss factor anisotropy in PZT-based piezoelectric ceramics (note the higher loss tangent value for s^E_{33} compared to that for s^E_{31})

Complex material constant	APC850 ("soft ceramic")			APC841 ("hard ceramic")		
	Real part	Imaginary part	Loss tangent	Real part	Imaginary part	Loss tangent
	Elastic compliance (10^{-12} m²/N)					
S11E*	16.9	−0.205	**0.0122**	11.4	−0.0105	**0.00092**†
S12E*	−6.4	0.0779	**0.0121**	−3.42	0.00383	**0.00112**
S13E*	−5.6	0.067	**0.0120**	−5.65	−0.032	**−0.00566**
S33E*	17.4	−0.347	**0.02**	15.35	−0.0692	**0.00451**†
S44E*	52.6	−3.82	**0.0726**	20.8	−0.24	**0.0115**
S66E*	46.7	−1.968	**0.042**	29.64	−0.028	**0.000945**
	Dielectric constant (10^{-9} F/m)					
ε11T*	10.01	0.346	−0.0346	12	−0.048	0.004
ε33T*	15.8	−0.34	0.022	18	−0.162	0.009

† If there is a difference between tan ϕ' for s^E_{33} and s^E_{11}, the final admitance curve changes.

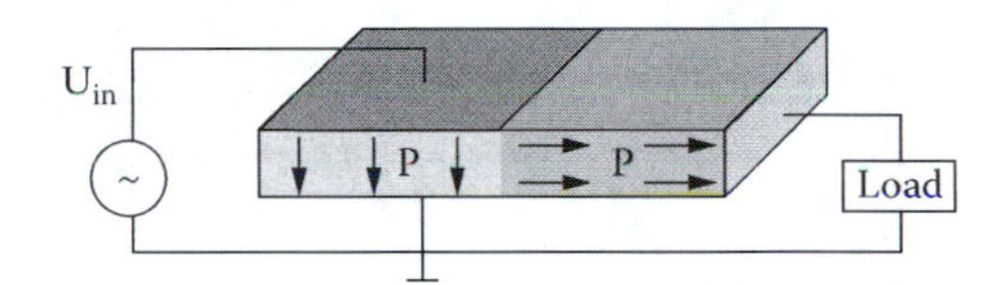

FIGURE 7.4 Rosen-type piezoelectric transformer with two polarization configurations, leading to the mode coupling with k_{31} and k_{33}.

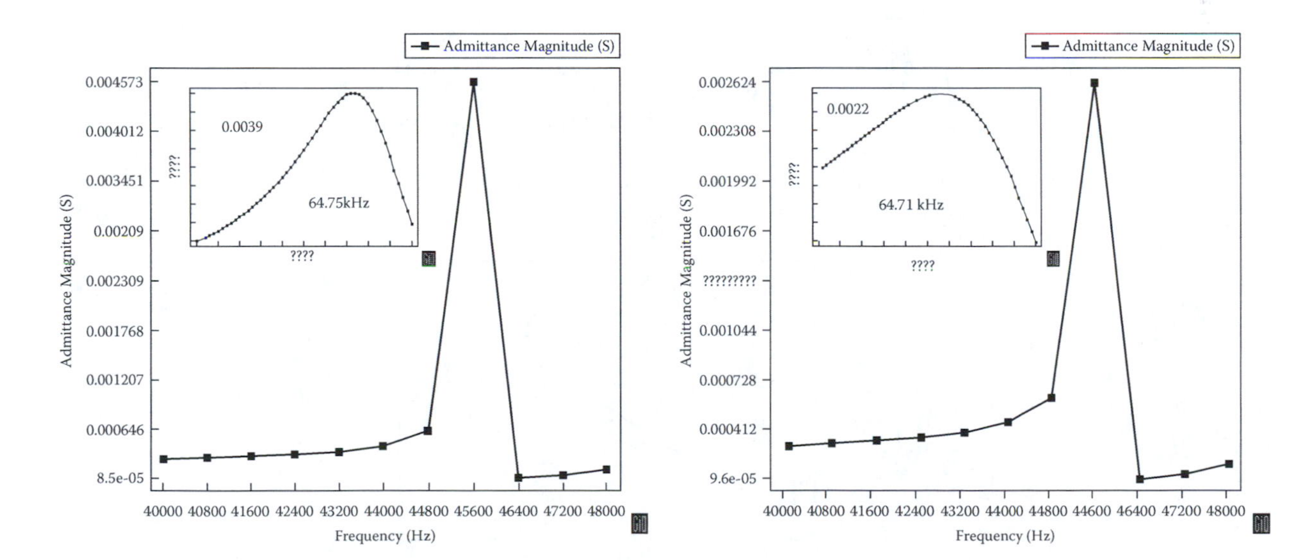

FIGURE 7.5 (a) Simulated admittance curves for an isotropic loss case (only one Q_m = 1087), and (b) anisotropic loss case (two Q_ms 1087 and 222), respectively. Note that the maximum admittance value is about double for case (a).

Zheng et al. reported the heat generation from various sizes of multilayer type piezoelectric ceramic actuators.[2] Figure 7.6 shows the temperature change with time in the actuators when driven at 3 kV/mm and 300 Hz, and Figure 7.7 plots the saturated temperature as a function of V_e/A, where V_e is the effective volume (electrode overlapped part) and A is the surface area. This linear relation is reasonable because the volume V_e generates the heat, and this heat is dissipated through the area A. Thus, if we need to suppress the temperature rise, a small V_e/A design is preferred.

According to the law of energy conservation, the rate of heat storage in the piezoelectric resulting from heat generation and dissipation effects can be expressed as

$$q_g - q_{out} = V \rho c (dT/dt), \tag{7.1}$$

assuming uniform temperature distribution in the sample. V, ρ, and c are total volume, density, and specific heat, respectively. The heat generation is considered to be caused by losses. Thus, the rate of heat generation (q_g) in the piezoelectric can be expressed as

$$q_g = u f V_e, \tag{7.2}$$

where u is the loss of the sample per driving cycle per unit volume; f, the driving frequency; and V_e, the effective volume where the ceramic is activated. According to the measuring condition, this u corresponds to the intensive dielectric loss w_e, which consists of the extensive dielectric loss tanδ and the electromechanical and piezoelectric combined loss ($\tan\phi - 2\tan\theta$):

$$u = w_e = \pi \varepsilon^X \varepsilon_0 E_0^2 \tan\delta'$$

$$= [1/(1 - k^2)][\tan\delta + k^2(\tan\phi - 2\tan\theta)]\, \pi \varepsilon^X \varepsilon_0 E_0^2. \tag{7.3}$$

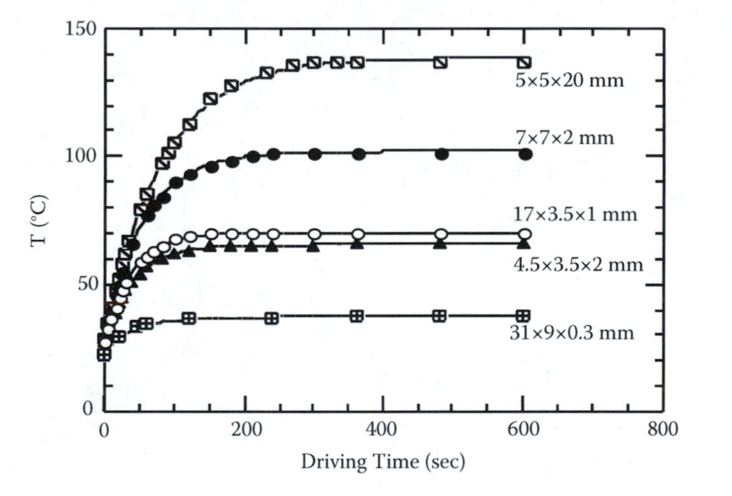

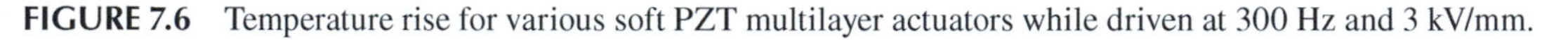
FIGURE 7.6 Temperature rise for various soft PZT multilayer actuators while driven at 300 Hz and 3 kV/mm.

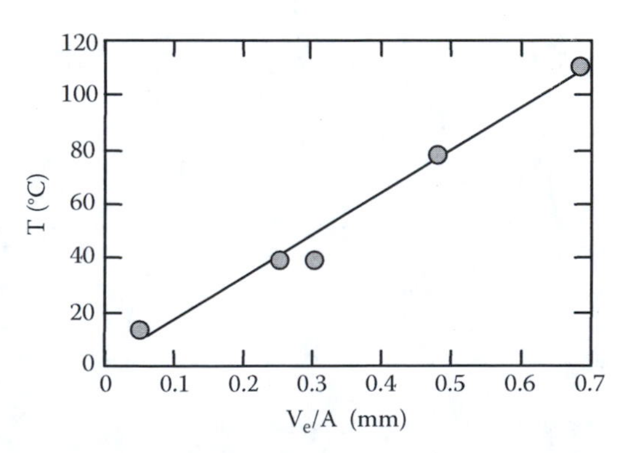

FIGURE 7.7 Temperature rise versus V_e/A (3 kV/mm, 300 Hz), where V_e is the effective volume generating the heat and A is the surface area dissipating the heat.

Note that we do not need to add w_{em} explicitly, because the corresponding electromechanical loss is already included implicitly in w_e.

When we neglect the conduction heat transfer, the rate of heat dissipation (q_{out}) from the sample is the sum of the rates of heat flow by radiation (q_r) and convection (q_c):

$$q_{out} = q_r + q_c$$
$$= \sigma e A(T^4 - T_0^4) + h_c A(T - T_0), \quad (7.4)$$

where σ is the Stefan-Boltzmann constant, e is the emissivity of the sample, h_c is the average convective heat transfer coefficient, and A is the sample surface area.

Thus, equation (7.1) can be written in the form:

$$u f V - A\, k(T)\, (T - T_0) = V \rho c\, (dT/dt), \quad (7.5)$$

where

$$k(T) = \sigma e(T^2 + T_0^2)(T + T_0) + h_c \quad (7.6)$$

is defined as the overall heat transfer coefficient. If we assume that $k(T)$ is relatively insensitive to temperature change, the solution to equation (7.5) for the piezoelectric sample temperature is given as a function of time (t):

$$T - T_0 = [u f V_e / k(T) A]\, [1 - e^{-t/\tau}], \quad (7.7)$$

where the time constant τ is expressed as

$$\tau = \rho c V / k(T) A. \quad (7.8)$$

Figures 7.8a and 7.8b show the dependence of $k(T)$ on applied electric field and frequency. Because $k(T)$ is not really constant, we can calculate the total loss u of the piezoelectric more precisely through equation (7.7). The calculated results are shown in Table 7.2. The experimental data of *P-E* hysteresis losses under a stress-free condition are also listed for comparison. It is seen that the *P-E* hysteresis-intensive loss agrees well with the total loss contributing to the heat generation under an off-resonance drive.

Temperature rise in the piezoelectric device and the successive property change should be included in the future calculation process of the FEM for smart transducer systems.

7.3 HYSTERESIS ESTIMATION PROGRAM

When using a statistical treatment of the domain-reversal mechanism, there are various computer simulation trials in polarization (*P-E*), magnetization (*M-H*), and strain (*x-X*) curves. We review here the calculation algorithm proposed by R. C. Smith.[3]

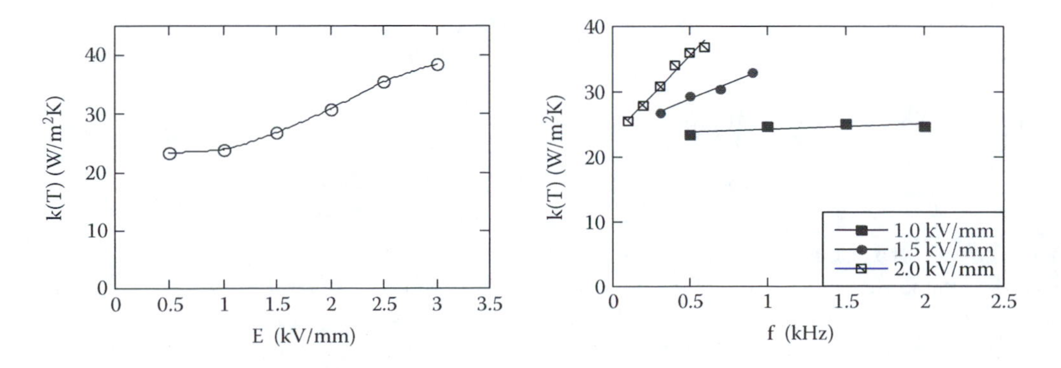

FIGURE 7.8 (a) Overall heat transfer coefficient $k(T)$ as a function of applied electric field (400 Hz) and (b) frequency (Data from the actuator with dimensions of 7 mm × 7 mm × 2 mm).

TABLE 7.2
Loss and overall heat transfer coefficient for PZT multilayer samples (E = 3 kV/mm, f = 300 Hz)

Actuator	4.5 × 3.5 × 2 mm	7 × 7 × 2 mm	17 × 3.5 × 1 mm
Total loss (×10³ J/m³) $u = \frac{\rho c v}{f v_e}\left(\frac{dT}{dt}\right)_{t\to 0}$	19.2	19.9	19.7
P-E hysteresis loss (×10³J/m³)	18.5	17.8	17.4
$k(T)$ (W/m^2K)	38.4	39.2	34.1

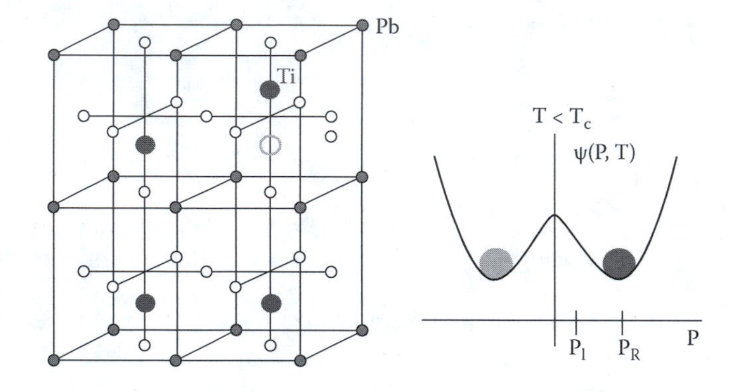

FIGURE 7.9 (a) $PbTiO_3$ crystal lattice model. There are two potential minima for the Ti ion position. (b) Double well model, showing the two potential minima.

We adopt here the so-called ***mesoscopic model*** to treat ferroelectric, ferromagnetic, and ferroelastic materials. Let us consider primarily a ferroelectric such as $PbTiO_3$, the crystallographic structure of which is illustrated in Figure 7.9a. There are basically two potential minima for the Ti ion position, so that we can assume the ***double well model*** shown in Figure 7.9b.

Introducing the Helmholtz energy (temperature stability of polarization) for this double-well polarization model at a finite temperature T:

$$\begin{aligned}\Psi(P,T) &= U - ST \\ &= \frac{\Phi_0 N}{4V}\left[1-(P/P_s)^2\right]+\frac{TkN}{2VP_s}\left[P\ln\left(\frac{P+P_s}{P_s-P}\right)+P_s\ln\left(1-(P/P_s)^2\right)\right]\end{aligned} \tag{7.9}$$

we obtain the following relationship:

$$\Psi(P)=\begin{cases}\frac{\eta}{2}(P\pm P_R)^2 & , |P|<P_I \\ \frac{\eta}{2}(P_I-P_R)\left[\frac{P^2}{P_I}-P_R\right] & , |P|<P_I\end{cases} \tag{7.10}$$

Then, we introduce the Gibbs energy for a ferroelectric:

$$G(E,P,T)=\Psi(P,T)-EP \qquad \text{(Ferroelectric)} \tag{7.11}$$

The Gibbs energy curves for various electric fields are depicted in Figure 7.10.

To introduce the temperature fluctuation of the polarization more explicitly, ***Boltzmann probability*** is introduced (refer to Figure 7.11):

$$\mu(G)=C_e^{-GV/kT} \qquad p_{+-}=\sqrt{\frac{kT}{2\pi m V^{2/3}}}\cdot\frac{e^{-G(E,P_0(T),T)V/kT}}{\int_{P_0}^{x} e^{-G(E,P,T)V/kT}\,dP} \tag{7.12}$$

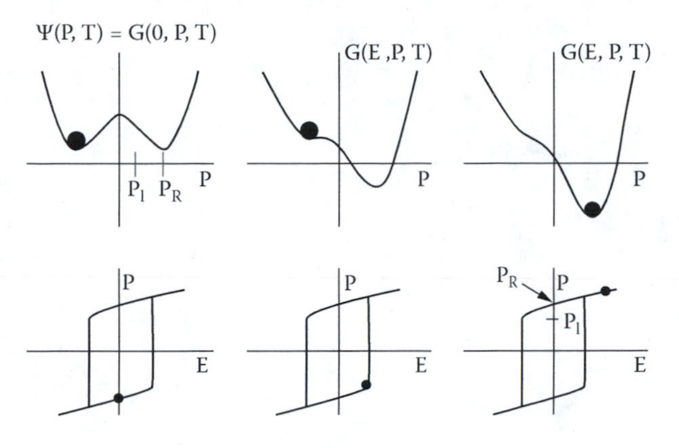

FIGURE 7.10 Gibbs energy curves for various electric field E levels.

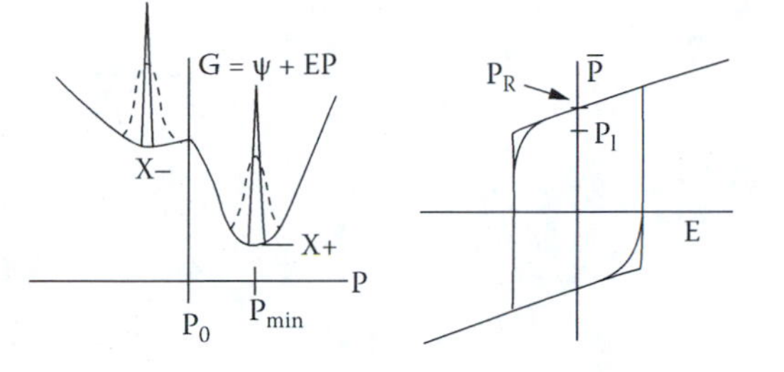

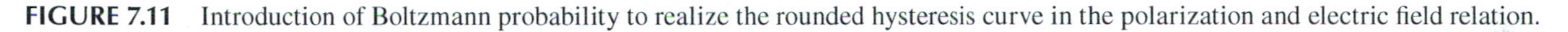

FIGURE 7.11 Introduction of Boltzmann probability to realize the rounded hysteresis curve in the polarization and electric field relation.

By taking the evolution relations

$$\frac{dx_+}{dt} = -p_{+-}x_+ + p_{-+}x_- \quad , \quad \frac{dx_-}{dt} = -p_{-+}x_- + p_{+-}x_+ \tag{7.13}$$

we finally obtain the polarization under a uniform lattice

$$P = x_+\langle P_+\rangle + x_-\langle P_-\rangle \tag{7.14}$$

where

$$\langle P_+\rangle = \int_{P_0}^{\infty} P\mu(G)dP \tag{7.15}$$

The Gibbs energy for a ferromagnetic and a ferroelastic can be obtained in a similar fashion:

$$\begin{aligned} G(H,M,T) &= \Psi(M,T) - \mu_0 HM \quad \text{(Ferromagnetic)} \\ G(\sigma,\varepsilon,T) &= \Psi(\varepsilon,T) - \sigma\varepsilon \quad \text{(Ferroelastic)} \end{aligned} \tag{7.16}$$

Then, the average magnetization can be obtained as

$$\bar{M} = x_+\langle M_+\rangle + x_-\langle M_-\rangle \tag{7.17}$$

Finally, Figure 7.12 demonstrates the *P-E*, *M-H*, and *x-X* hysteresis curves for PZT 5A piezoelectric, Terfenol-D magnetostrictive, and NiTi film shape memory materials, respectively. Note that the hysteresis curve starting from any intermediate external parameter (E, H, or X) exhibits a reasonable agreement with the experimental curves.

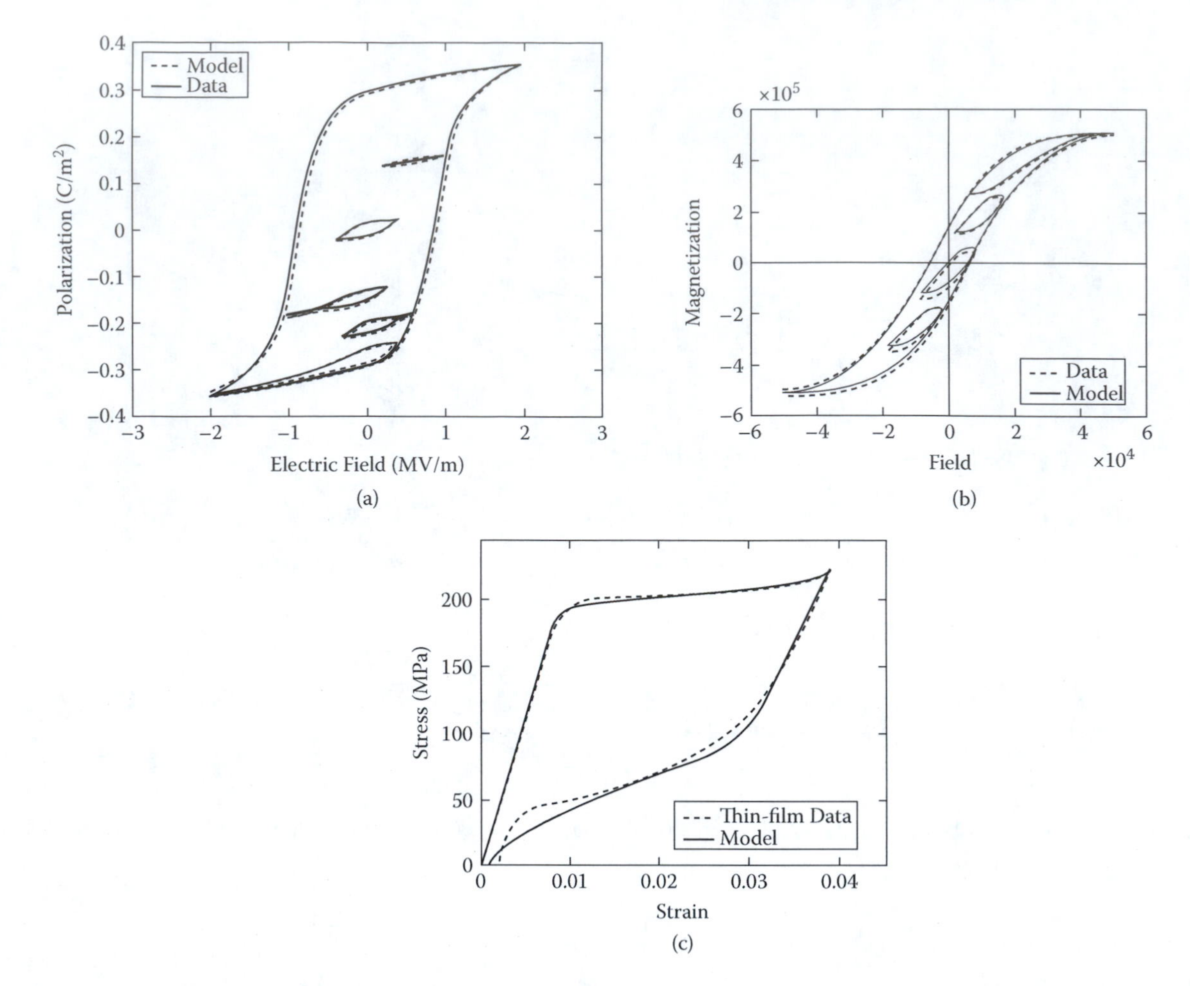

FIGURE 7.12 Hysteresis curves obtained from the mesoscopic approach in comparison with the experimental data: (a) polarization versus electric field in piezoelectric PZT 5A, (b) magnetization versus magnetic field in magnetostrictive Terfenol-D, and (c) strain versus stress in shape memory NiTi film. (From R. C. Smith, *Smart Materials Systems: Model Development*, Soc. Industrial and Appl. Math., Philadelphia [2005]. With permission.)

We will seek the combination of this hysteresis estimation program with ATILA FEM software in the future to introduce nonlinear and hysteretic behavior in the vibration calculation, as well as heat generation evaluation.

REFERENCES

1. N. Aurella, D. Guyomar, C. Richard, P. Gonrard, and L. Eyraud: *Ultrasonics*, **34**, 187–191 (1996).
2. J. Zheng, S. Takahashi, S. Yoshikawa, K. Uchino, and J. W. C. de Vries: *J. Am. Ceram. Soc.*, **79**, 3193–3198 (1996).
3. R. C. Smith: *Smart Materials Systems: Model Development*, Soc. Industrial and Appl. Math., Philadelphia (2005).

<<DO YOU KNOW? >>

BIMORPH FAMILY

You should distinguish between the following device terms: *monomorph*, *unimorph*, *bimorph*, and *multimorph*. All are bending devices; however, the definitions are as follows:

Monomorph—single actuator ceramic plate
Unimorph—single actuator plate and an elastic shim
Bimorph—double actuator plates bonded together with or without an elastic shim
Multimorph—multiple actuator plates bonded together with or without multiple elastic shims

Part II

How to Use ATILA

0 Preparation

II. HOW TO USE ATILA 

Part II. HOW TO USE ATILA: List of Contents

Π-SHAPED LINEAR MOTOR 2D FINITE ELEMENT ANALYSIS

Atila-Light Software Instalation

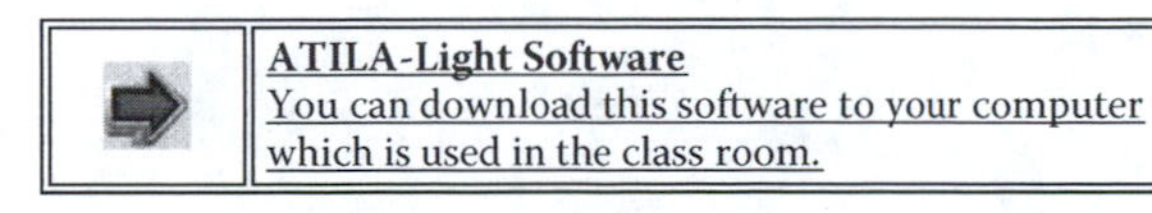
ATILA-Light Software
You can download this software to your computer which is used in the class room.

General Simulation Process

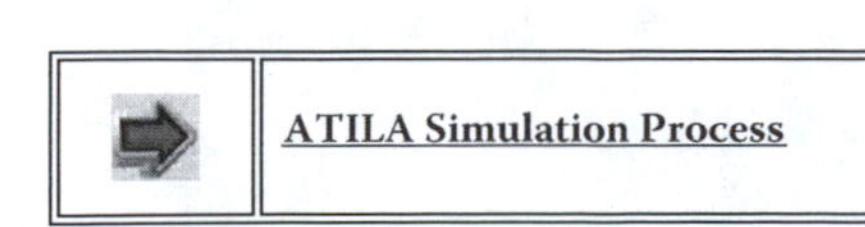
ATILA Simulation Process

Useful References

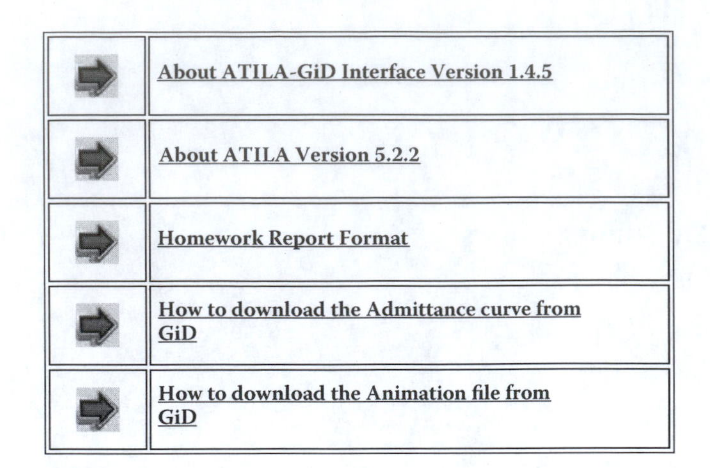

➡	About ATILA-GiD Interface Version 1.4.5
➡	About ATILA Version 5.2.2
➡	Homework Report Format
➡	How to download the Admittance curve from GiD
➡	How to download the Animation file from GiD

Final Examination Report

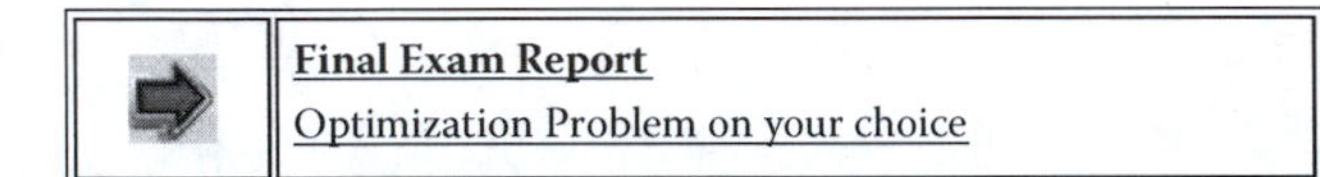

➡	Final Exam Report Optimization Problem on your choice

Back to Top

ATILA Demo Version Installation

Introduction

1. **Install ATILA Demo Version**
2. **Install GiD**
3. **Install ATILA–GiD Interface**

This sequence is essential.
Test
Problem During Installation?

Introduction ATILA GiD ATILA-GiD Test Problem
(The most recent ATILA Demo Version is available at http://www.mmech.com. Mecromechatronics, Inc. [200 Innovation Blvd. State College, PA 16803] is an official distributor of ATILA FEM software.)

0.1 GENERAL SIMULATION PROCESS/LEARN GID

GEOMETRY/DRAWING

Draw the structure, using the layers in the Utilities.

MATERIALS ASSIGMENT

Assign the materials, Piezoelectric, Elastic, Magnetostrictive, Magnetic, and Acoustic Medium.

BOUNDARY CONDITIONS

Assign Polarization (dielectric or magnetic), Potential, Magnetic Field, Mechanical Stress, and Electrical Load.

PROBLEM DATA

Assign the Problem Type; 3D, 2D (Axisymmetric, Planer Stress, Planer Strain).
Assign the Calculation Type; Static, Modal, Harmonic, and Transient.

Static—At one frequency lower than 1 kHz
Modal—At all the resonance frequencies
Harmonic—Frequency dependence of Admittance/Impedance (magnitude and phase)
Transient—Unavailable for the Demo Version

FREQUENCY DATA

Assign the calculation Frequency range for the Harmonic case.

MESHING

Learn the smart meshing with Structured Mesh.

CALCULATION

Calculation Window

POST PROCESS

Admittance/Impedance Result
View Results—Vibration Mode Animation at the resonance (or other) frequency

Displacement, Stress, Potential, Medium Pressure

Back to Top

Back to Home

0.2 ANIMATION, ADMITTANCE CURVE, AND REPORT FORMAT

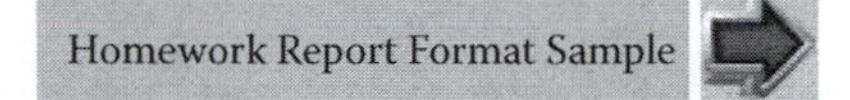

How to Copy the Admittance Curve from GiD

- Use the "Camera Icon" on the GiD Toolbar.
- Choose GIF (or others)

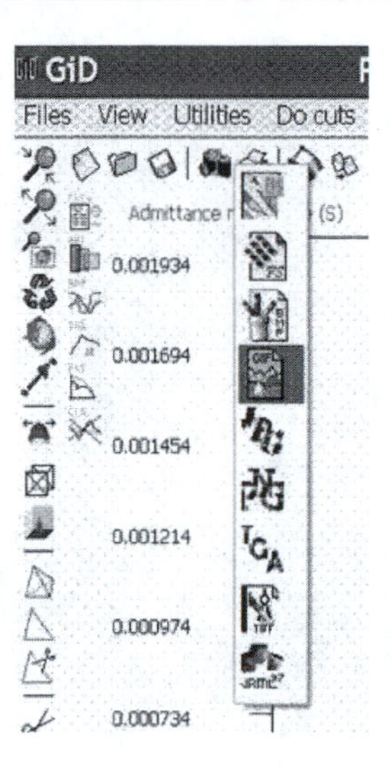

How to Copy the Animation from GiD

- Use the Animation Toolbar.
- Click "SAVE" on either MPEG or TIFF.
- Choose "MPEG": choose the download file position: Then, click the triangle mark to download the movie file.
- Or choose "GIF" on the "TIFF" portion: choose the download file position. Then click the triangle mark to download the movie file. In this case, as shown below, 19 GIF pictures are successively downloaded. Use ULEAD Animation Creator or an equivalent animation maker to make as a movie file. Different from the MPEG movie, the GIF movie can be directly inserted on a Power Point File, which is easier for presentation purposes.
- Or, if you are using the most recent version of ATILA, you can find a "GIF" movie file selection button.
- Remember: Unclick "Endless" item. Click to start recording.

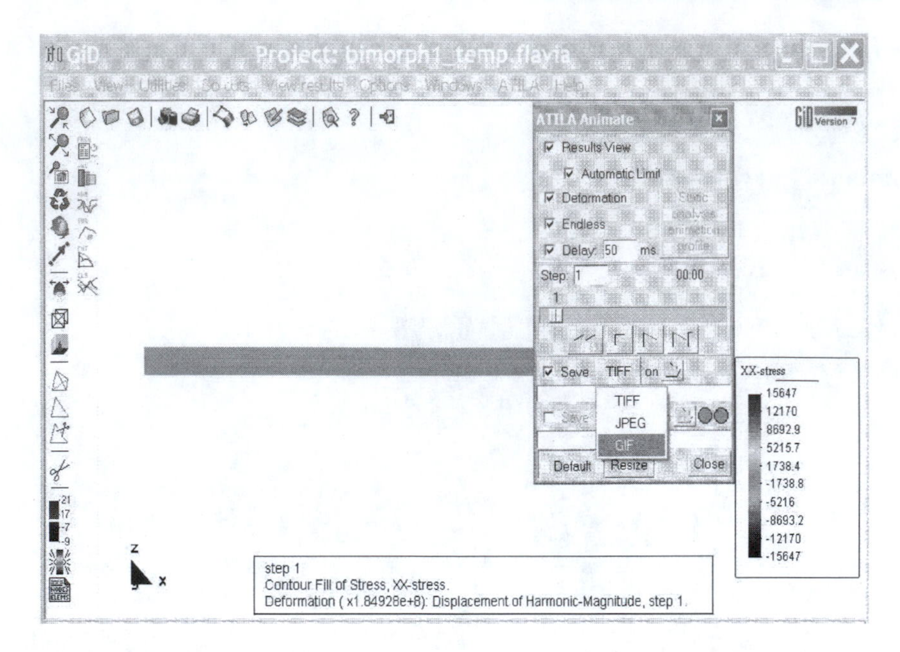

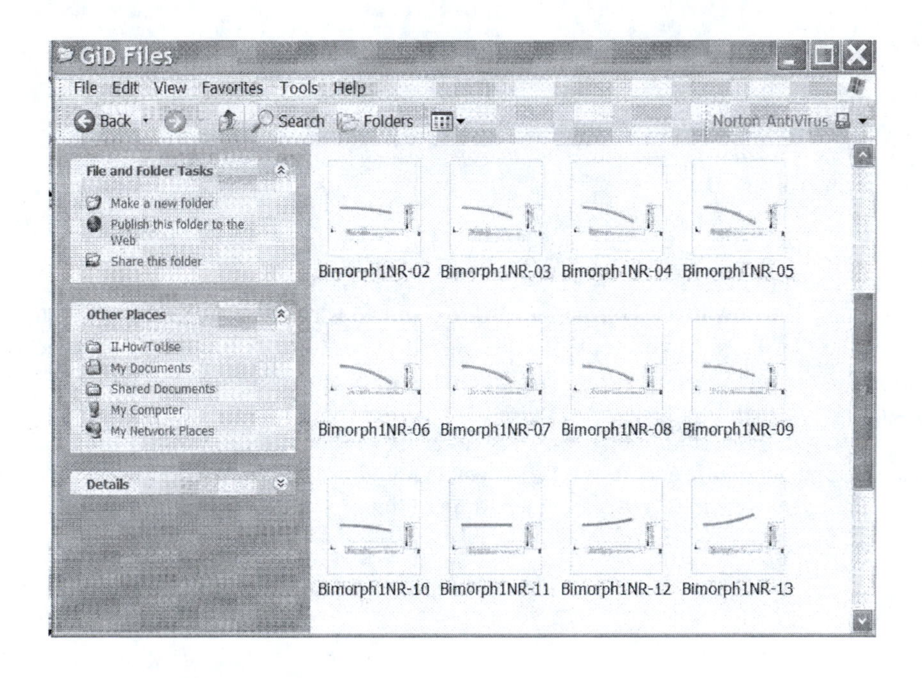

Back to Top

Back to Home

0.3 HOW TO USE THE ATTACHED GID FILE IN THIS CD

1) Click and open "GiD 7.2" file.

(2) Click "Files" on the Toolbar, then "Open," "Read Project" will show up.

(3) Find your desired ....GiD, in CD-FEMappl\II.HowToUse\GiD Files. This CD includes six demonstration GiD files as shown in the right picture.

(4) Click the desired file; then you can upload the calculated results, and enjoy "Impedance Spectrum" and "Animation."

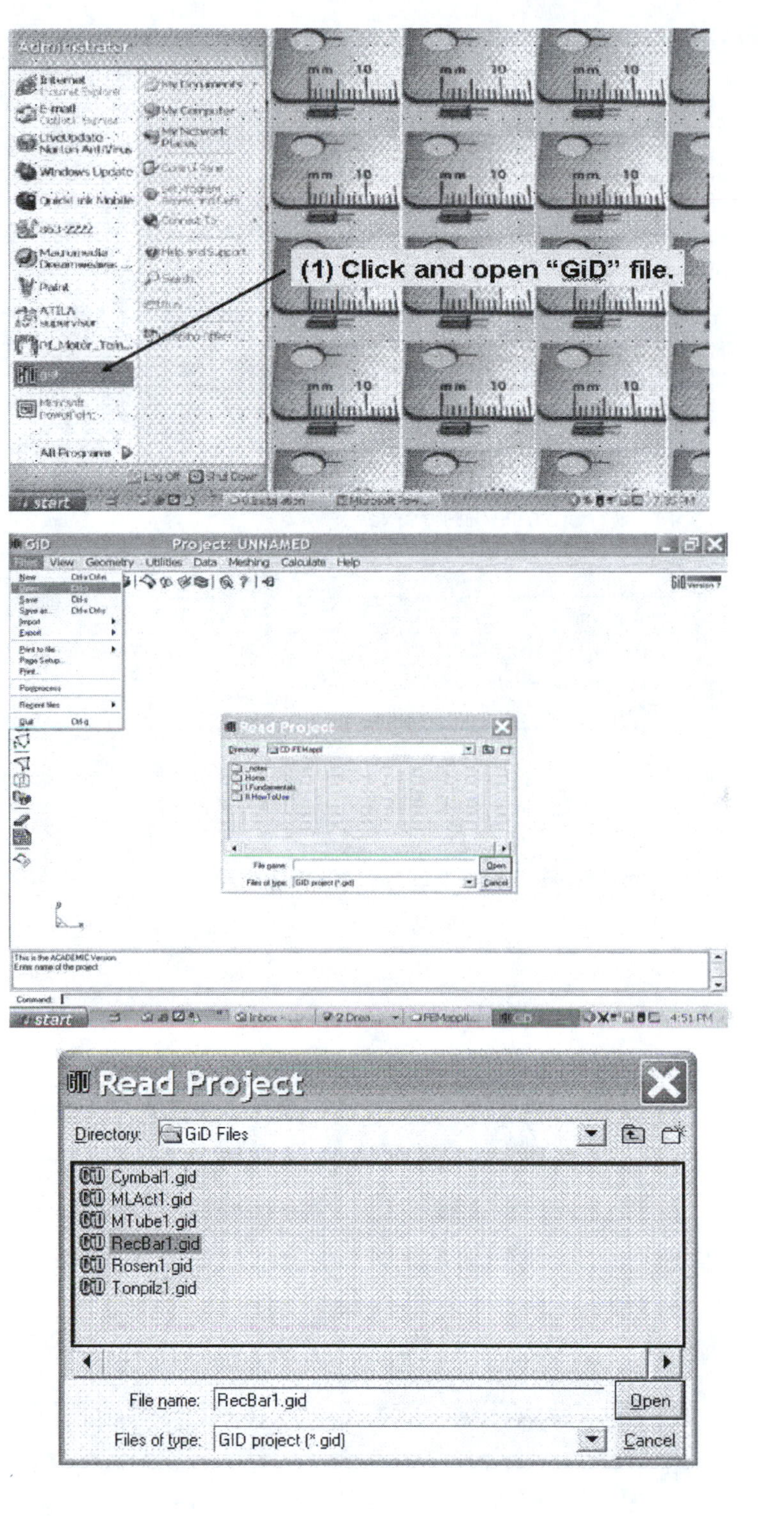

Go to How-To-Use Menu

I. FUNDAMENTALS

CONTENTS	WHAT WILL BE LEARNED
1. Trend of Micromechatronics and Computer Simulation	• Why computer simulation?
2. Review of Piezoelectricity and Magnetostriction	
2.1 Piezoelectric Materials: Overview 2.2 Magnetostrictive Materials 2.2Mathematical Treatment	• Principle of piezoelectricity and magnetostriction • PZT, Terfenol-DTensor/matrix notation • Loss mechanisms and heat generation
3. Structures of Smart Transducers	
3.1 Multilayer 3.2 Bimorph/Ringmorph/Cymbal 3.3 Magnetostrictive Actuator 3.4 Ultrasonic Motor	• Operating principles of multilayer, bimorph, and other composite structures • Magnetic coil • Prestress • Operating principles of ultrasonic motors
4. Drive/Control Techniques of Smart Transducers	
4.1 Pulse Drive 4.2Resonance Drive	• Overshoot, vibration ringing • Resonance and antiresonance • Mechanical quality factor
5. Finite Element Analysis	
5.1 Fundamentals of FEM 5.2 Combining with Piezoelectric Equations 5.3 Examples of ATILA Simulation	• Discretization, mesh • Piezoelectric equations
6. Design Optimization	
6.1 Optimization Concept 6.2 Genetic Optimization 6.3 Case Studies	• Genetic optimization theory
7. Future of FEM in Smart Structures	

II. HOW TO USE ATILA

CONTENTS	WHAT WILL BE LEARNED	ADVANCED LEARNING
0. Preparation		
0.1 ATILA Download 0.2 Simulation Process	• ATILA Download • General simulation process	• ATILA-GiD Interface 1.4.5 • ATILA Version 5.2.2
1. Piezoelectric Plate		
1.1 Rectangular Plate 1.2 Circular Disk	• How to draw the structure • How to assign the "Polarization" and "Electric field" • "Structured mesh" • How to take the admittance curve • How to make the animation • Homework report format • 3D and 2D simulation	• First and second harmonics • PZT ring
2. Magnetostrictive Rod	• How to assign the "Polarization" and "Magnetic coil/field"	
3. Composite Structure		

3.1 Bimorph 3.2 Unimorph 3.3 Multilayer 3.4 Cymbal	• How to couple two materials • How to set the mechanical boundary condition • Difference between the resonance and off-resonance vibration modes • Various problem types—Static, Modal, Modal Resantires, Harmonic, and Transient • How to use "Layer" • Stress concentration	• Cantilever and both-end clamp conditions • Optimization of metal/PZT thickness • Stress concentration at the internal electrode edge • MEMS membrane
4. Piezoelectric Transformer		
4.1 Rosen Type 4.2 Ring-Dot Type	• How to set the "Electrical load"	• Step-up ratio versus electrical load
5. Ultrasonic Motor		
5.1 L-Shaped Motor 5.2 π-Shaped Motor 5.3 Metal-Tube Motor	• How to apply sine and cosine voltages • Difference between 2D and 3D analysis	• Surface-wave-type motor • Design optimization (thickness, length, angle)
6. Underwater Transducer		
6.1 Langevin Type 6.2 Underwater Cymbal 6.3 Tonpilz Sonar	• How to assign "Water" • How to set the boundary condition • How to calculate the medium pressure	• Unipole and Dipole sound source • 1-3 Composite
7. Acoustic Lens	• Acoustic impedance	

Back to Top

1 Piezoelectric Plate

- How to draw the structure
- How to assign the "Polarization" and "Electric field"
- "Structured mesh"
- How to take the admittance curve
- How to make the animation
- Homework report format
- 3D and 2D simulation

TEXT CONTENT

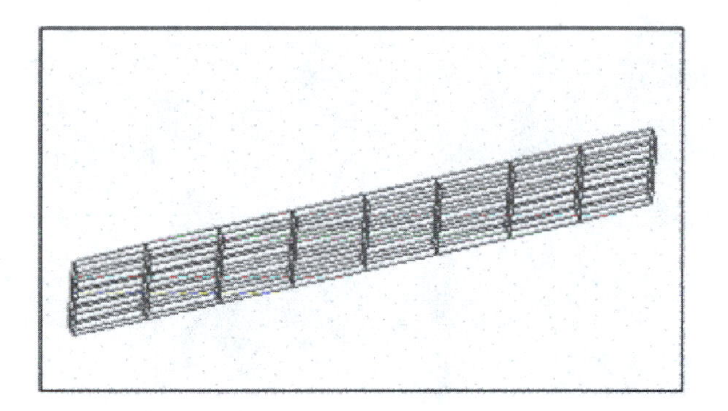

Rectangular Plate Mesh

➡	**1.1 Rectangular Plate**
➡	**1.2 Circular Disk**
➡	**Homework 1** Rectangular Bar Second Harmonic Mode
➡	**Homework 2** PZT Ring Vibration
How to Use this GID file	**Example GiD File attached in this CD** RecPlate.gid, 3DDisk.gid

Back to Top

1.1 RECTANGULAR PLATE (1)

GEOMETRY/DRAWING

PROBLEM—3D RECTANGULAR PLATE, K31 MODE, PZT4 = 40 MM × 6 MM × 1 MM

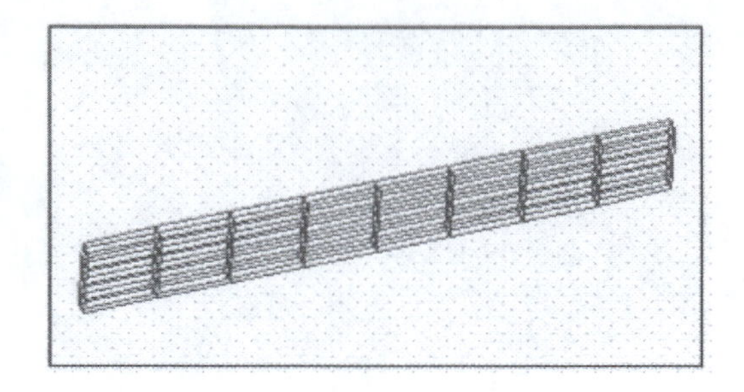

GEOMETRY/DRAWING

CREATE SURFACE SIMPLE STRUCTURE LIST

Geometry -> Create -> Object -> Rectangle -> Enter first corner point -> 0,0 -> Enter second corner point -> 40,6 -> Enter.

Remember that Point = Black color, Line = Blue color, Surface = Magenta, Volume = Cyan.

CREATE VOLUMES COPY MODE

Utilities -> Copy -> Surfaces -> Second point z=1.0 -> Do Extrude Surfaces -> Select.
After selecting the rectangular plate (click on the surface line) -> Finish.

When you select the surface (click on the Magenta line), the surface color Magenta changes to Red.

If Extrude is chosen, the four side surfaces are also assigned. If you do not choose Extrude, two separate surfaces are created.

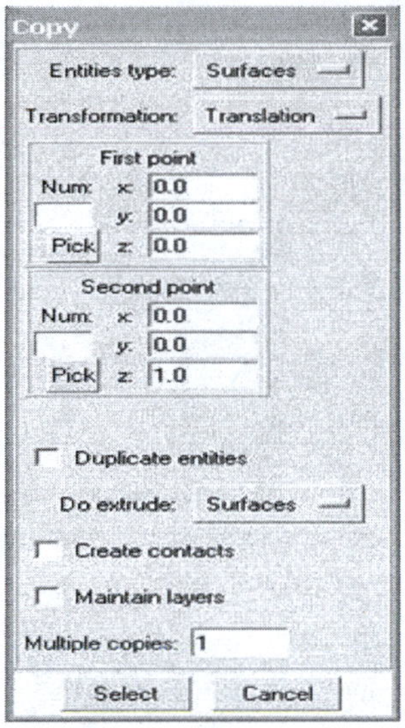

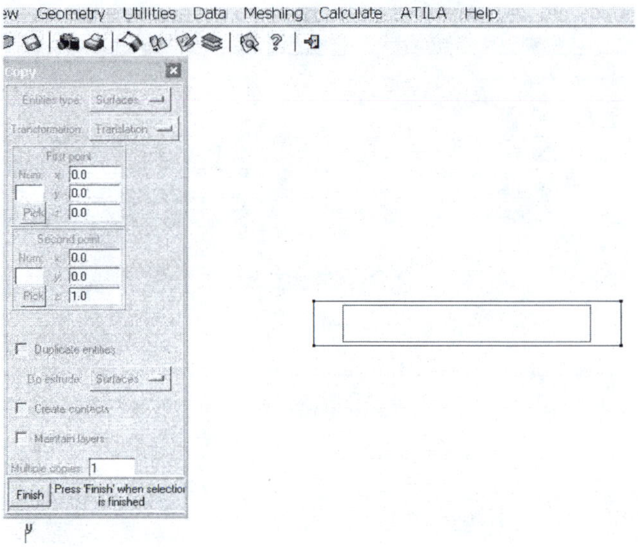

View the Structure and Learn Trackball and Shift Key

View -> Rotate and/or Shift -> Trackball or Shift ("Escape" after using).

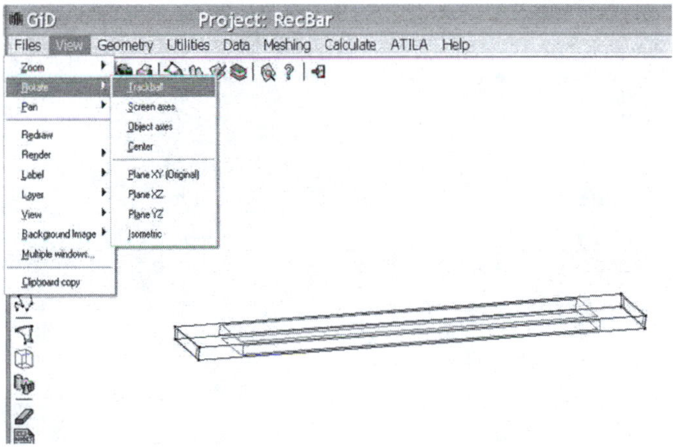

Create Volumes 2

Geometry -> Create -> Volume -> Automatic 6-sided volumes.

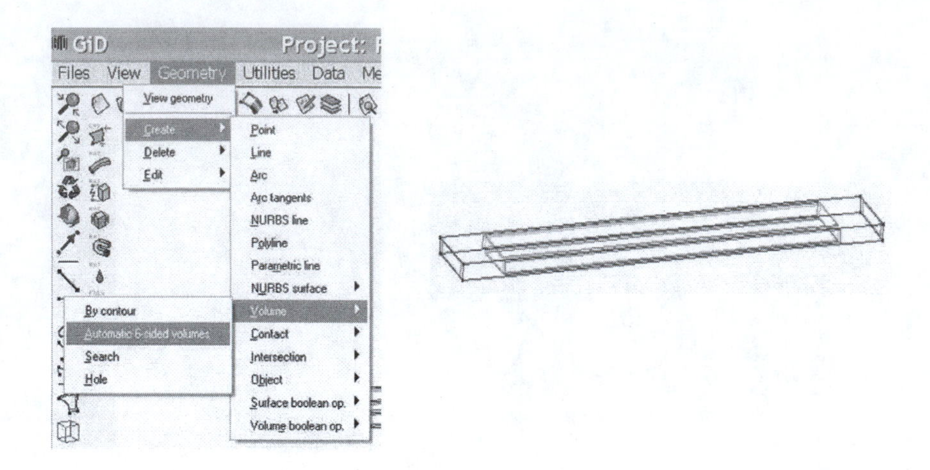

When you make the volume, you will find the Cyan color (which is different from the previous drawing).

NB: When you choose Extrude Volume, you can create Volume directly by skipping Create Volume.

Back to Top

Back to Home

1.1 RECTANGULAR PLATE (2)

MATERIAL ASSIGNMENT 

PROBLEM—3D RECTANGULAR PLATE, K31 MODE, PZT4 = 40 MM × 6 MM × 1 MM

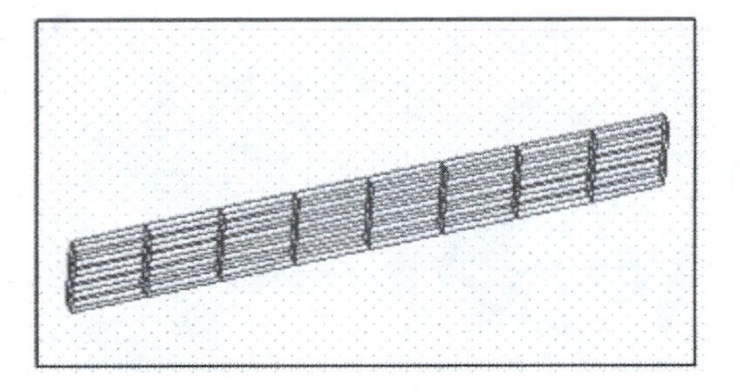

MATERIAL ASSIGMENT

PIEZOELECTRIC MATERIALS 

Data -> Materials -> Piezoelectric -> PZT4 -> Assign -> Pick Volumes.
Click Draw -> All materials -> Check the material is assigned as PZT4.

Back to Top

Back to Home

1.1 RECTANGULAR PLATE (3)

BOUNDARY CONDITIONS

PROBLEM—3D Rectangular Plate, k31 Mode, PZT4 = 40 mm × 6 mm × 1 mm

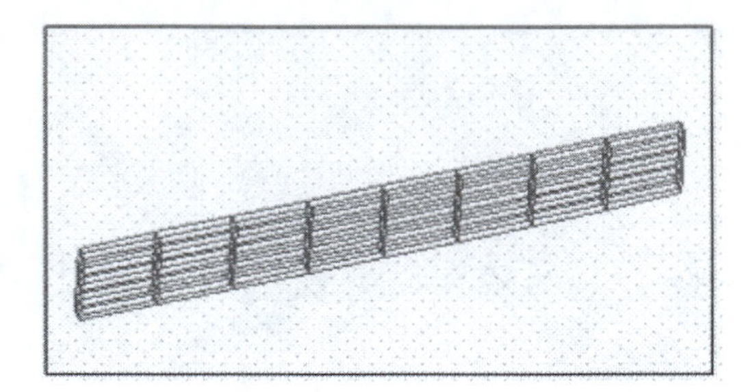

BOUNDARY CONDITIONS

Potential

Data -> Conditions -> Surfaces -> Potential 0.0 (or Ground in ATILA 5.2.4 or higher version) -> Assign -> Potential 1.0 -> Assign. To check the assignment -> Draw -> All conditions -> Include local axes.

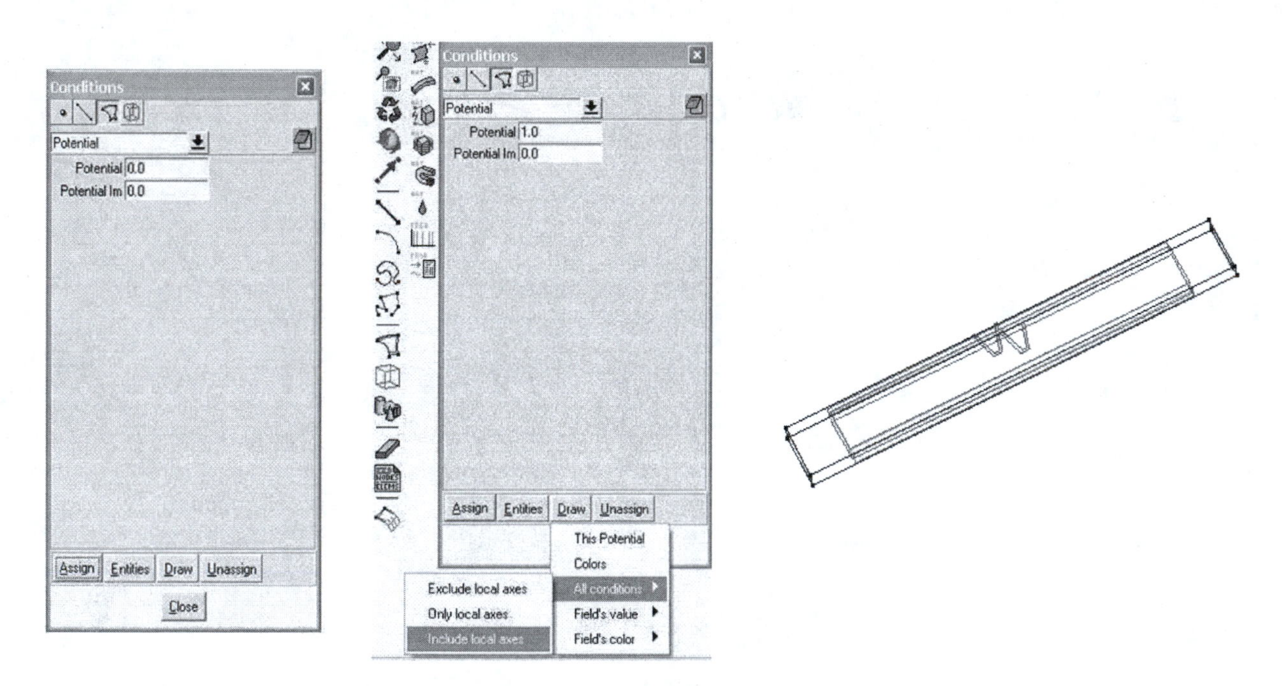

NB: 1V is applied in most cases in this textbook because other voltage merely provides a proportional result in displacement, strain, stress and so forth.

Polarization

Data -> Conditions -> Volumes -> Geometry -> Polarization (Cartesian) -> Define Polarization (P1).

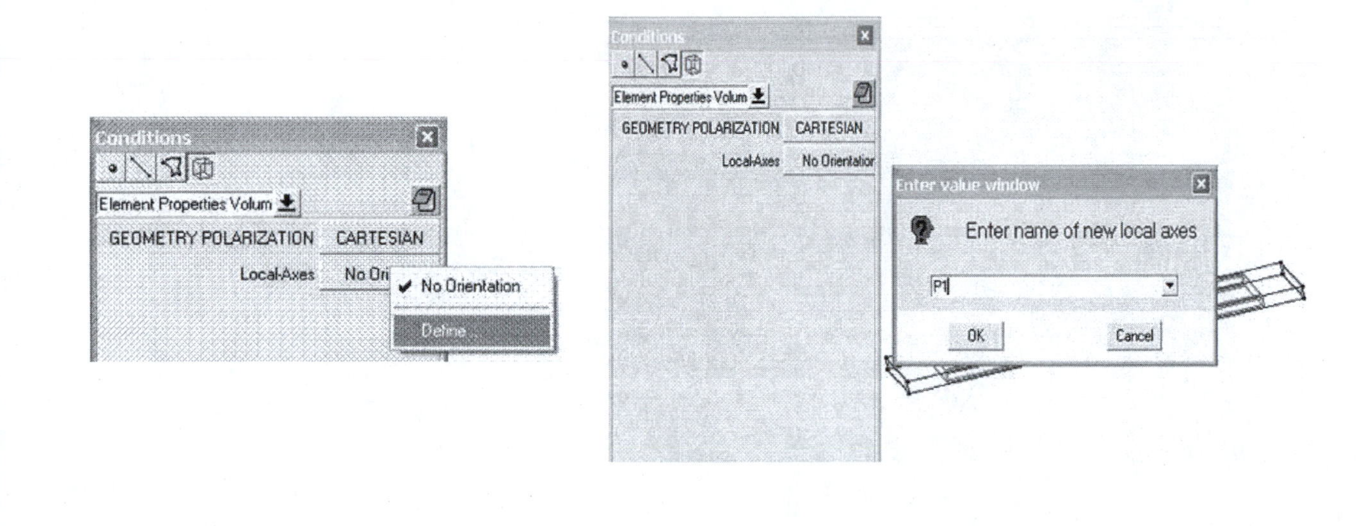

Polarization/Local Axis

Choose X and Angle -> Define X axis -> Click two points to define X axis -> Choose the suitable Cartesian P1 direction (Z axis) -> Escape.

Or: Choose 3PointsXZ -> Define X axis by putting (1,0,0) -> Then define Z axis by putting (0,0,1) -> Escape.

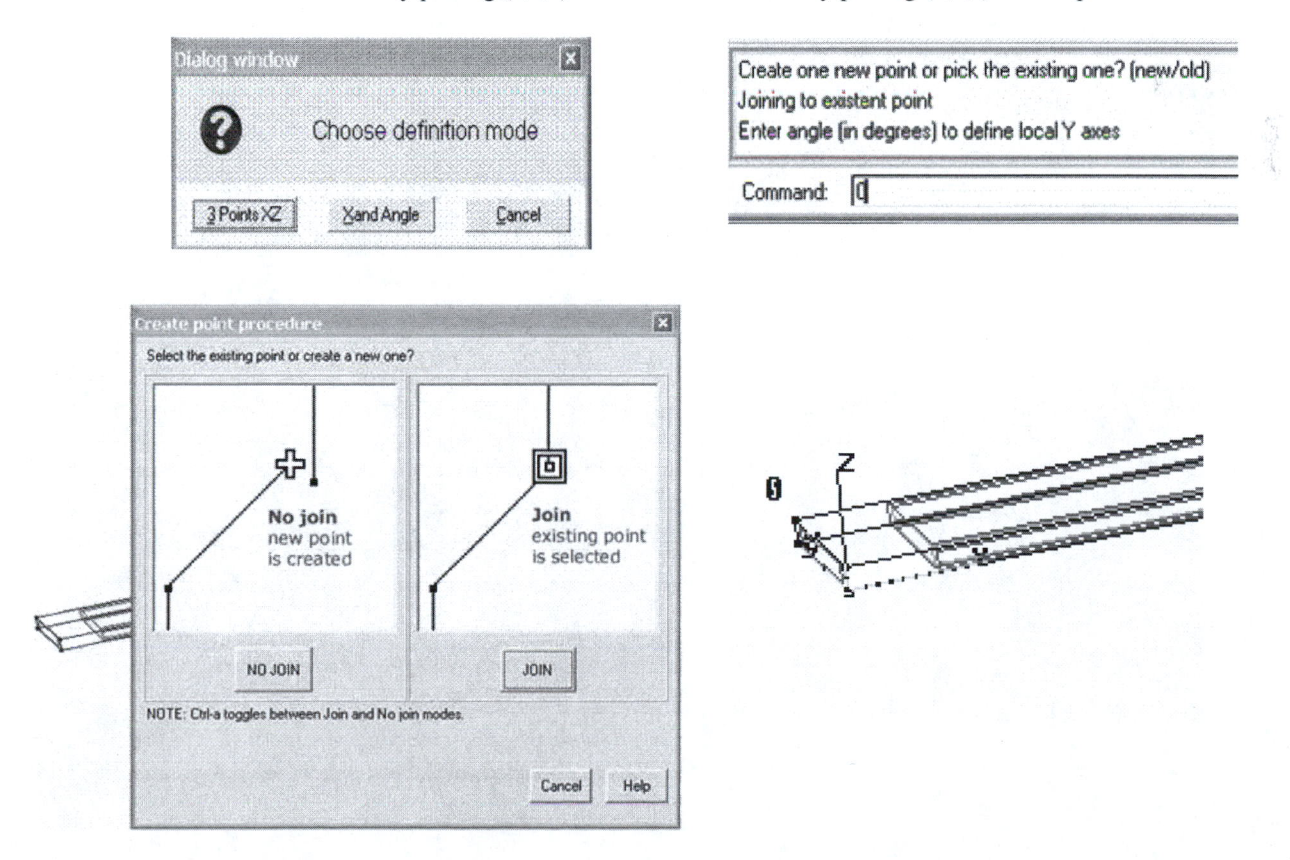

Assign Polarization (Local-Axes P1) -> Choose the volume -> Finish.

To check the assignment -> Draw -> All the conditions -> Include Local axes.

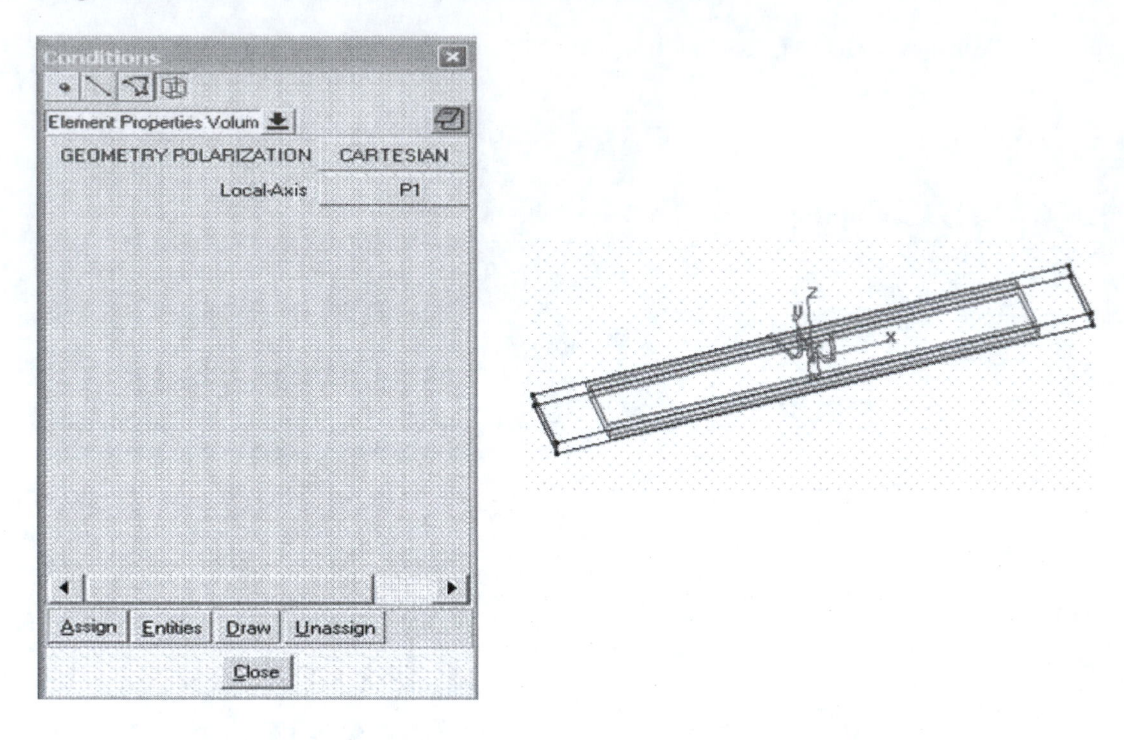

NB: The polerization is defined in terms of "local axis" (z should be the positive P direction), which should not be confused with the Cartesian drawing map axis.

Back to Top

Back to Home

1.1 RECTANGULAR PLATE (4)

MESHING 

PROBLEM—3D Rectangular Plate, k31 Mode, PZT4 = 40 mm × 6 mm × 1 mm

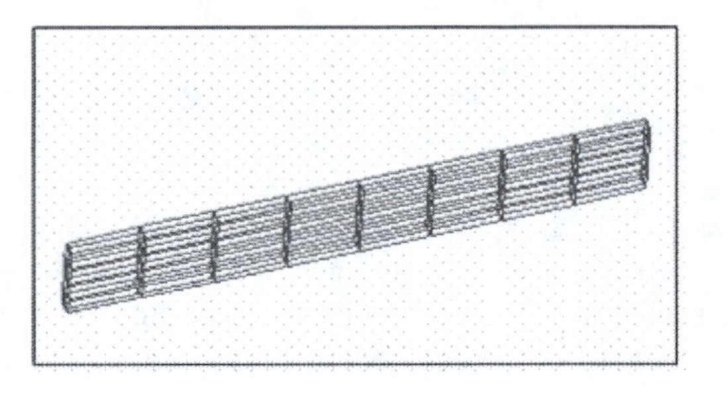

MESHING 

Free Mesh

Meshing -> Generate -> Enter size of elements (4) -> Mesh Generated ->
Number of nodes should not exceed 500 for ATILA Light Version

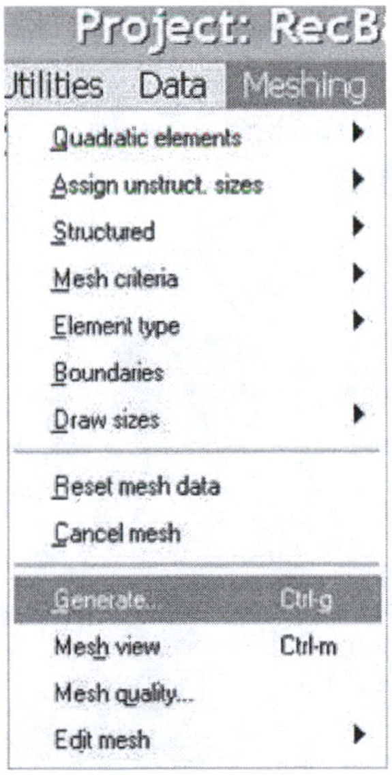

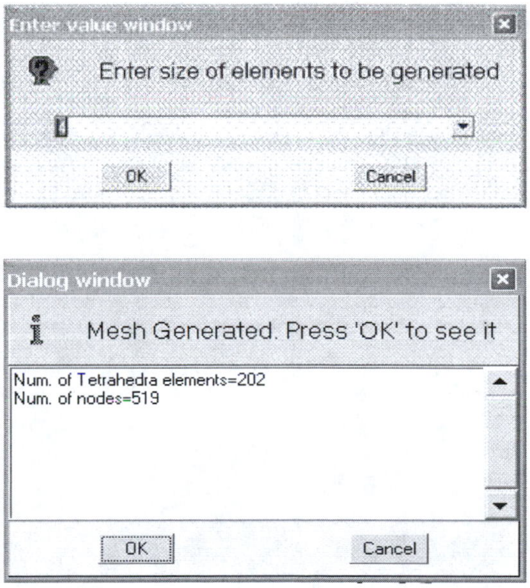

Mapped Mesh

Meshing -> Structured -> Volumes -> Select all volumes -> Then Cancel the further process.
Meshing -> Structured -> Lines -> Enter number to divide line -> 8 and 4 are assigned to X and Y axes.

In practice, for ATILA Light Version, assign 1 to Z axis, 1 or 2 to Y axis, and 4 or 8 to X axis. A value of 8 is required to show the second resonance vibration so that a better position resolution is obtained.

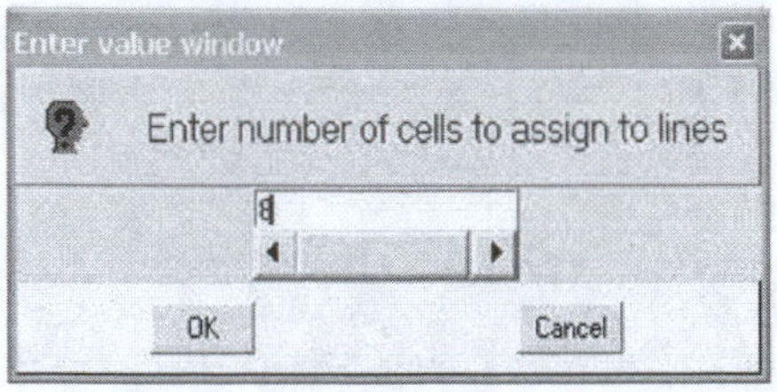

Enter value window

Enter number of cells to assign to lines

8

OK Cancel

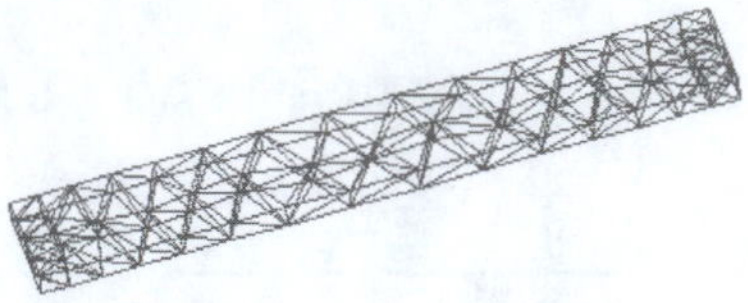

Free mesh (Not good)

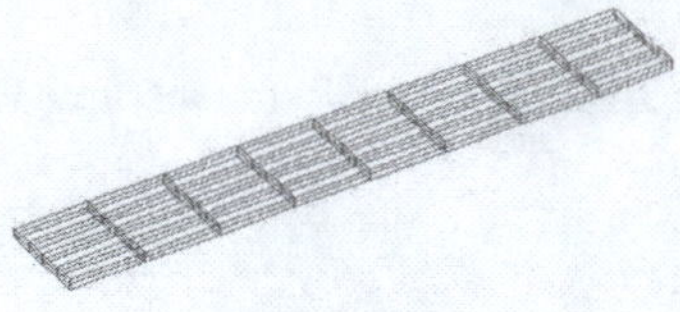

Mapped mesh (good)

NB: Though the free mesh can simulate the mode, the structured mapped mesh with your intuitive sense provides better results with faster calculation speed. In particular, because the demo version has a node number limitation, the structured mesh is highly recommended.

Back to Top

Back to Home

1.1 RECTANGULAR PLATE (5)

CALCULATION

PROBLEM—3D RECTANGULAR PLATE, K31 MODE, PZT4 = 40 MM × 6 MM × 1 MM

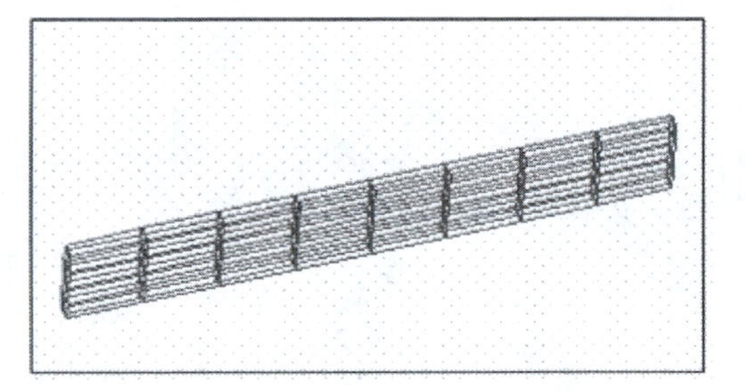

PROBLEM TYPE

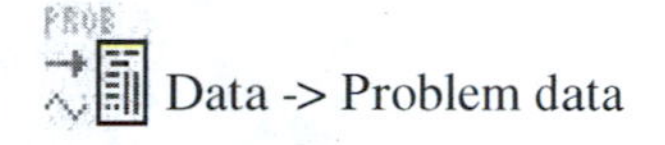
Data -> Problem data

Choose: 3D and Harmonic.

FREQUENCY

Data -> Interval data

Number of frequencies: "11" is chosen to get 10 frequency interval.

CALCULATION

Calculate -> Calculate window -> Start

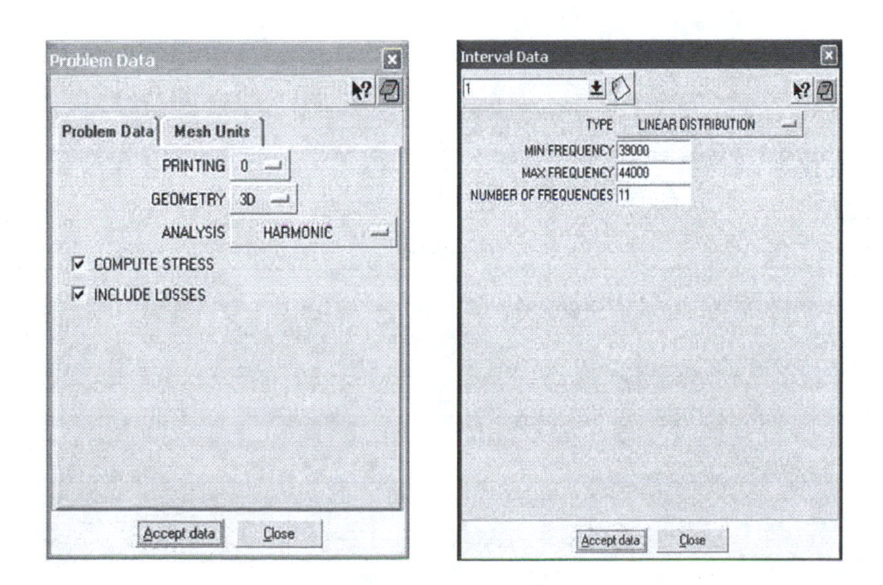

Meshing Calculate ATILA ATILA Help
Cancel process
View process info
Calculate window...
Calculate
Calculate remote
Process window
Project Start time UID
Output view Kill
Start Start remote Remote... Close

1.1 RECTANGULAR PLATE (6)

POST PROCESS/VIEW RESULTS

PROBLEM—3D RECTANGULAR PLATE, K31 MODE, PZT4 = 40 MM × 6 MM × 1 MM

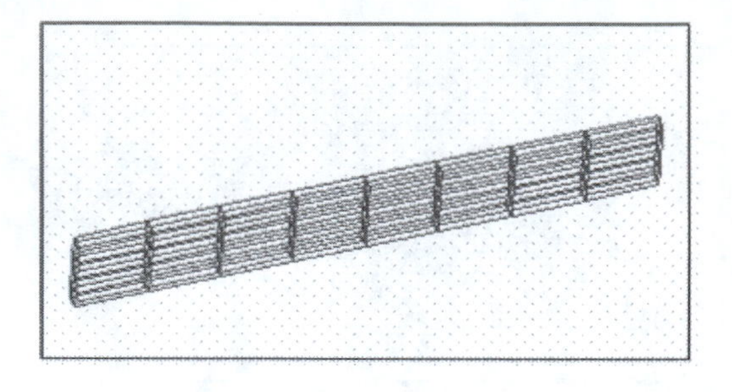

POST PROCESS/VIEW RESULTS

Click to Open the Post Calculation (Flavia) Domain.

Note the new Toolbar appearance.

Click ADM **Admittance Result** (Mesh Dependence)

Peak Point: 41,000 Hz.

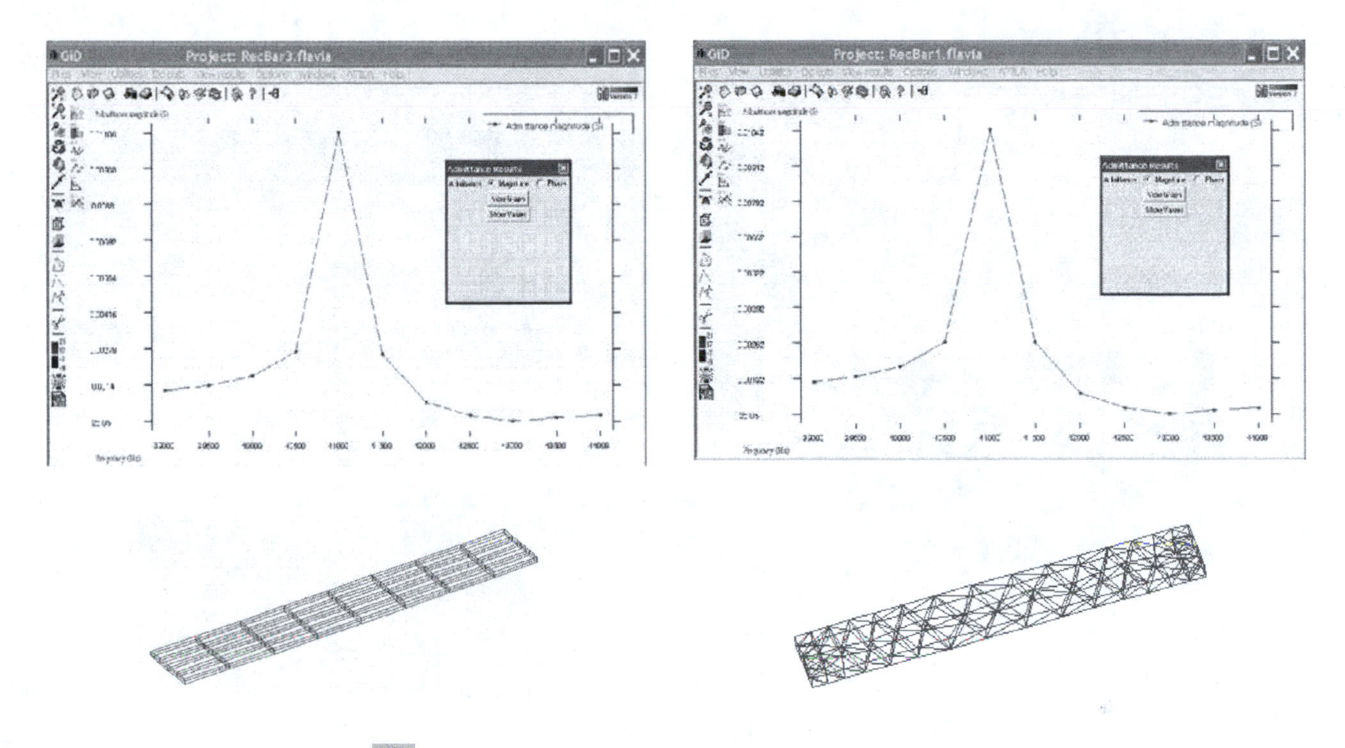

To take this graph out -> Go To

View Results (1)—Admittance/Impedance Curve

View results -> Default Analysis/Step -> Harmonic-Magnitude -> Set a Frequency (Peak Point: 41,000 Hz for Resonance Mode).

GiD
Project: RecBar3.flavia
Files View Utilities Do cuts View results Options Windows ATILA Help
No Results
No Graphs
Default Analysis/Step
Harmonic-Magnitude
Harmonic-Phase
Contour Fill
Contour Lines
Contour Ranges
Show Min Max
Display Vectors
Iso Surfaces
Stream Lines
Graphs
Deformation
Line Diagram
39000
39500
40000
40500
41000
41500
42000
42500
43000
43500
44000
Admittance magnitude (S)
Admittance Results
Admittance Magnitude Phase
View Graph
Show Values
GiD Version 7
0.00416
Frequency (Hz)
39000 39500 40000 40500 41000 41500 42000 42500 43000 43500 44000

View Results (2)—Vibration Mode at a Particular Frequency

Double-Click  Loading frequency: 41,000 Hz Harmonic Case -> Click OK (single-click and wait for a while).

Set Deformation -> Displacement

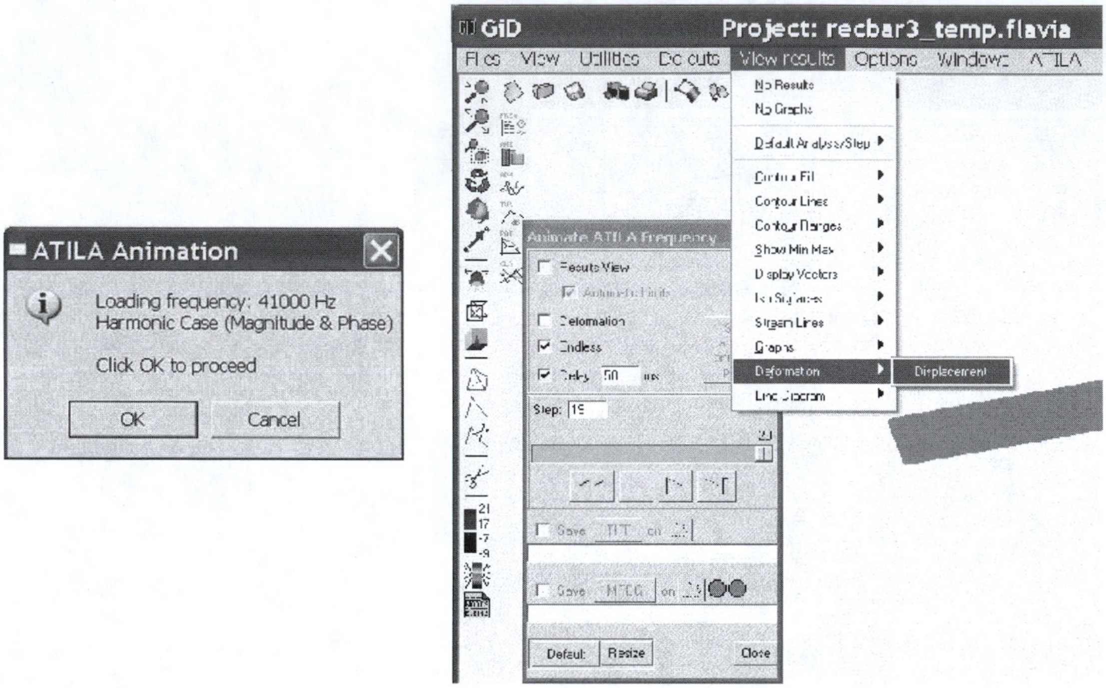

Deformation Scale Change

Click Windows -> Deform mesh -> Change the Factor in Mesh Deformation.

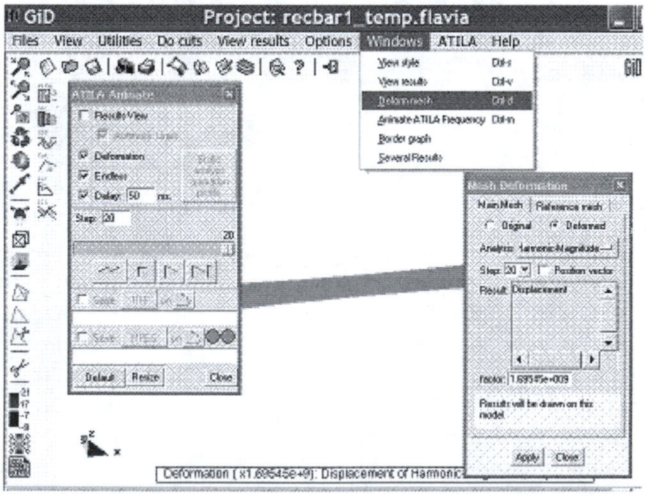

View Results (3)—Parameter other than Displacement

Set Results View -> View results -> Contour Fill -> Stress -> XX-stress.

After setting both "Results view" and "Deformation" -> Click to see the Animation.

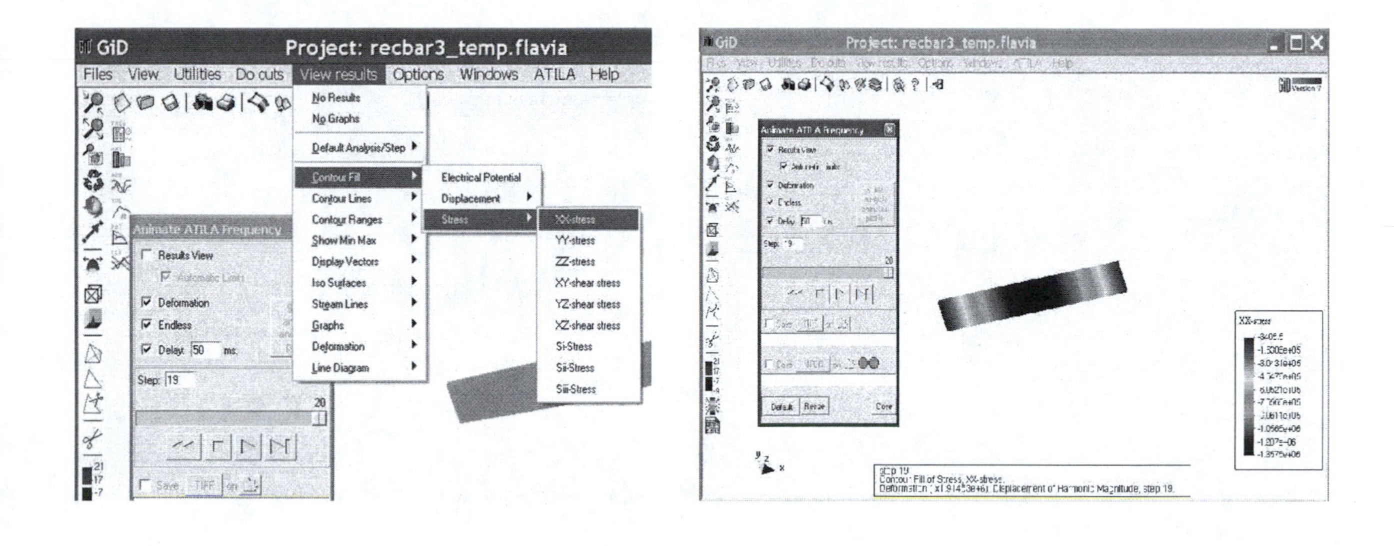

To take this graph out -> Go To

Mesh and Calculation Time

# of unit (z/y/x)	Hexahedra Element	# of nodes	Calculation Time
1/2/4	8	89	13 sec
2/2/4	16	141	20 sec
2/2/8	32	261	90 sec
2/2/16	64	501	6.5 min
2/4/8	64	453	6 min
4/8/8	256	1449	60 min
4/8/16	512	2777	220 min

Calculation Time α (# of nodes)^2

Back to Top

Back to Home

1.2 CIRCULAR DISK (1)

GEOMETRY/DRAWING

PROBLEM—3D CIRCULAR PLATE, KP MODE, PZT4 = 50 MM Φ × 1 MM T

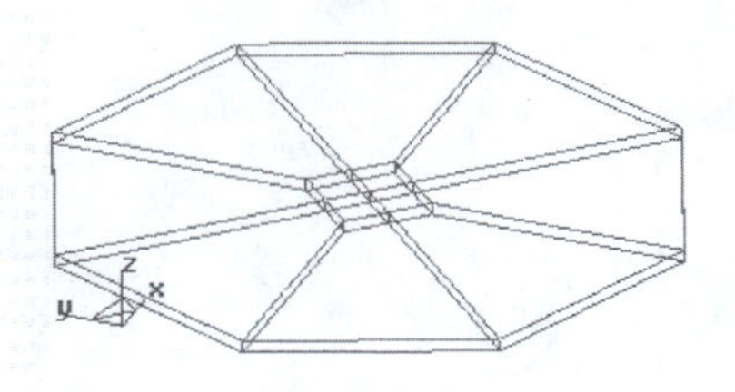

and **(2D Calcualtion)**

- Difference from the rectangular plate lies in how the center portion is treated in meshing (quadrilateral element cannot be made in the center)

GEOMETRY/DRAWING

CREATE SURFACE **SIMPLE STRUCTURE LIST**

Geometry -> Create -> Object -> Circle -> Enter a center for the circle -> 0,0 -> Enter a normal for the circle -> Positive Z -> OK -> Enter a radius for the circle -> 25 -> Enter.

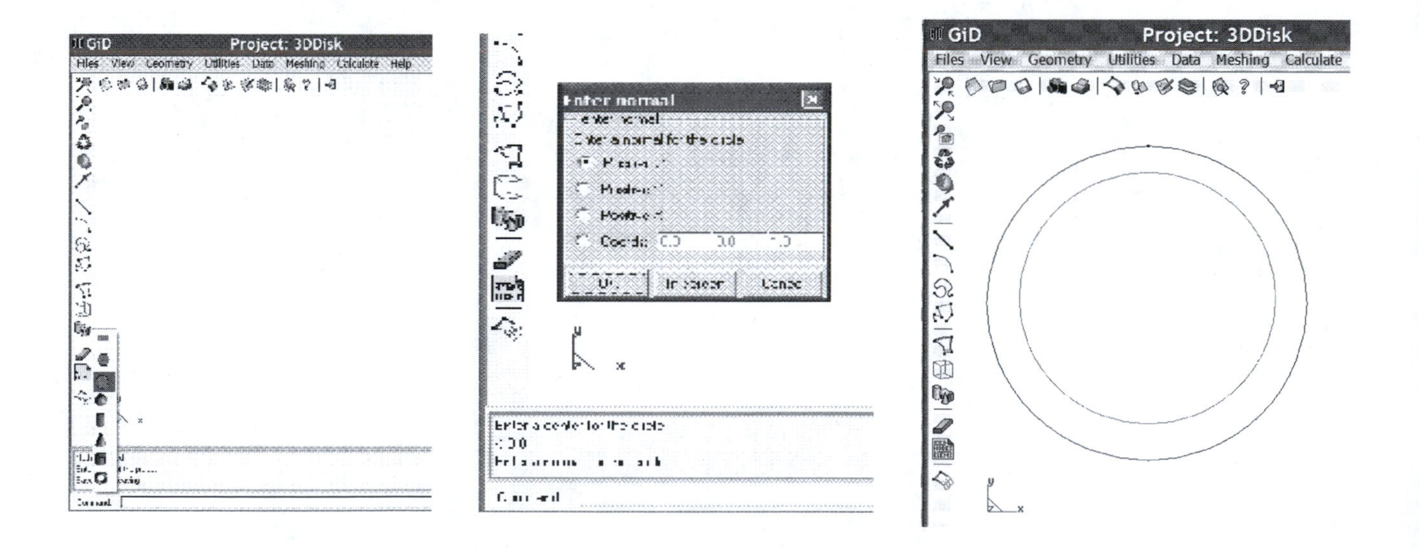

Remember that Point = Black color, Line = Blue color, Surface = Magenta, Volume = Cyan.

Create the Center Quadrilateral Element—To obtain the better Meshing

Geometry -> Delete -> Surface -> Select the surface (click the magenta circle) -> Escape.

Geometry -> Create -> Line -> Enter points to define line -> 5,0; 0,5; -5,0; 0,-5 and finally 5,0, successively -> Create point procedure -> JOIN -> Escape.

Geometry -> Edit -> Divide -> Num Divisions -> Select line to divide -> If you click the line, the line color change to Red -> Enter Value Window -> Enter "4" -> You will find four points with black color.

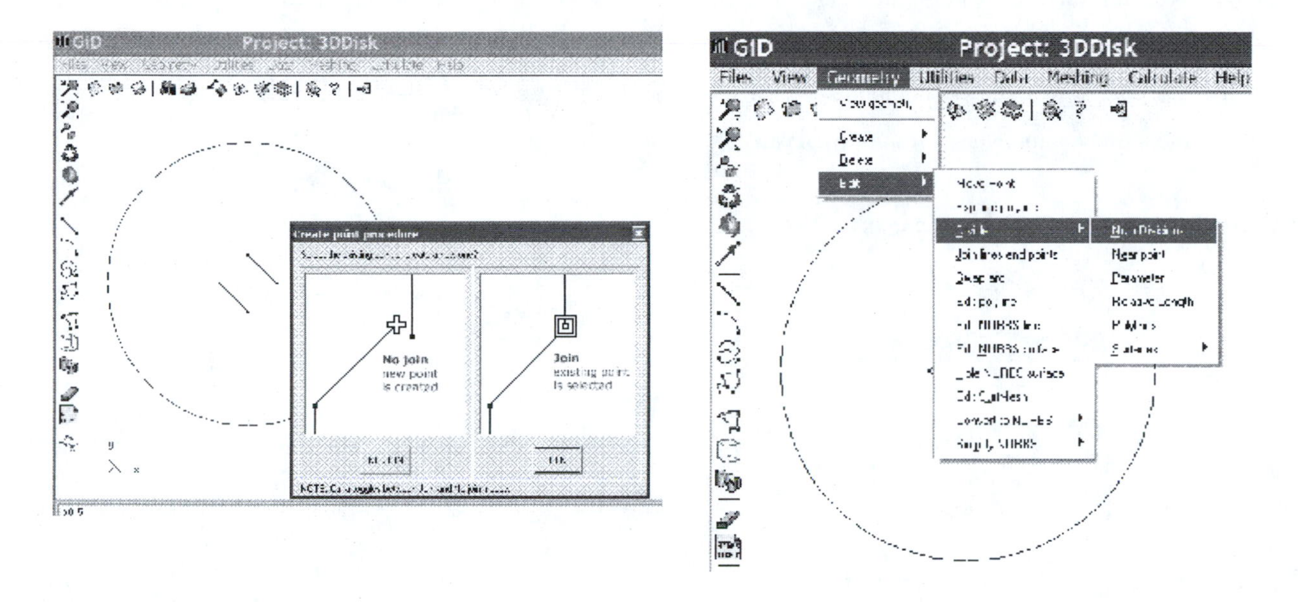

Geometry -> Create -> Line -> Enter points to define line -> Click with Thick Cross Cursor successively -> Create point procedure -> JOIN -> Escape.

Geometry -> Create -> NURBS Surface -> Automatic -> Enter Value Window -> Enter number "4" -> You will create five new surfaces.

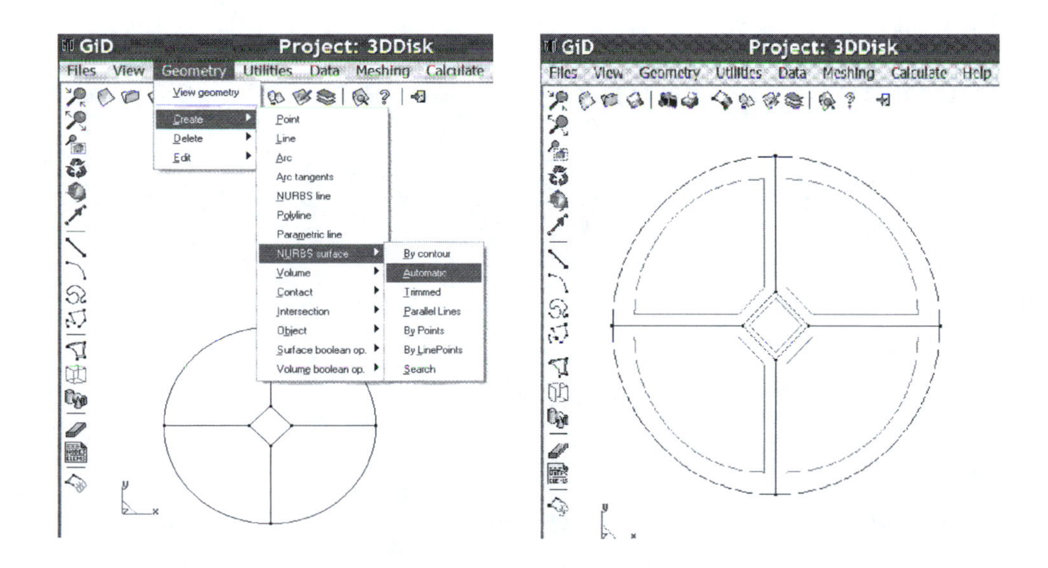

Create Volumes — Copy mode

Utilities -> Copy -> Surfaces -> Second point z=1.0 -> Do Extrude Surfaces -> Select After selecting the circle (click on the magenta color circle line) -> Finish.

When you select the surface (click on the Magenta line), the surface color Magenta changes to Red.

If Extrude is chosen, the four side surfaces are also assigned. If you do not choose Extrude, two separate surfaces are created.

Geometry -> Create -> Volume -> Automatic 6-sided volumes.

When you make the volume, you will find five Cyan color volumes.

View the Structure—Use Trackball and Shift

Copy
Entities type: Surfaces
Transformation: Translation
First point
Num: x: 0.0
y: 0.0
Pick z: 0.0
Second point
Num: x: 0.0
y: 0.0
Pick z: 1.0
Duplicate entities
Do extrude: Surfaces
Create contacts
Maintain layers
Multiple copies 1
Select Cancel

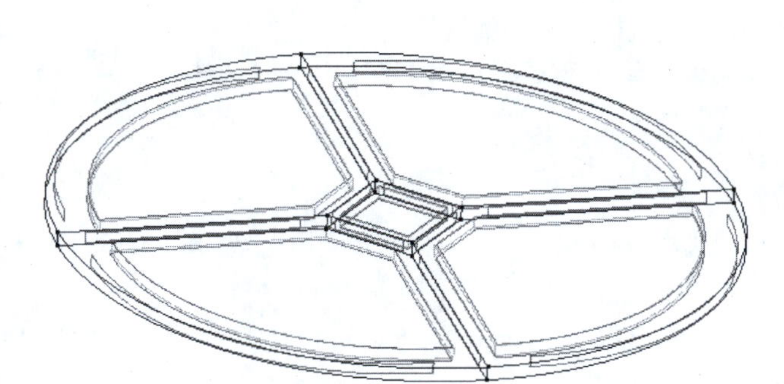

Back to Top

Back to Home

1.2 CIRCULAR DISK (2)

MATERIAL ASSIGNMENT

PROBLEM—3D Circular Plate, kp Mode, PZT4 = 50 mm Φ × 1 mm t

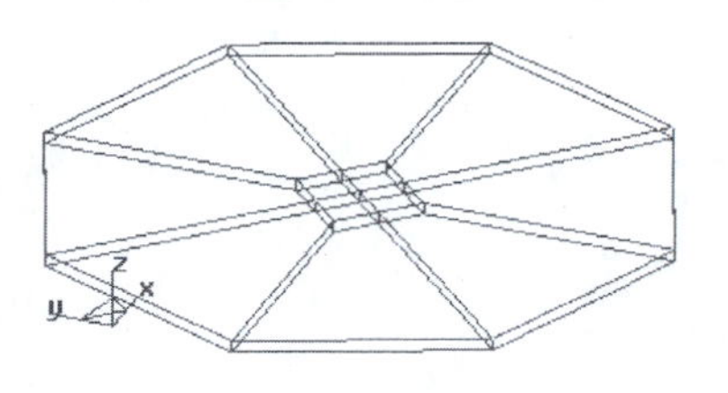

MATERIAL ASSIGNMENT

Piezoelectric Materials

Data -> Materials -> Piezoelectric -> PZT4 -> Assign -> Pick five Volumes.
Click Draw -> All materials -> Check that the material is assigned as PZT4.

Piezoelectric
PZT4
General | Mechanical | Coupling | Dielectric | Losses
DELTA M 0.002
DELTA P 0.
DELTA D 0.004
Assign | Draw | Unassign | Import/Export
Close

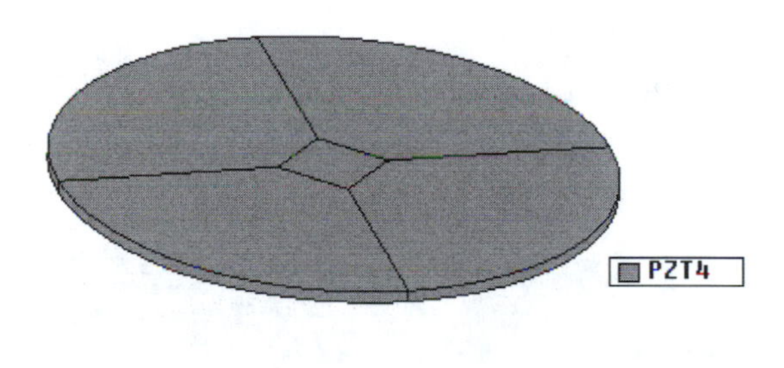

Back to Top

Back to Home

1.2 CIRCULAR DISK (3)

BOUNDARY CONDITIONS

PROBLEM—3D CIRCULAR PLATE, KP MODE, PZT4 = 50 MM Φ × 1 MM T

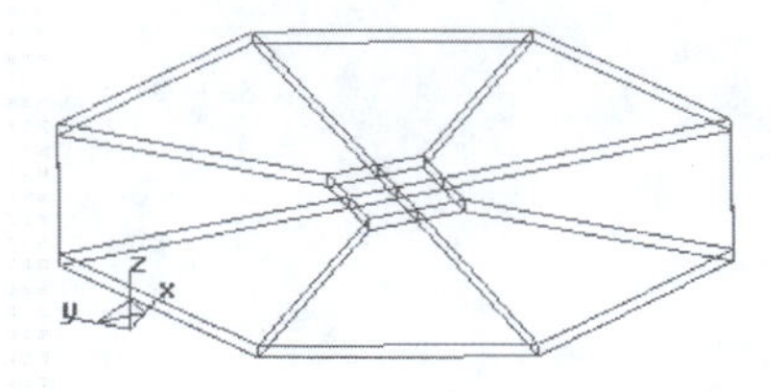

BOUNDARY CONDITIONS

POTENTIAL

Data -> Conditions -> Surfaces -> Potential 0.0 (or Ground in ATILA 5.2.4 or higher version) -> Assign -> Potential 1.0 -> Assign.

Because the gap between the top and bottom surfaces is narrow, it is recommended that the high magnification be used, as shown in the following text.

Focus-in and Focus-out.

To check the assignment -> Draw -> Color.
(As shown below, the color seems to be better than [All conditions -> Include local axes] in some cases.)

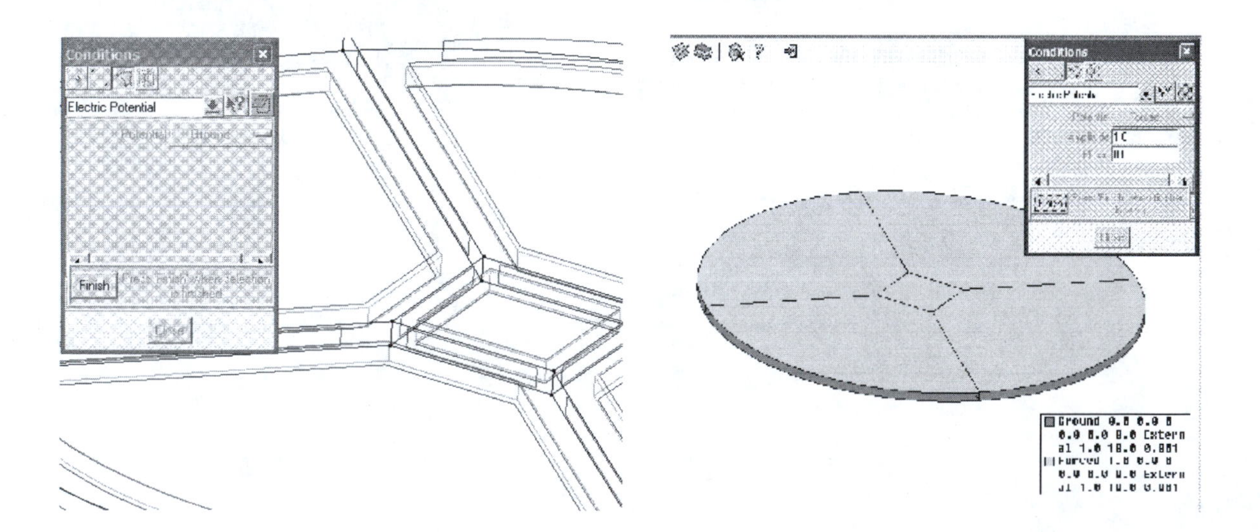

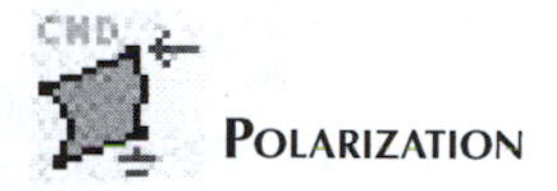

Polarization

Data -> Conditions -> Volumes -> Geometry -> Polarization (Cartesian) -> Define Polarization (P1).

We take the P1 downward along Z [Local axis (0,0,–1)].

Choose 3PointsXZ -> Define X axis by putting (1,0,0) -> Then define Z axis by putting (0,0,–1) -> Escape.

Polarization/Local Axis

Choose 3PointsXZ -> Enter local axes center -> (0,0,0) -> Define X axis by putting (1,0,0) -> Then define Z axis by putting (0,0,-1) -> Escape.

Assign Polarization (Local-Axes P1) -> Choose the volume -> Finish.

To check the assignment -> Draw -> All the conditions -> Include Local axes.

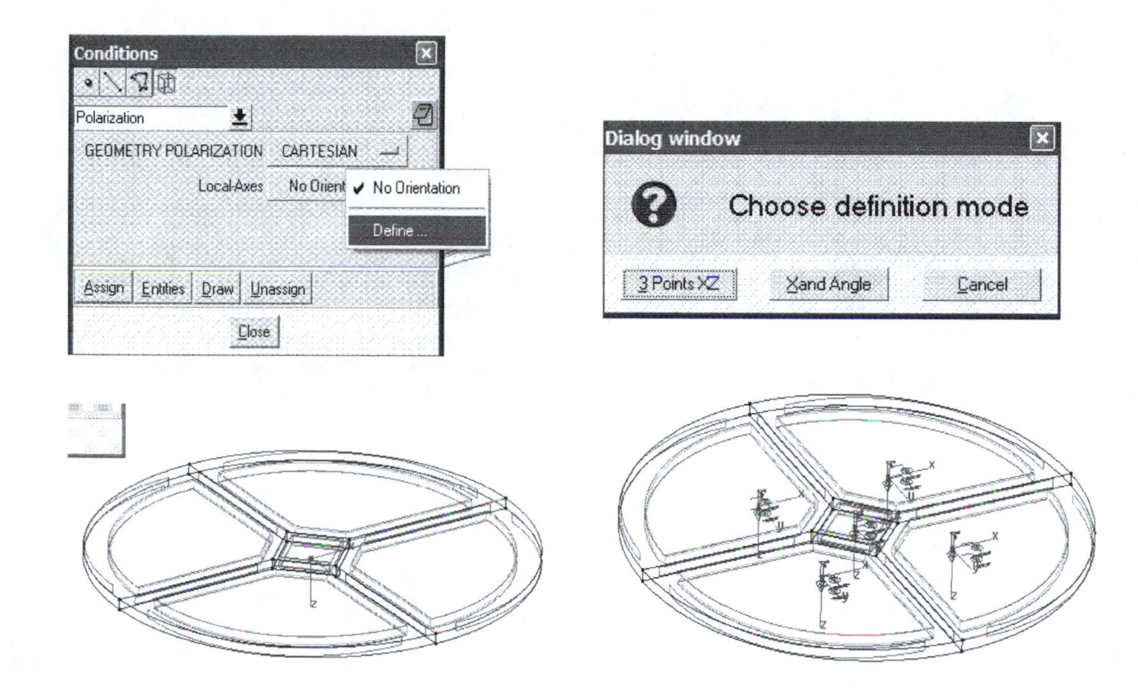

Back to Top

Back to Home

1.2 CIRCULAR DISK (4)

MESHING

PROBLEM—3D CIRCULAR PLATE, KP MODE, PZT4 = 50 MM Φ × 1 MM T

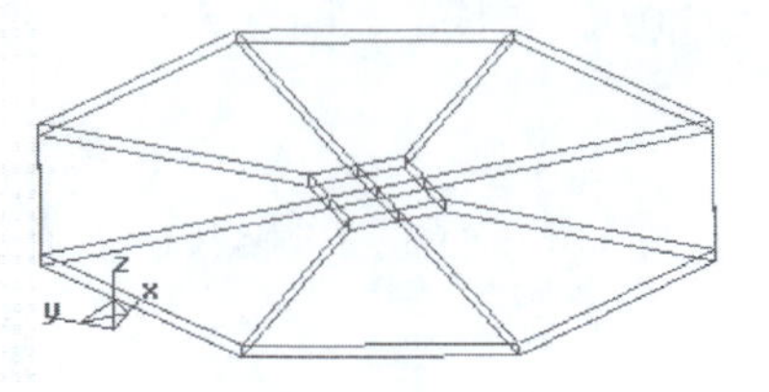

MESHING

Cant view is not very suitable for Meshing -> View -> Rotate -> Plane xy (original).

MAPPED MESH

Meshing -> Structured -> Volumes -> Select all volumes -> Then Cancel the further process.
Meshing -> Structured -> Lines -> Enter number to divide line -> Choose "1" for all lines (which makes the thickness line no division) -> Choose 2 for only the top and bottom lines.

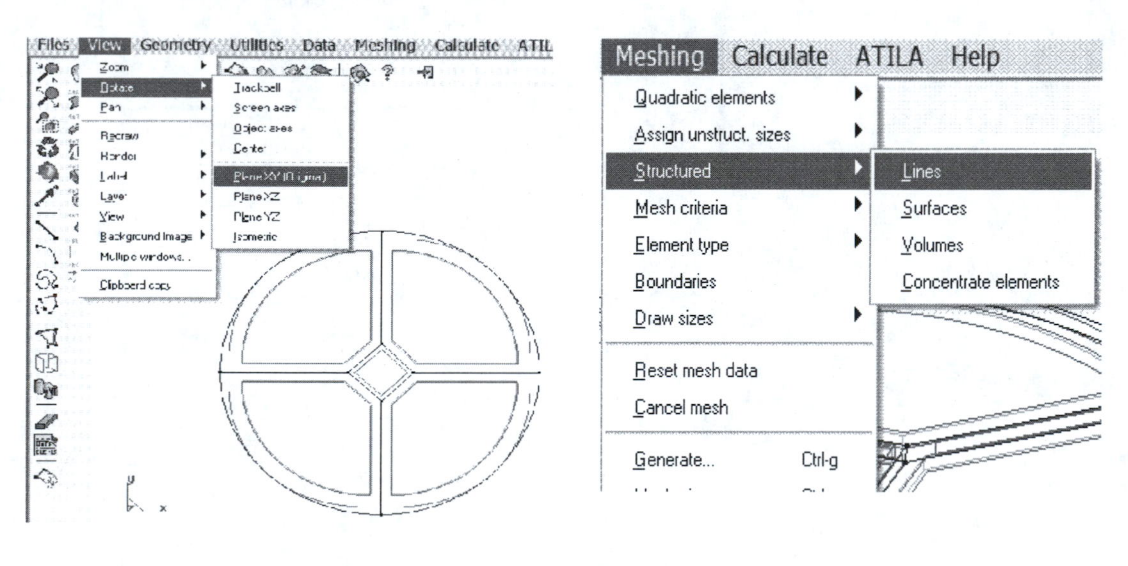

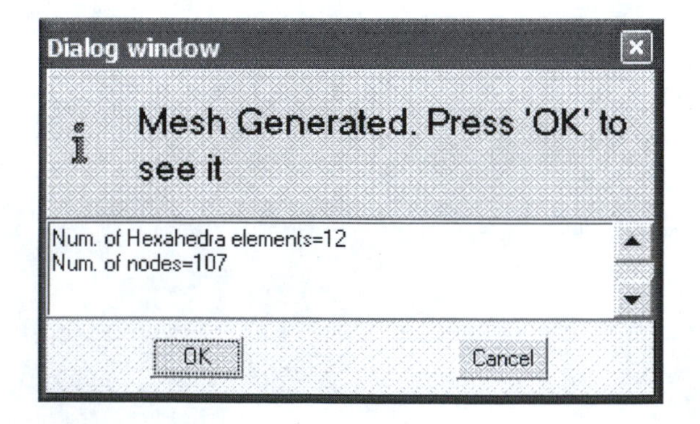

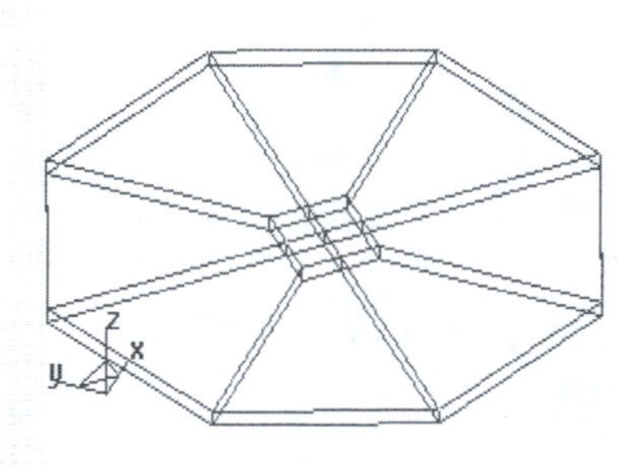

Mapped mesh (good)

Back to Top

Back to Home

1.2 CIRCULAR DISK (5)

PROBLEM TYPE PROB FREQ **VIEW RESULTS [9.83]**

PROBLEM—3D Circular Plate, kp Mode, PZT4 = 50 mm Φ × 1 mm t

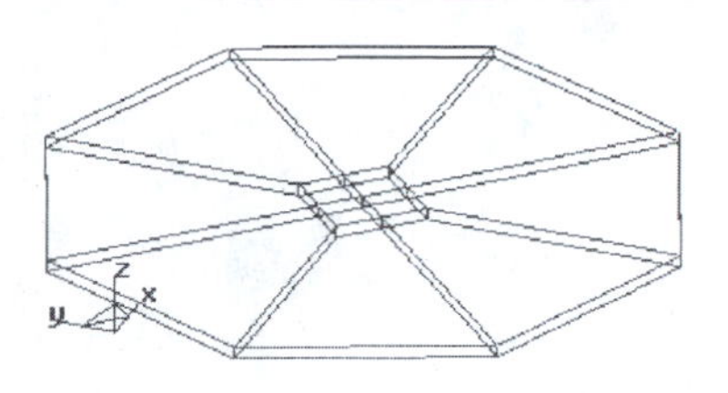

PROBLEM TYPE

Data -> Problem data -> 3D, Harmonic, Click both Compute Stress and Include Losses.

FREQUENCY

Data -> Interval data (40,000–55,000 Hz).

CALCULATION

Calculate -> Calculate window -> Start.

Problem Data
Problem Data | Mesh Units
PRINTING 0
GEOMETRY 3D
ANALYSIS HARMONIC
COMPUTE STRESS
INCLUDE LOSSES
Accept data | Close

Interval Data
1
TYPE LINEAR DISTRIBUTION
MIN FREQUENCY 40000.
MAX FREQUENCY 55000.
NUMBER OF FREQUENCIES 11
Accept data | Close

Process info
Process '3DDisk' started at Tue May 31 21:18:33 has finished.
OK | Postprocess

POST PROCESS/VIEW RESULTS

Click to Open the Post Calculation (Flavia) Domain.

Note the new Toolbar appearance.

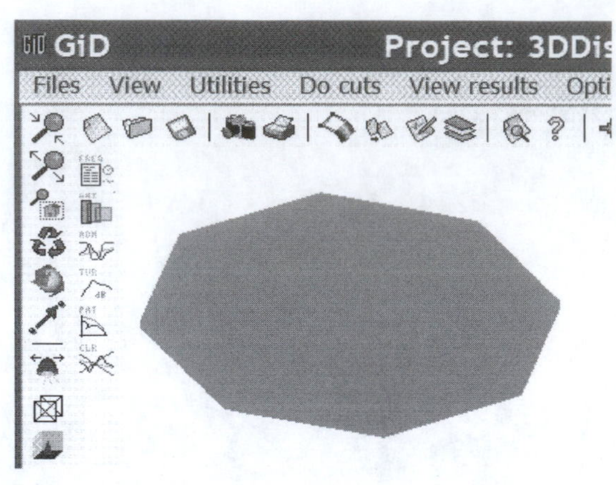

CLICK ADN **ADMITTANCE/IMPEDANCE RESULT**

Admittance Peak Point: 46,000 Hz -> This peak corresponds to the First Resonance.
Impedance Peak Point: 53,500 Hz -> This peak corresponds to the First Anti-Resonance.

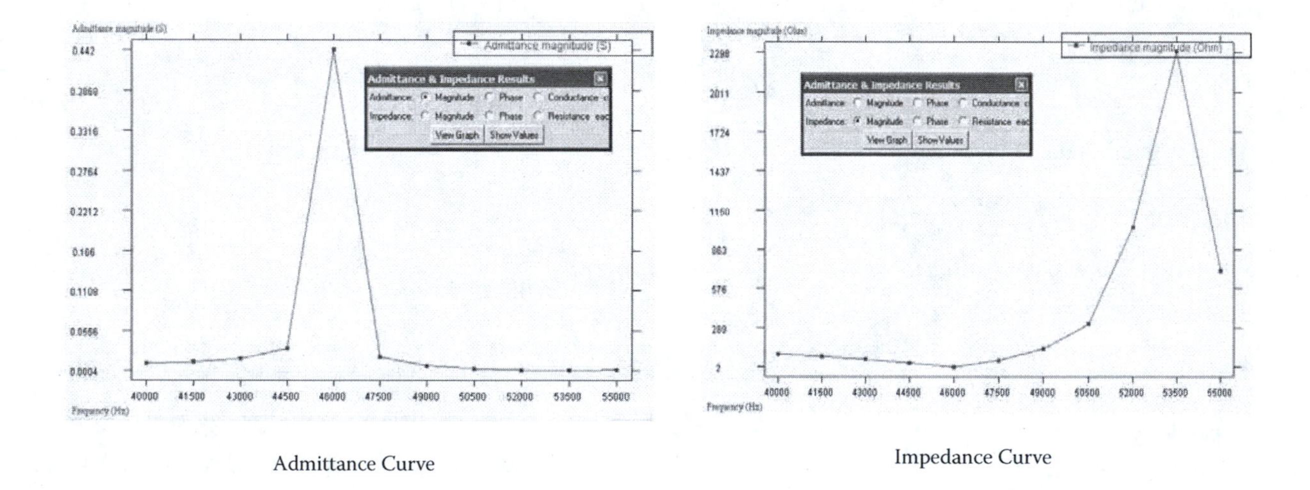

Admittance Curve

Impedance Curve

To take this graph out -> <u>Go To</u>

TO OBTAIN THE MORE PRECISE FREQUENCY DEPENDENCE

Repeat FREQ Data -> Interval data -> Use New Interval -> Copy condition's entities from interval 1 -> Yes -> Create interval 2 -> 45,000–47,000 Hz.

Interval Data
1 New interval
TYPE LINEAR DISTRIBUTION
MIN FREQUENCY 40000.
MAX FREQUENCY 55000.
NUMBER OF FREQUENCIES 11
Accept data Close

Confirm
Copy conditions' entities from interval 1?
Yes No Cancel

Interval Data
2
TYPE LINEAR DISTRIBUTION
MIN FREQUENCY 45000.
MAX FREQUENCY 47000.
NUMBER OF FREQUENCIES 11
Accept data Close

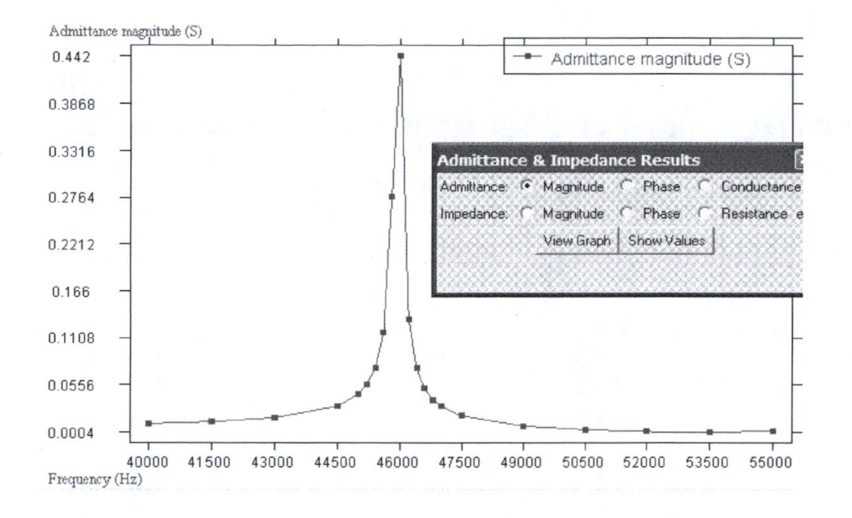

View Animation—Vibration Mode at a Particular Frequency

View results -> Default Analysis/Step -> Harmonic-Magnitude -> Set a Frequency (Peak Point: 46,000 Hz for Resonance Mode).

Double-Click Loading frequency: 46,000 Hz Harmonic Case -> Click OK (single-click and wait for a while).

Set Deformation -> Displacement

View results -> Deformation -> Displacement.

Deformation Scale Change

Click Windows -> Deform mesh -> Change the Factor in Mesh Deformation.

Parameter other than Displacement

Set Results View -> View results -> Contour Fill -> Von Mises Stress.

After setting both "Results view" and "Deformation" -> Click to see the Animation.

To save the GIF Animation Movie, use Save GIF, then Click after setting the file saving file.

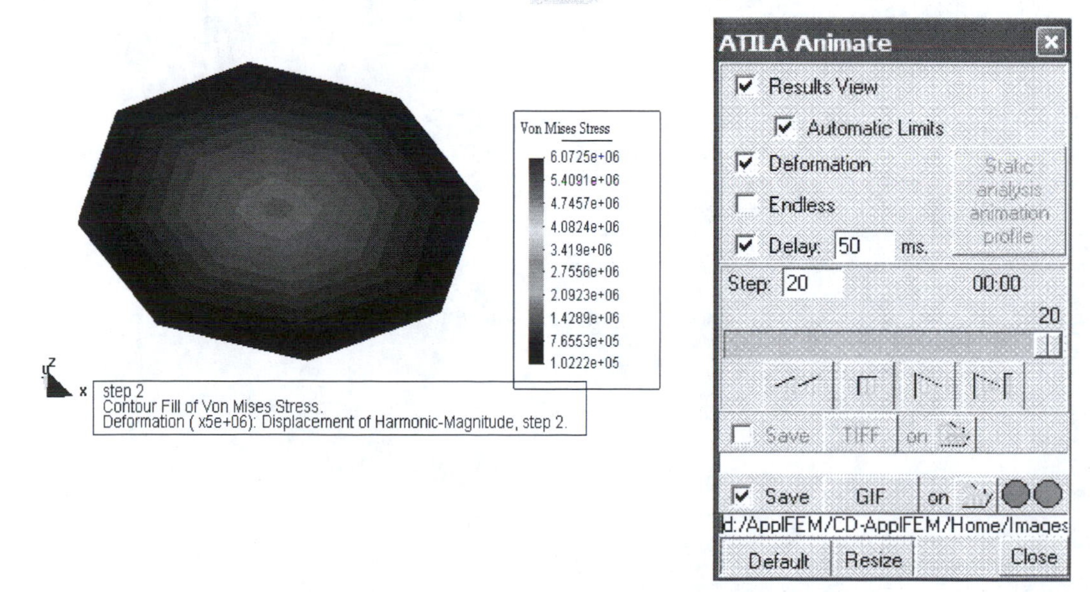

Back to Top

Back to Home

1.2 CIRCULAR DISK (6)

GEOMETRY MATERIAL ASSIGMENT BOUNDARY CONDITIONS

PROBLEM—2D CIRCULAR PLATE, KP MODE, PZT4 = 50 MM Φ × 1 MM T

PROBLEM -2D Circular Plate, kp Mode, PZT4 = 50mm Φ× 1mmt

- Axisymmetrical shape can be solved by using 2D calculation. The symmetrical axis should be X axis.

GEOMETRY/DRAWING

CREATE SURFACE SIMPLE STRUCTURE LIST

Geometry -> Create -> Object -> Rectangle -> Enter first corner point-> 0,0 -> Enter second corner point-> 1,25 -> Enter.

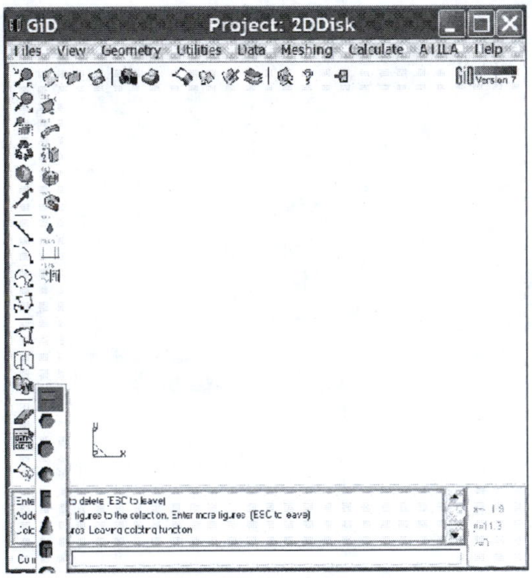

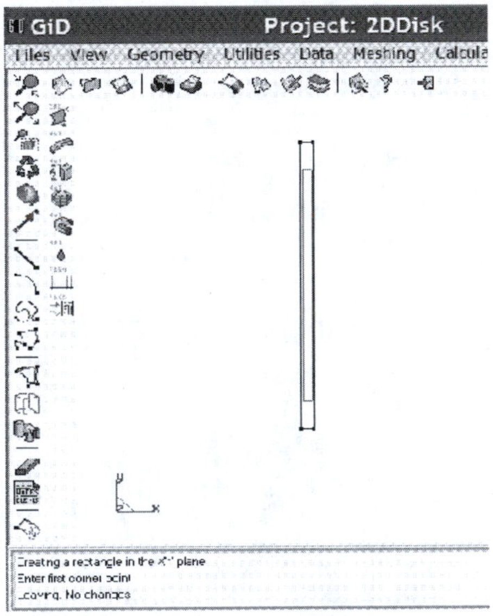

Remember that Point = Black color, Line = Blue color, Surface = Magenta, Volume = Cyan.

MATERIAL ASSIGMENT

Piezoelectric Materials

Data -> Materials -> Piezoelectric -> PZT4 -> Assign -> Surfaces -> Pick a Surface Click Draw -> All materials -> Check the material is assigned as PZT4.

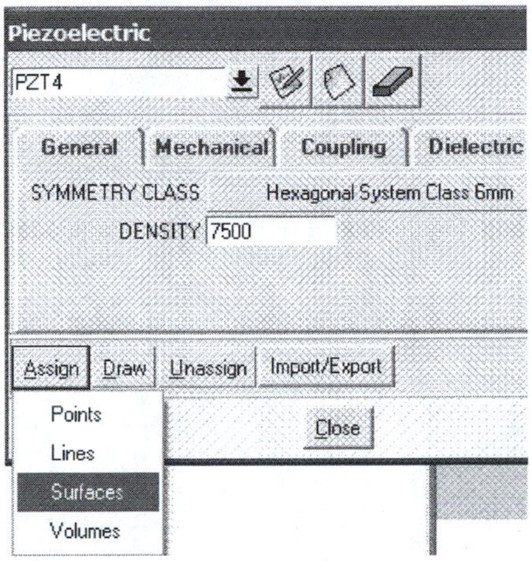

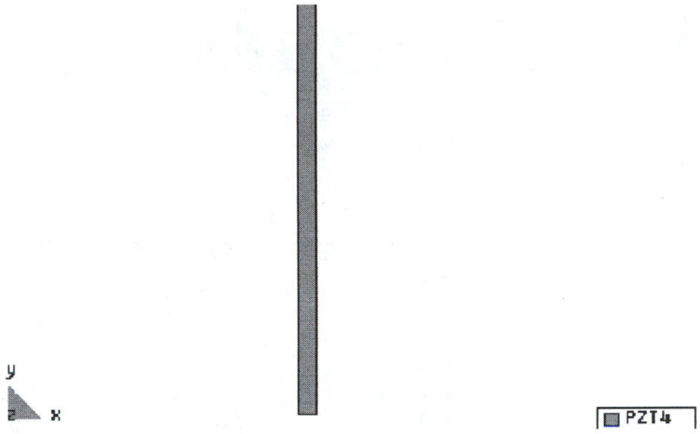

BOUNDARY CONDITIONS

Potential on the Line

Data -> Conditions -> Lines -> Potential 0.0 (or Ground in ATILA 5.2.4 or higher version) -> Assign the right line-> Potential 1.0 -> Assign the left line.

Polarization on the Surface

Data -> Conditions -> Surfaces -> Geometry -> Polarization (Cartesian) -> Define Polarization (P1)

We take the P1 rightward along X [Local axis (1,0,0)].

Choose 3PointsXZ -> Define X axis by putting (0,1,0) -> Then define Z axis by putting (1,0,0) -> Escape

Assign Polarization (Local-Axes P1) -> Choose the surface -> Finish.

To check the assignment -> Draw -> All the conditions -> Include Local axes.

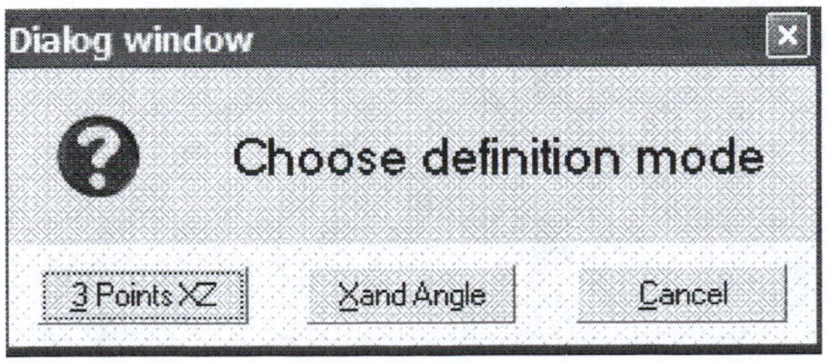

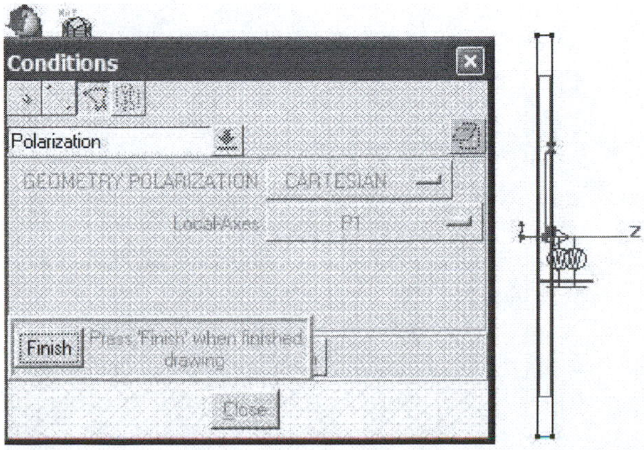

Mechanical Constraints on the Symmetrical Axis

This is the important technique for 2D modeling.

Data -> Conditions -> Lines -> Line-Constraints -> Choose X-component "None," Y and Z components "Clamped" -> Assign the bottom line (the symmetrical line should be taken on the X axis).

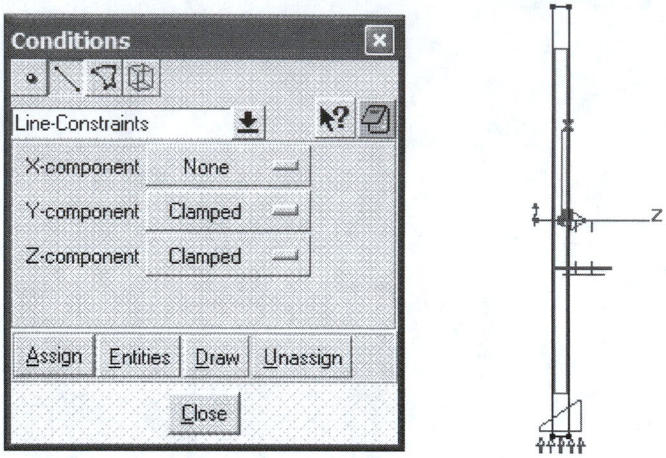

NB: When you choose the Axisymmetric 2D simulation, the X-axis should be the symmetric axis. Also, do not forget the line-constraint (Y and Z components = clamped).

Back to Top

Back to Home

1.2 CIRCULAR DISK (7)

MESHING VIEW RESULTS [9.114]

PROBLEM—2D CIRCULAR PLATE, KP MODE, PZT4 = 50 MM Φ × 1 MM T

PROBLEM - 2D Circular Plate, kp Mode, PZT4 = 50mm Φ×1mm t

MESHING

MAPPED MESH

Meshing -> Structured -> Surfaces -> Select the surface-> Then Cancel the further process.
Meshing -> Structured -> Lines -> Enter number to divide line -> Choose "1" for all lines (which makes the thickness line no division) -> Choose 8 for only the right and left lines.

Dialog window

Mesh Generated. Press 'OK' to see it

Num. of Quadrilateral elements=16
Num. of nodes=69

OK Cancel

Mapped mesh (good)

PROBLEM TYPE

Data-> Problem data -> 2D, Axisymmetric, Harmonic, Click both Compute Stress and Include Losses.

FREQUENCY

FREQ Data -> Interval data (40,000–55,000 Hz).

CALCULATION

Calculate -> Calculate window -> Start.

Problem Data
Problem Data | Mesh Units
PRINTING 0
GEOMETRY 2D
CLASS AXISYMMETRIC
ANALYSIS HARMONIC
COMPUTE STRESS
INCLUDE LOSSES
Accept data | Close

Interval Data
1
TYPE LINEAR DISTRIBUTION
MIN FREQUENCY 45000.
MAX FREQUENCY 47000.
NUMBER OF FREQUENCIES 11
Accept data | Close

POST PROCESS/VIEW RESULTS

Click to Open the Post Calculation (Flavia) Domain.

Click ADM Admittance/Impedance Result

Admittance Peak Point: 46,000 Hz -> This peak corresponds to the First Resonance.
Impedance Peak Point: 53,000 Hz -> This peak corresponds to the First Anti-Resonance.

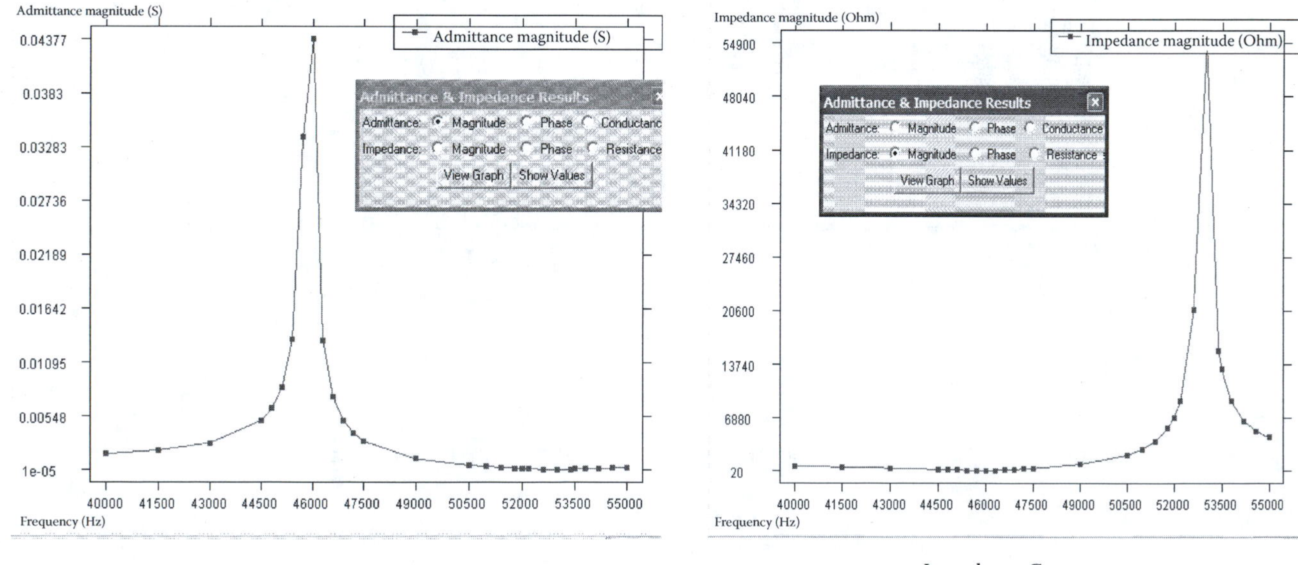

Admittance Curve

Impedance Curve

To find the exact frequency for the maximum Impedance, use "Show Values."

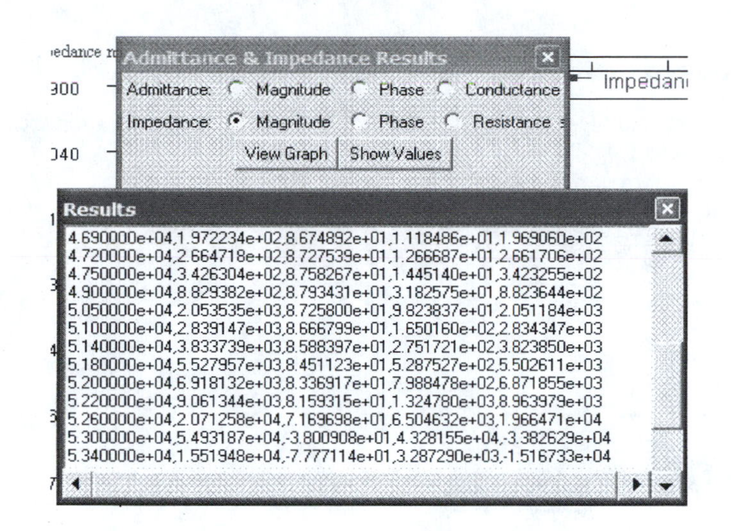

NB: An alternate path to obtain the peak frequency: Right click on the Impedance/Admittance curve. Choose "Label," "Select," then "Res. •" successively. Then choose the peak "dot," left click and "escape." You will find the numerical data (frequency and impedance).

View Animation—Vibration Mode at a Particular Frequency

View results -> Default Analysis/Step -> Harmonic-Magnitude -> Set a Frequency (Peak Point: 46,000 Hz for Resonance Mode).

Double-Click Loading frequency: 46,000 Hz Harmonic Case -> Click OK (single-click and wait for a while).

Set Deformation -> Displacement

View results -> Deformation -> Displacement.

Deformation Scale Change

Click Windows -> Deform mesh -> Change the Factor in Mesh Deformation.

Parameter other than Displacement

Set Results View -> View results -> Contour Fill -> Stress -> YY stress.

After setting both "Results view" and "Deformation" -> Click [icon] to see the Animation.

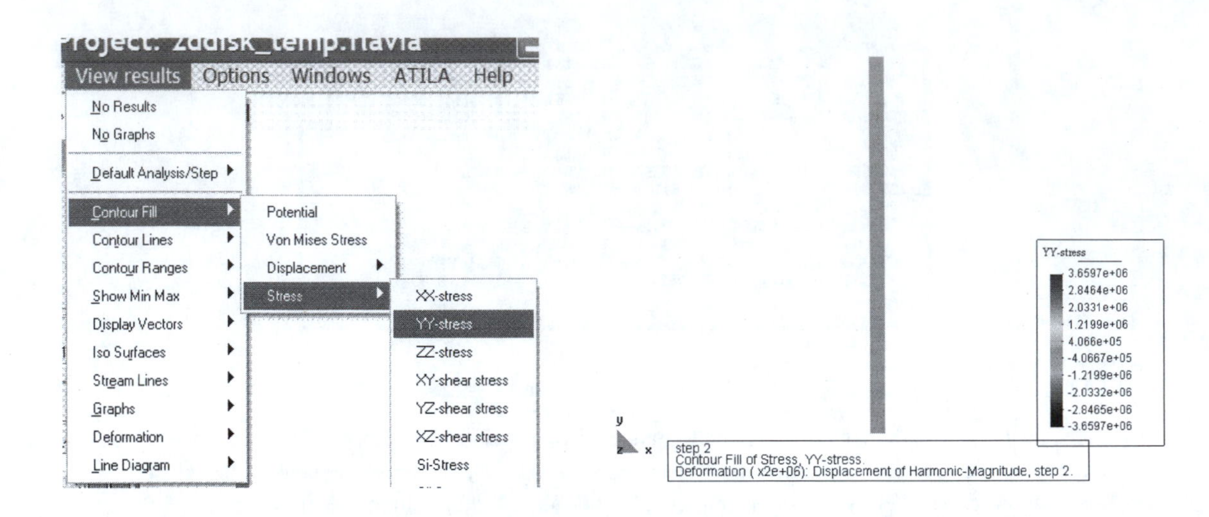

Note that the resonance frequency, resonating deformation, and the stresses are the same as the ones obtained in the 3D calculation within the calculation resolution.

Back to Top

Back to Home

2 Magnetostrictive Rod

- How to assign the "Polarization" and "Magnetic coil/field."

TEXT CONTENT

Rectangular Plate Mesh

Back to Top

	Magnetostrictive Rod
How To Use this GiD file	**Example GiD File** JMRod.gid

NB: The demo version (ATILA Light) has limitations in magneto-strictive simulation. Only the registered version can do this calculation.

2.1 MAGNETOSTRICTIVE ROD (1)

GEOMETRY MATERIAL ASSIGNMENT

PROBLEM—2D CYLINDER, JM = 50 MM L × 6.5 MM DIA

This example problem illustrates a coil energized with a current source inducing movement in a rod of magnetostrictive material. The magnetostrictive rod has a plane of symmetry normal to its axis. The rod can also be represented using an axisymmetric model. Therefore, only a quarter of the cross-sectional area of the rod needs to be used.

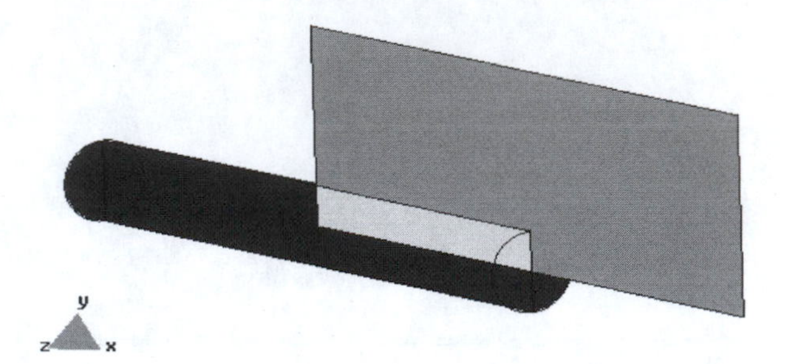

GEOMETRY/DRAWING

CREATE SURFACE SIMPLE STRUCTURE LIST

Geometry -> Create -> Object -> Rectangle -> Enter first corner point-> 0,0 -> Enter second corner point-> 50,16.25-> Enter.

Geometry -> Edit -> Divide -> Surfaces -> Near point -> Select the surface -> Choose NURBS sense -> V-sense -> 25,0 -> Select now the two surfaces -> Choose NURBS sense -> U-sense -> 0,3.25.

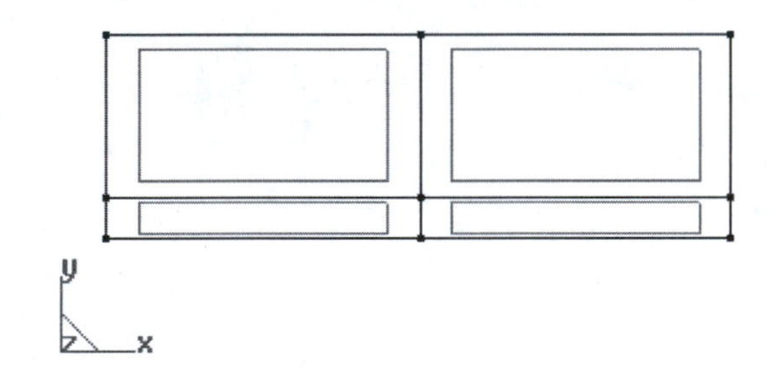

MATERIAL ASSIGNMENT

MAGNETOSTRICTIVE MATERIALS

Data -> Materials -> Magnetostrictive -> JM -> Assign -> Surfaces -> Pick a surface.

MAGNETIC MATERIALS

Data -> Materials -> Magnetic -> Air Mag -> Assign -> Surfaces -> Pick three outer regions.
Click Draw -> All materials -> Check that the material is assigned as JM and Air Mag.

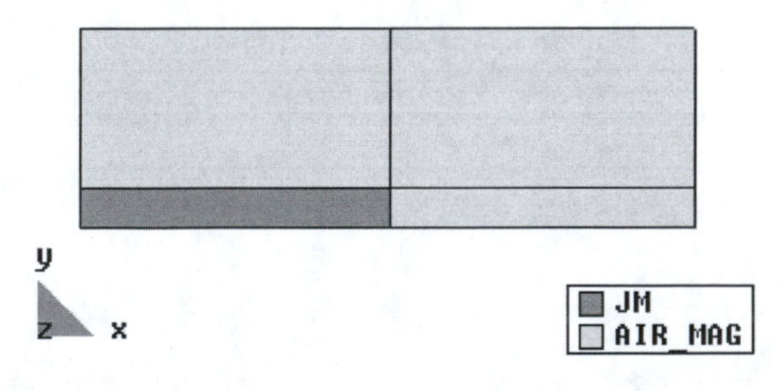

Back to Top

Back to Home

2.1 MAGNETOSTRICTIVE ROD (2)

BOUNDARY CONDITIONS

PROBLEM—2D CYLINDER, JM = 50 MM L × 6.5 MM DIA

This example problem illustrates a coil energized with a current source inducing movement in a rod of magnetostrictive material. The magnetostrictive rod has a plane of symmetry normal to its axis. The rod can also be represented using an axisymmetric model. Therefore, only a quarter of the cross-sectional area of the rod needs to be used.

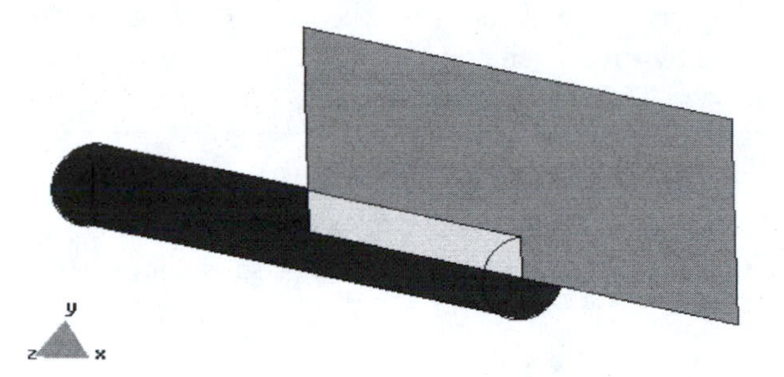

BOUNDARY CONDITIONS

INDUCTOR CONDITION

The ATILA–GiD interface uses a point boundary condition to define the inductor.

The inductor used is a cylindrical one with the following dimensions:

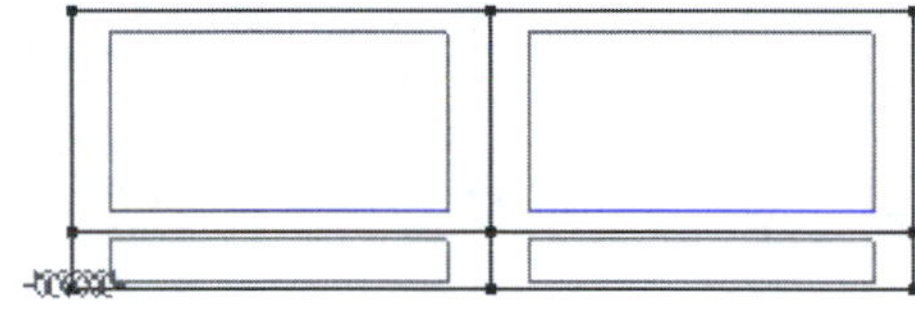

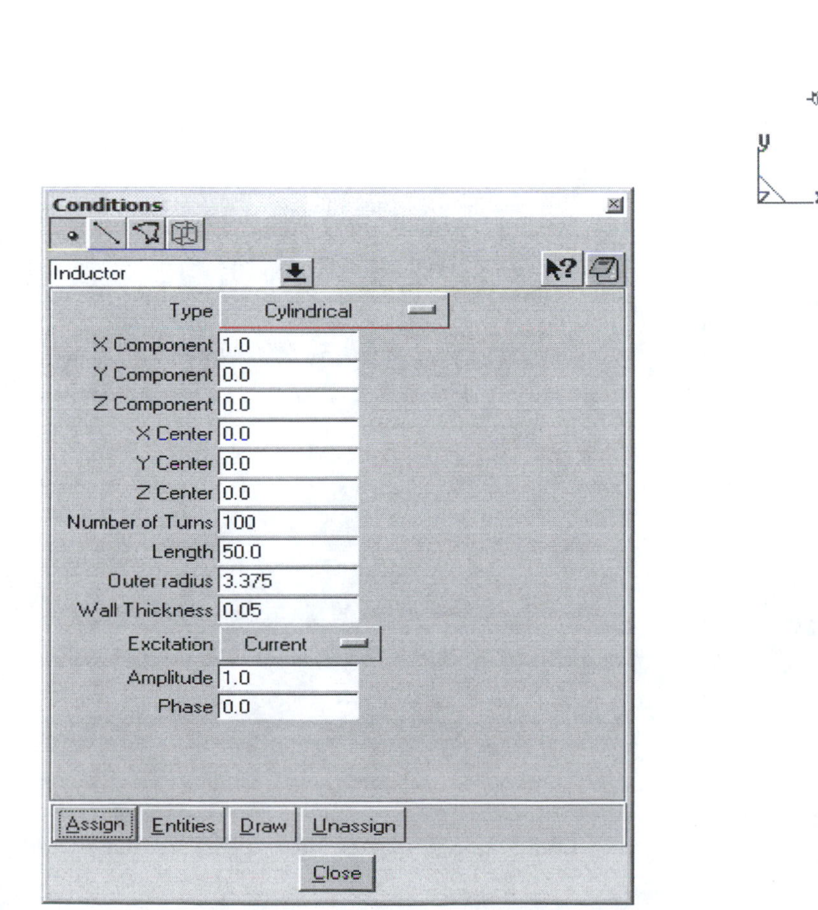

NB: The length of the inductor does not account for symmetry. You must enter the full length of the inductor and set the center point at the origin.

For clarity, apply the inductor condition to the point at the center of the magnetostrictive rod. ATILA will use the center point as defined in the dialog, not the point where the inductor is applied in GiD. Placing the inductor at the correct point in the geometry provides a visual consistency with the information in the dialog.

The external source energizing the coil is also defined in this dialog. The ATILA–GiD interface will supply the correct excitation based on the values entered in the dialog. This coil is energized with a per-unit current source.

Magnetic Potential

The entire magnetic field is assumed to be contained within the air region surrounding the rod. Force the magnetic potential to zero on the lines surrounding the problem.

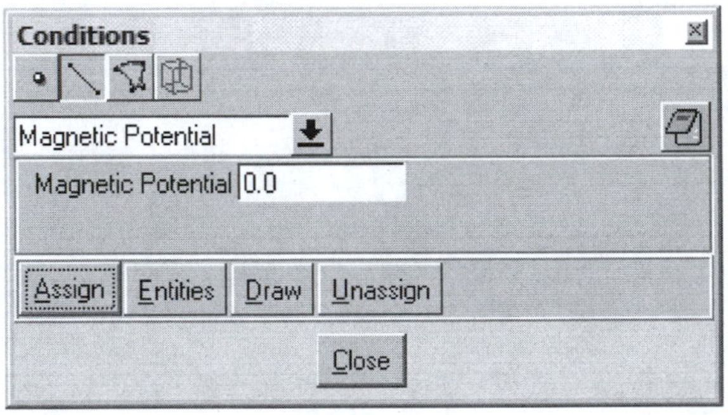

Conditions
Magnetic Potential
Magnetic Potential 0.0
Assign Entities Draw Unassign
Close

y
x

Symmetry Conditions

To enforce the axisymmetric condition, apply a line constraint clamping the *y* displacement along the centerline of the rod.

To enforce the symmetry condition, apply a line constraint clamping the *x* displacement along the middle of the rod.

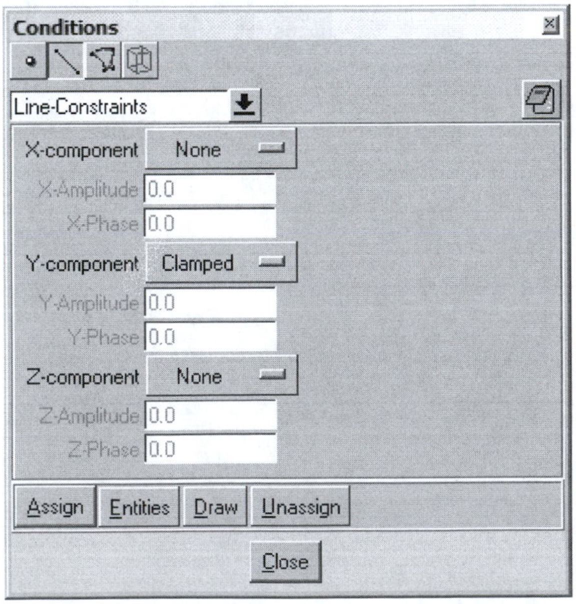

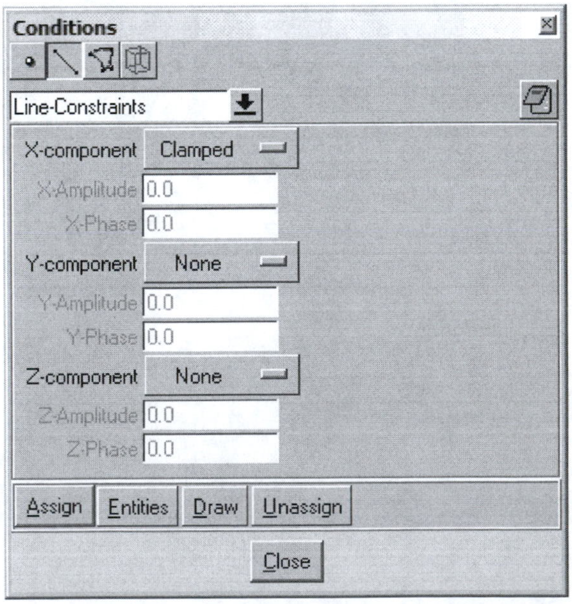

Conditions
Line-Constraints
X-component None
X-Amplitude 0.0
X-Phase 0.0
Y-component Clamped
Y-Amplitude 0.0
Y-Phase 0.0
Z-component None
Z-Amplitude 0.0
Z-Phase 0.0
Assign Entities Draw Unassign
Close

Conditions
Line-Constraints
X-component Clamped
X-Amplitude 0.0
X-Phase 0.0
Y-component None
Y-Amplitude 0.0
Y-Phase 0.0
Z-component None
Z-Amplitude 0.0
Z-Phase 0.0
Assign Entities Draw Unassign
Close

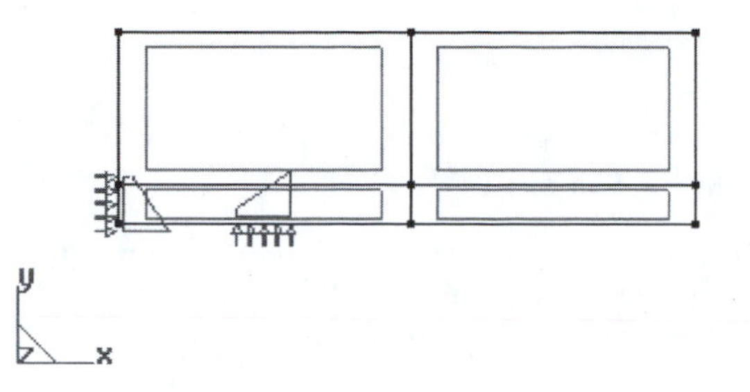

Polarization of Magnetostrictive Material

The polarization direction of the magnetostrictive material is set using a surface condition.

Define the local axis using the X and Angle option.

Pick the center point of the rod as the origin of the axis. Then, enter the point (0, 0, −1) on the command line to indicate the point in the positive x direction. Finally, enter −90° to place the Y axis. The local Z axis is then oriented along the global X axis, as shown.

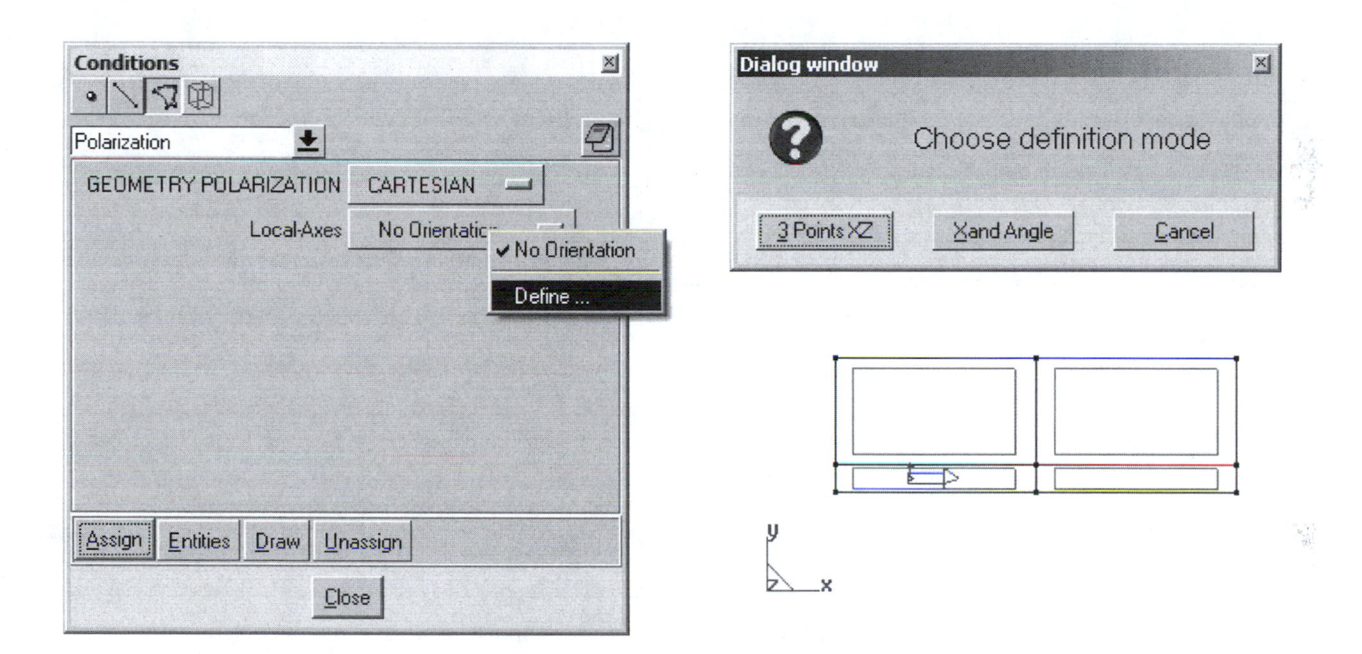

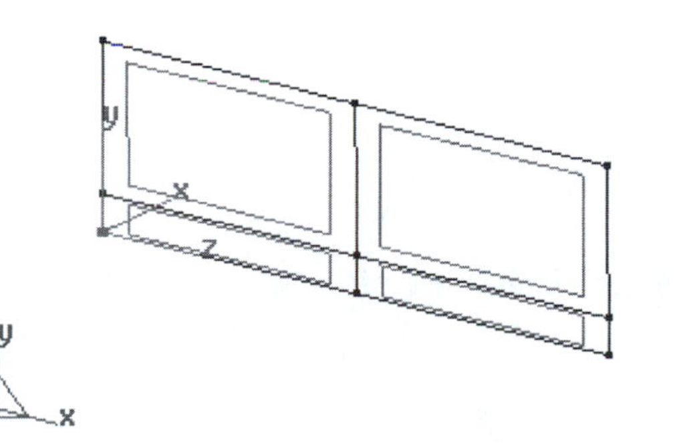

Back to Top

Back to Home

2.1 MAGNETOSTRICTIVE ROD (3)

PROBLEM TYPE 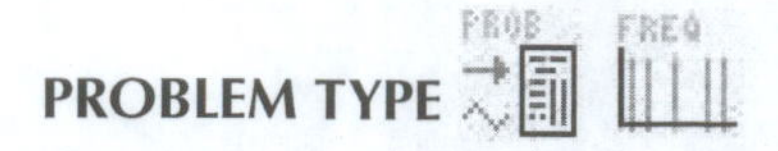

PROBLEM—2D CYLINDER, JM = 50 MM L × 6.5 MM DIA

This example problem illustrates a coil energized with a current source inducing movement in a rod of magnetostrictive material. The magnetostrictive rod has a plane of symmetry normal to its axis. The rod can also be represented using an axisymmetric model. Therefore, only a quarter of the cross-sectional area of the rod needs to be used.

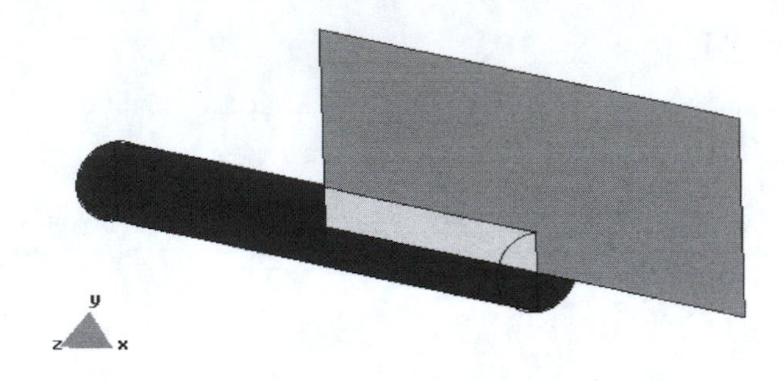

MESH GENERATION

Use a structured, quadrilateral mesh for the rod and the surrounding air.

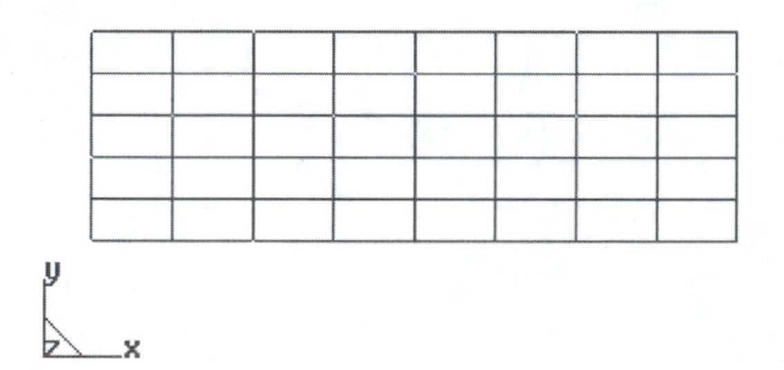

PROBLEM TYPE

Data -> Problem data -> 2D, Axisymmetric, Harmonic, Click both Compute Stress and Include Losses.

FREQUENCY

Data -> Interval data (10,000–30,000 Hz).

To obtain the more precise frequency dependence

Repeat Data -> Interval data -> Use New Interval -> Copy condition's entities from Interval 1 -> Yes -> Create Interval 2 -> 20,000–24,000 Hz -> Create Interval 3 -> 15,000–17,000 Hz.

CALCULATION

Calculate -> Calculate window -> Start.

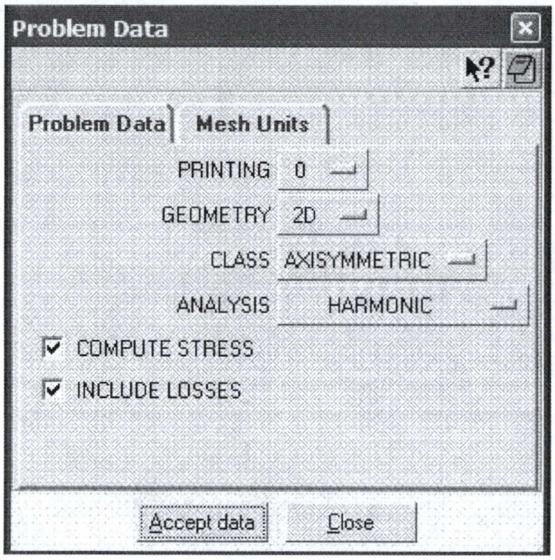

Problem Data

Problem Data | Mesh Units

PRINTING 0

GEOMETRY 2D

CLASS AXISYMMETRIC

ANALYSIS HARMONIC

COMPUTE STRESS

INCLUDE LOSSES

Accept data | Close

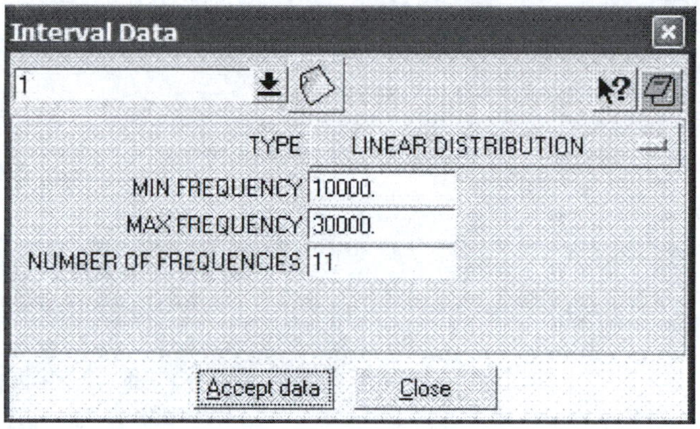

Interval Data

1

TYPE LINEAR DISTRIBUTION

MIN FREQUENCY 10000.

MAX FREQUENCY 30000.

NUMBER OF FREQUENCIES 11

Accept data | Close

Back to Top

Back to Home

2.1 MAGNETOSTRICTIVE ROD (4)

VIEW RESULTS

PROBLEM—2D CYLINDER, JM = 50 MM L × 6.5 MM DIA

This example problem illustrates a coil energized with a current source inducing movement in a rod of magnetostrictive material. The magnetostrictive rod has a plane of symmetry normal to its axis. The rod can also be represented using an axisymmetric model. Therefore, only a quarter of the cross-sectional area of the rod needs to be used.

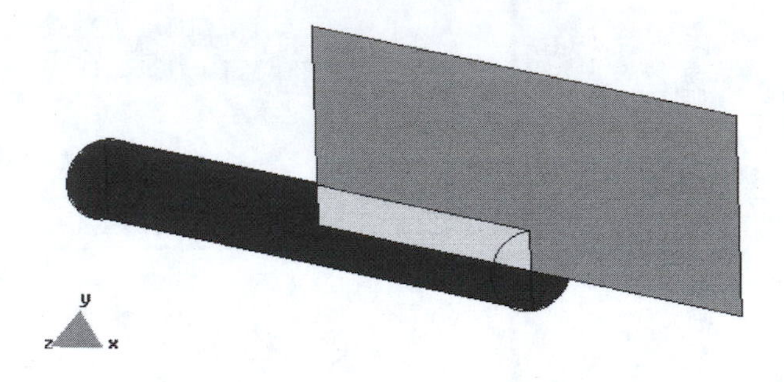

POST PROCESS/VIEW RESULTS

Click to Open the Post Calculation (Flavia) Domain.

Note the new Toolbar appearance.

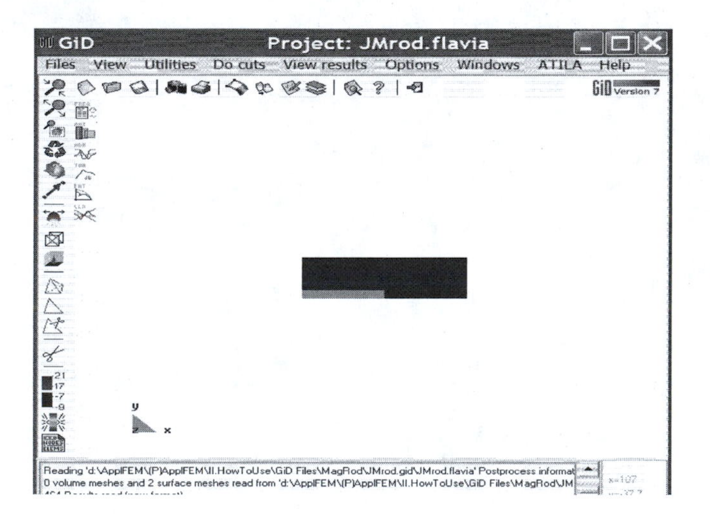

Click Admittance/Impedance Result

Impedance Peak Point: 17,000 Hz -> This peak corresponds to the first resonance.
Impedance Peak Point: 21,600 Hz -> This peak corresponds to the first antiresonance.

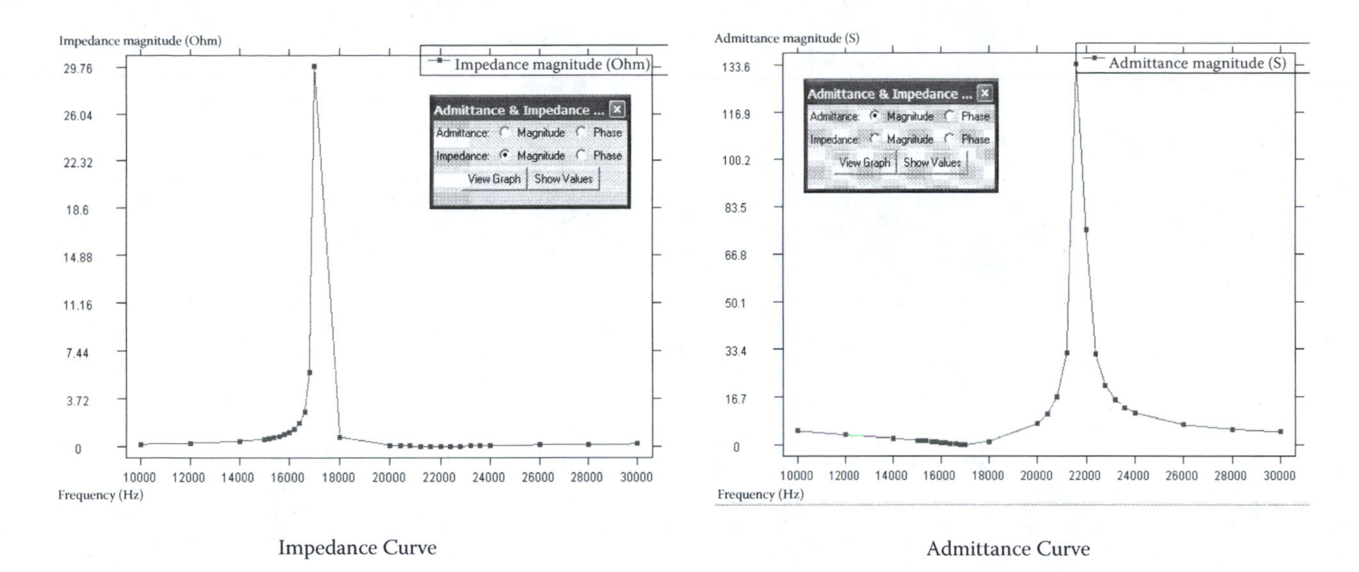

Impedance Curve Admittance Curve

NB: Note that the resonance in a piezoelectric shows the maximum admittance, which is reversed in a magnetostrictor.

To take this graph out -> Go To

View Animation—Vibration Mode at a Particular Frequency

View results -> Default Analysis/Step -> Harmonic-Magnitude -> Set a frequency (Peak Point: 17,000 Hz for Resonance Mode).

Double Click Loading frequency: 17,000 Hz Harmonic Case -> Click OK (single-click and wait for a while).

Set Deformation -> Displacement

View results -> Deformation -> Displacement.

Deformation Scale Change

Click Windows -> Deform mesh -> Change the factor in Mesh Deformation.

Parameter other than Displacement

Set Results View -> View results -> Contour Fill -> Von Mises Stress.

After setting both "Results view" and "Deformation" -> Click to see the Animation.

To save the GIF Animation Movie, use Save GIF, then Click] after setting the file-saving file.

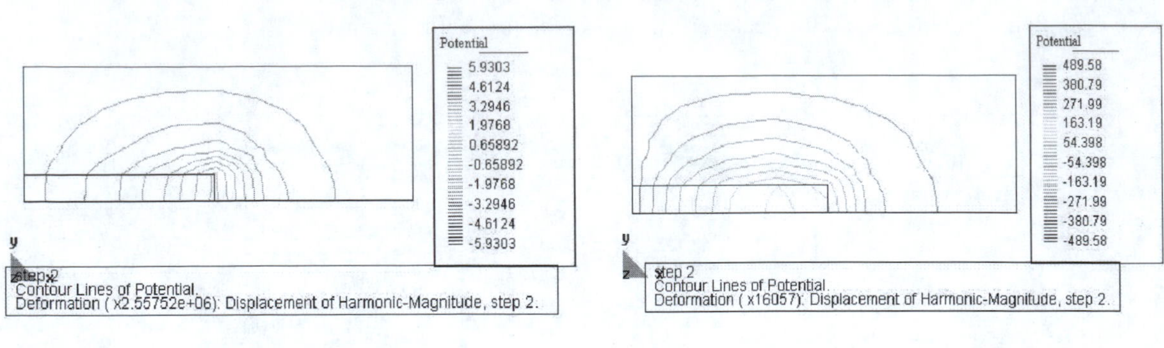

10,000 Hz (Off resonance)

17,000 Hz (Resonance) (Notice the potential max point shift from the off resonance)

Back to Top

Back to Home

3 Composite Structure

- How to couple two materials
- How to set the mechanical boundary condition
- Difference between the resonance and off-resonance vibration modes
- How to use "Layer"
- Stress concentration

TEXT CONTENT

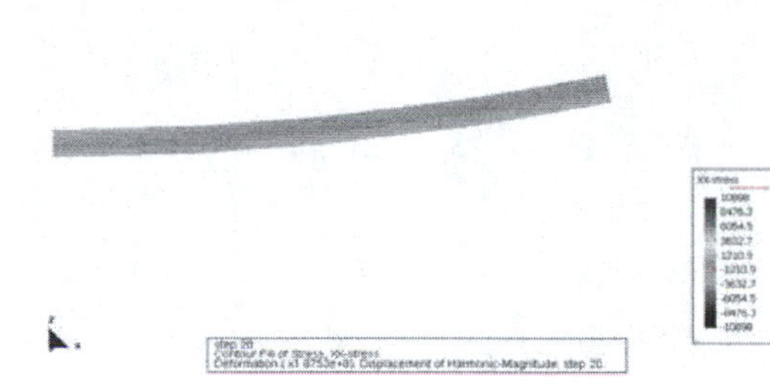

Bimorph Cantilever Motion

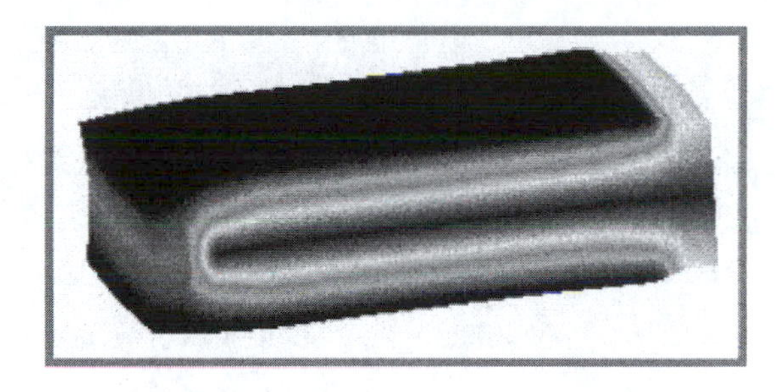

Potential Distribution of a ML Actuator

	3.1 Bimorph
	3.2 Unimorph
	3.3 Multilayer
	3.4 Cymbal
	Homework 3 Cantilever or both-clamped unimorph
	Homework 4 Metal/ceramic thickness optimization
	Homework 5 Stress concentration in 2D ML Actuator
How To Use this GiD file	**Example GiD File** Bimorph.gid, Unimorph.gid, MLAct.gid, Cymbal.gid

3.1 BIMORPH (1)

GEOMETRY/DRAWING MATERIAL ASSIGNMENT

PROBLEM—3D RECTANGULAR PLATE, PZT4 = 40 MM × 6 MM × 1 MM × 2 PIECES, CANTILEVER SUPPORT

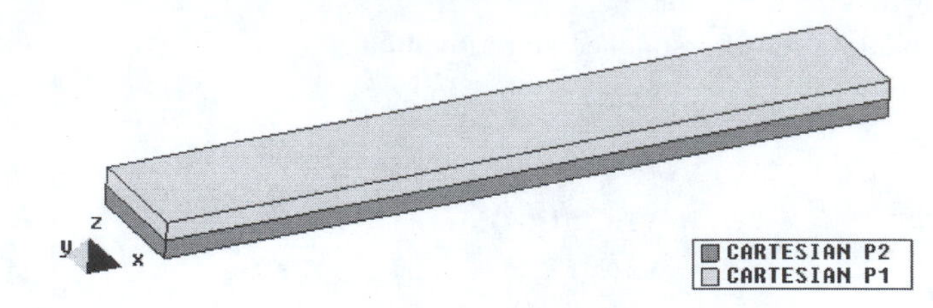

Composite of two different polarization layers, with cantilever (one-end mechanical clamp) configuration

GEOMETRY/DRAWING

CREATE SURFACE SIMPLE STRUCTURE LIST

Geometry -> Create -> Object -> Rectangle -> Enter first corner point -> 0, 0 -> Enter second corner point -> 40,6 -> Enter.

CREATE VOLUMES COPY MODE

Utilities -> Copy -> Surfaces -> Second point z=1.0 -> Do Extrude Surfaces -> Select -> Multiple copies -> 2. After selecting the rectangular (click on the surface line) -> Finish.

When you select the surface (click on the Magenta line), the surface color Magenta changes to Red.

Geometry -> Create -> Volume -> Automatic 6-sided volumes -> Two new volumes are created.

When you make the volume, you will find the Cyan color.

> NB: When you choose "Extrude Volume," you can create Volume directly.

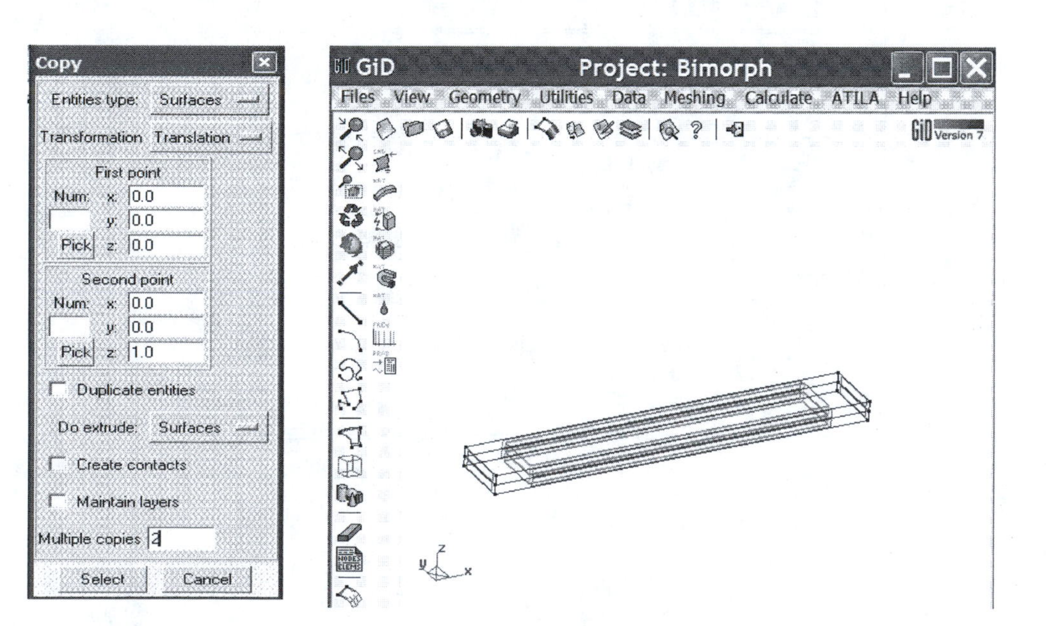

MATERIAL ASSIGNMENT

Piezoelectric Materials 

Data -> Materials -> Piezoelectric -> PZT4 -> Assign -> Pick Volumes
Click Draw -> All materials -> Check the material is assigned as PZT4.

Back to Top

Back to Home

3.1 BIMORPH (2)

BOUNDARY CONDITIONS

PROBLEM—3D RECTANGULAR PLATE, PZT4 = 40 MM × 6 MM × 1 MM × 2 PIECES, CANTILEVER SUPPORT

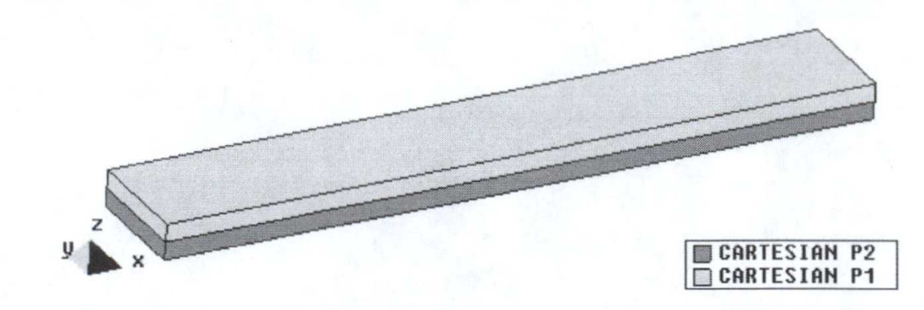

BOUNDARY CONDITIONS

POLARIZATION/LOCAL AXES

Data -> Conditions -> Volumes -> Geometry -> Polarization (Cartesian) -> Define Polarization (P1) (0,0,1) direction.

Choose X and Angle -> Define X axis -> Click two points to define X axis -> Choose the suitable Cartesian P1 direction (Z axis) -> Escape.

Or, choose 3PointsXZ -> Define X axis by putting (1, 0, 0) -> Then define Z axis by putting (0,0,1) -> Escape.

Data -> Conditions -> Volumes -> Geometry -> Polarization (Cartesian) -> Define Polarization (P2) (0,0,-1) direction.

Assign Polarization (Local-Axes P1 and P2) -> Choose the volume -> Finish.

Enter value window
Enter name of new local axes
P1
OK Cancel

Dialog window
Choose definition mode or delete
3 Points XZ Xand Angle Delete
Cancel

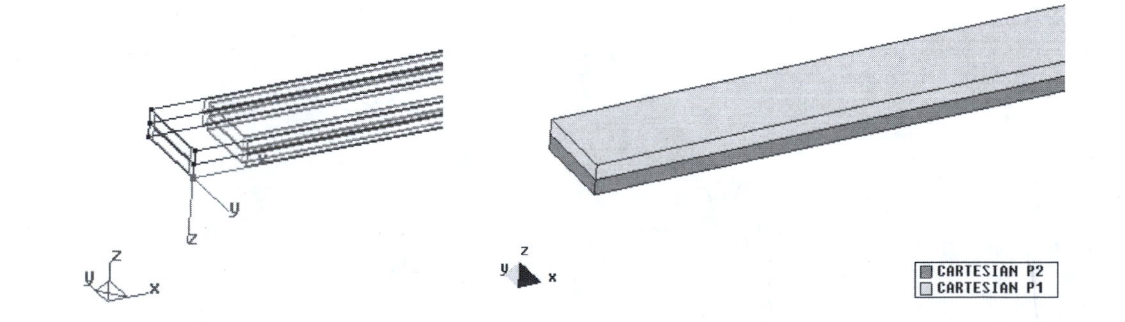

Potential

Data -> Conditions -> Surfaces -> Potential 0.0 (or Ground in ATILA 5.2.4 or higher version) -> Assign the bottom surface-> Forced Potential 1.0 -> Assign the top surface -> Floating -> Assign the center surface.

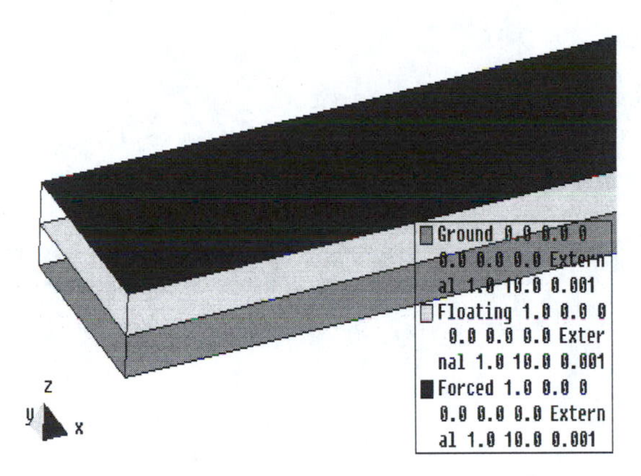

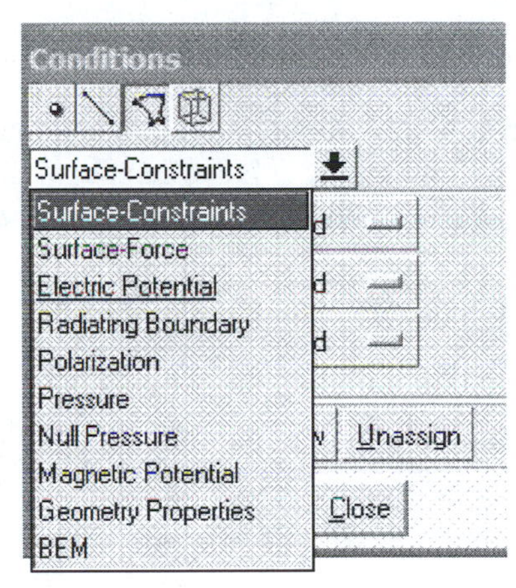

Surface Constraints (Cantilever: One-End Clamp)

Data -> Conditions -> Surfaces -> Surface Constraints -> X, Y, and Z components -> Clamped -> Assign left two surfaces.

To check the assignment -> Draw -> All conditions -> Include local axes.

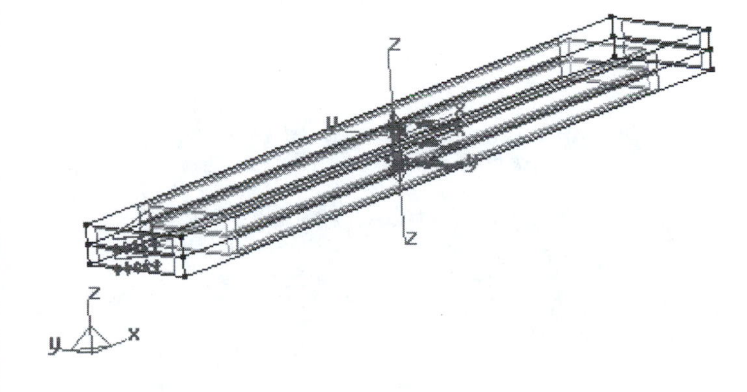

Back to Top

Back to Home

3.1 BIMORPH (3)

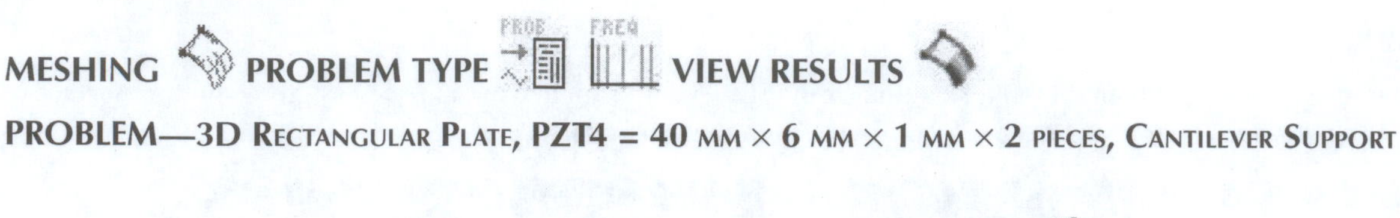

PROBLEM—3D Rectangular Plate, PZT4 = 40 mm × 6 mm × 1 mm × 2 pieces, Cantilever Support

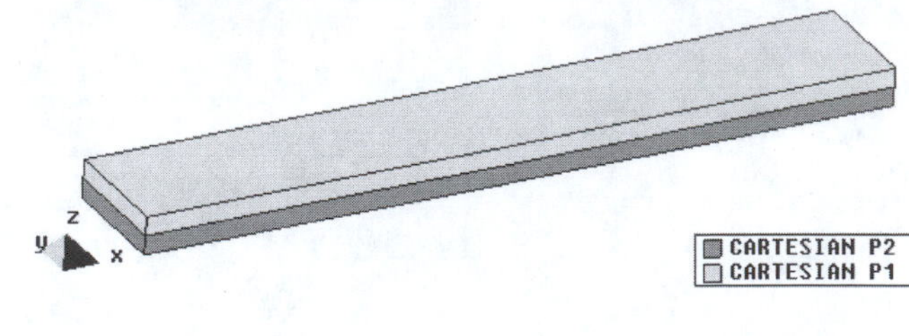

MESHING

Cant view is not very suitable for Meshing -> View -> Rotate -> Plane xy (original).

Mapped Mesh

Meshing -> Structured -> Volumes -> Select all volumes -> Then Cancel the further process.
Meshing -> Structured -> Lines -> Enter number to divide line -> Choose “1” for all lines (which makes the thickness line no division) -> Choose 2 for only the right and left lines -> Choose 4 for only the top and bottom lines -> Generate.

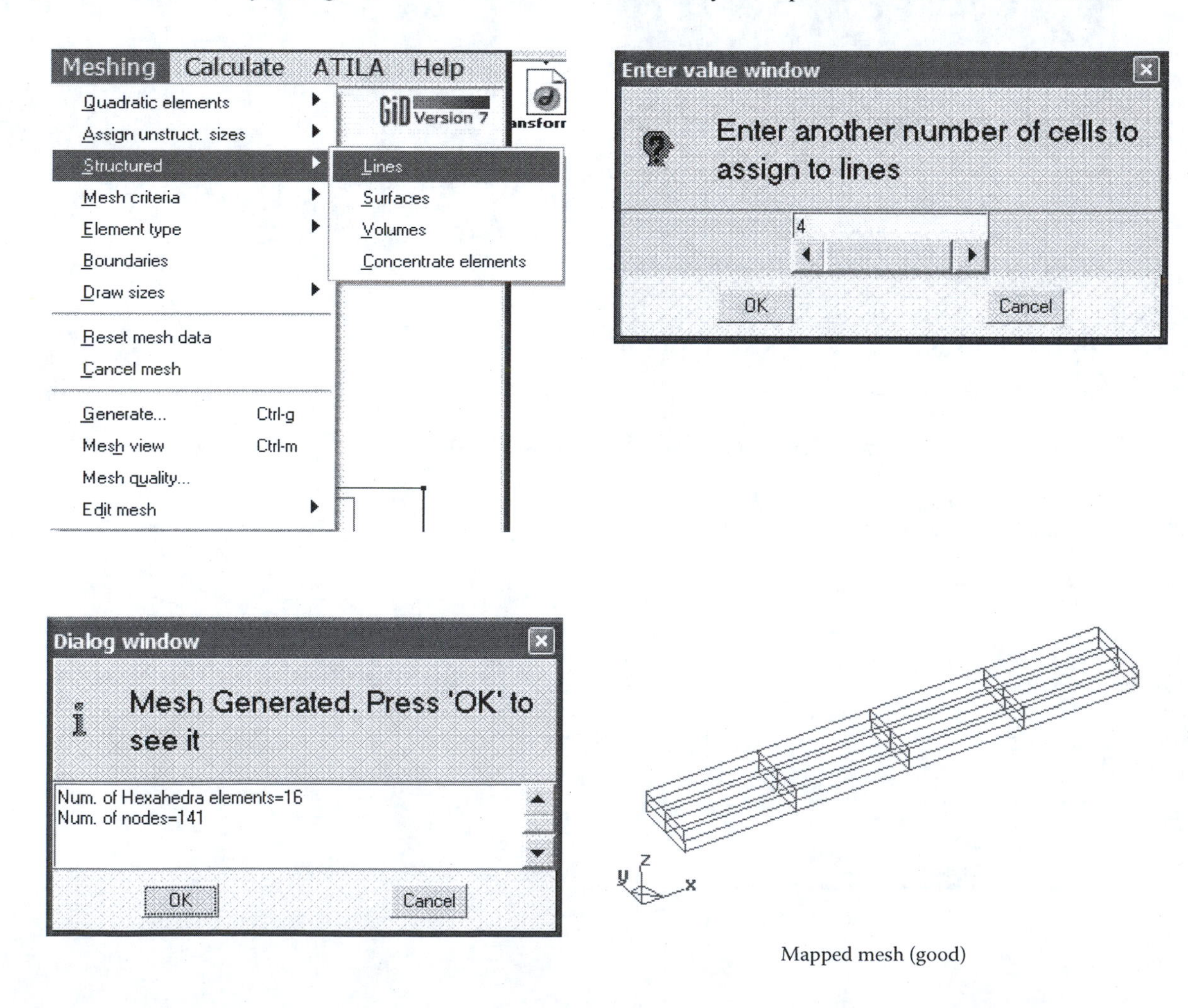

Mapped mesh (good)

PROBLEM TYPE

PROB Data -> Problem data -> 3D, Harmonic, Click both Compute Stress and Include Losses.

FREQUENCY

FREQ Data -> Interval data Interval 1 (100 Hz), Interval 2 (4,000–5,000 Hz)

CALCULATION

Calculate -> Calculate window -> Start.

Problem Data
Problem Data | Mesh Units
PRINTING 0
GEOMETRY 3D
ANALYSIS HARMONIC
COMPUTE STRESS
INCLUDE LOSSES
Accept data | Close

Interval Data
2
TYPE LINEAR DISTRIBUTION
MIN FREQUENCY 4000.
MAX FREQUENCY 5000.
NUMBER OF FREQUENCIES 11
Accept data | Close

POST PROCESS/VIEW RESULTS

Click to Open the Post Calculation (Flavia) Domain.
Note the new Toolbar appearance.

CLICK ADM ADMITTANCE/IMPEDANCE RESULT

Admittance Peak Point: 4,500 Hz -> This peak corresponds to the First Resonance.

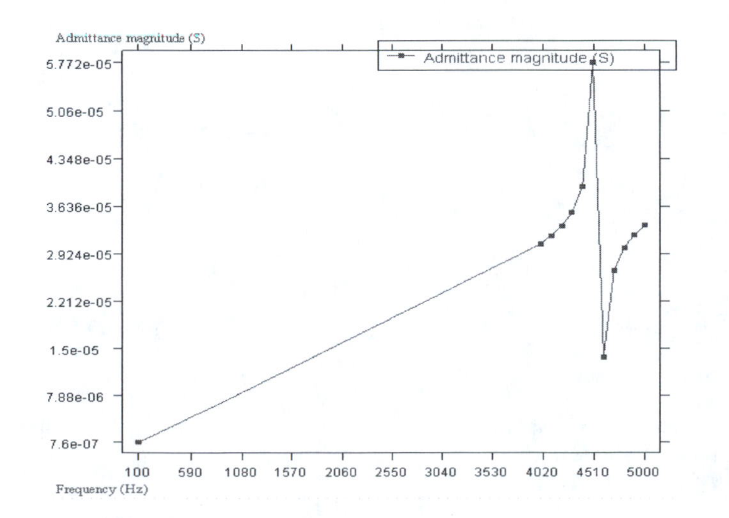

Admittance Curve

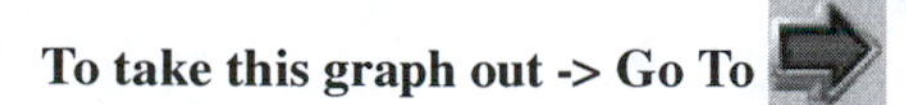

To take this graph out -> Go To

View Animation—Vibration Mode at a Particular Frequency

View results -> Default Analysis/Step -> Harmonic-Magnitude -> Set a Frequency
(100 Hz for Off-Resonance Mode; Peak Point: 4,500 Hz for Resonance Mode).

Double Click [ANI] Loading frequency: 4,500 Hz Harmonic Case -> Click OK (single-click, and wait for a while).

Set Deformation -> Displacement

View results -> Deformation -> Displacement.

Deformation Scale Change

Click Windows -> Deform mesh -> Change the Factor in Mesh Deformation.

Parameter other than Displacement

Set Results View -> View results -> Contour Fill -> Von Mises Stress.

After setting both "Results view" and "Deformation" -> Click [ ] to see the Animation.

To save the GIF Animation Movie, use Save GIF, then click [ ] after setting the file-saving file.

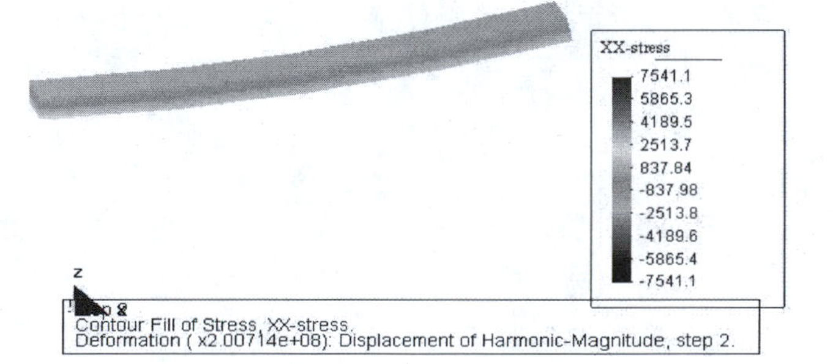

100 Hz - Off Resonance

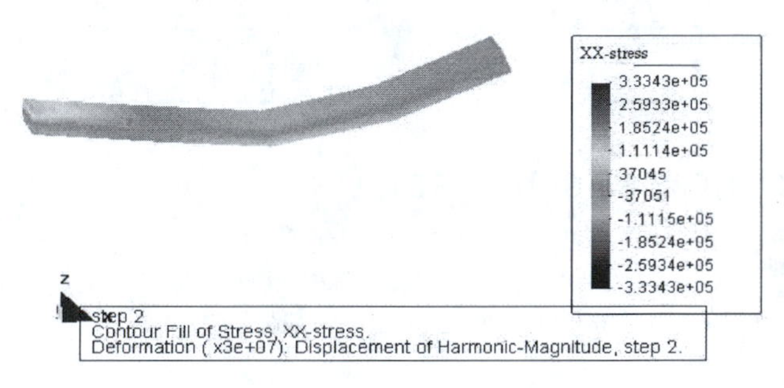

4,500 Hz - Resonance

Comparison of FEM Calculation with Beam Theory

Beam theory is applied for a very thin and narrow (width) beam.

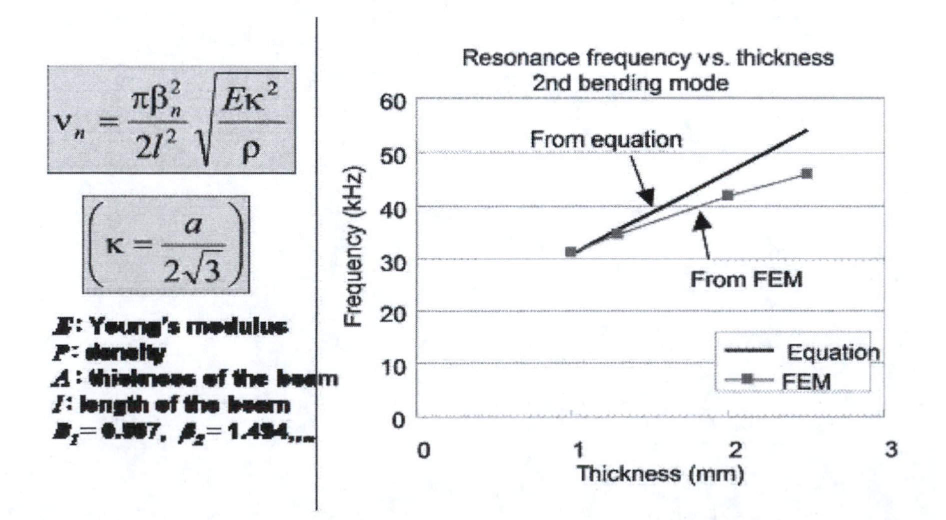

Back to Top

Back Home

3.1 BIMORPH (4)

VARIOUS PROBLEM TYPES VIEW RESULTS

PROBLEM—3D Rectangular Plate, PZT4 = 40 mm × 6 mm × 1 mm × 2 pieces, Cantilever Support

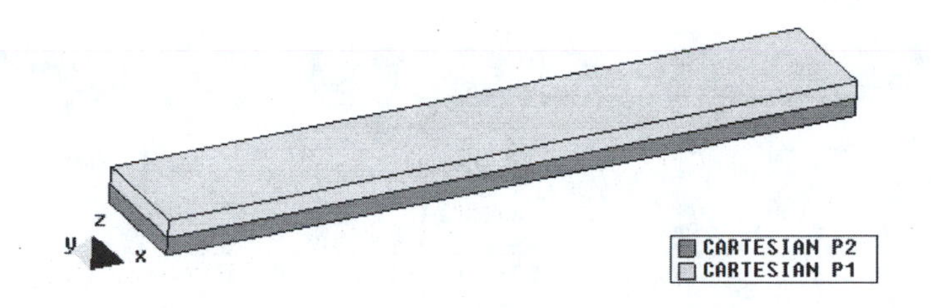

PROBLEM TYPE

We consider here various problem types:

Data -> Problem data -> 3D.

- **Static**—Click Compute Stress -> Pseudostatic mode is obtained.

Problem Data
Problem Data | Mesh Units
PRINTING 0
GEOMETRY 3D
ANALYSIS STATIC
☑ COMPUTE STRESS
Accept data | Close

- **Modal**—Put the number of modes (10) -> Only the resonance modes are obtained.

Problem Data
Problem Data | Mesh Units
PRINTING 0
GEOMETRY 3D
ANALYSIS MODAL
NUMBER OF MODES 10
Accept data | Close

Modal Analysis

Results
Title, Normal frequencies
Mode, Frequency (Hz)
1, 7.198307e+02
2, 2.096654e+03
3, 4.454962e+03
4, 7.346000e+03
5, 1.190077e+04
6, 1.228769e+04
7, 2.195769e+04
8, 2.218885e+04
9, 2.366692e+04
10, 2.959000e+04

Resonance Modes

- **Modal Resantires**—Put the number of modes (10)
    - -> Both the resonance and antiresonance modes are obtained.

NB: "Resantires" means "Resonance-Antiresonance."

Modal Resantires Analysis

Title, Resonance & antiresionance frequencies,,
Mode, Resonance Frequency (Hz), Antiresonance Frequency (Hz),Coupling coefficient (%)
1, 6.914116e+02, 7.083269e+02, 2.172352e+01
2, 1.994885e+03, 1.994885e+03, 0.000000e+00
3, 4.294269e+03, 4.324615e+03, 1.182579e+01
4, 7.340730e+03, 7.340730e+03, 0.000000e+00
5, 1.139385e+04, 1.139385e+04, 0.000000e+00
6, 1.189923e+04, 1.193192e+04, 7.397507e+00
7, 2.171192e+04, 2.171192e+04, 0.000000e+00
8, 2.217038e+04, 2.217038e+04, 0.000000e+00
9, 2.304654e+04, 2.307538e+04, 4.998734e+00
10, 2.849885e+04, 2.849885e+04, 0.000000e+00

Resonance and Antiresonance Modes with Coupling coefficient

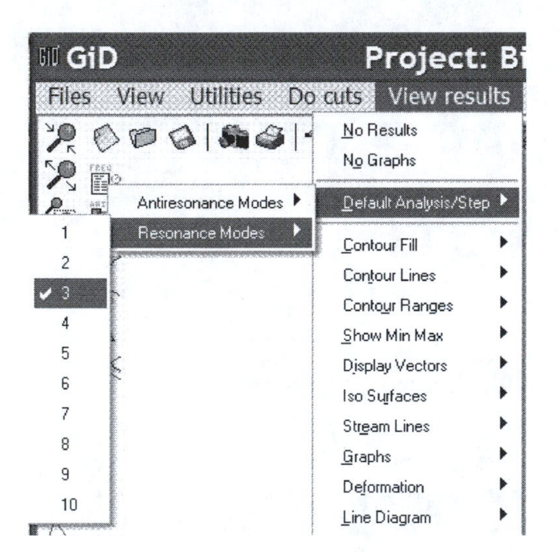

Assign the first resonance mode (Mode 3)

The animation for the second resonance mode (Mode 6)

- **Harmonic**—Click both Compute Stress and Include Losses -> Admittance curve can be obtained.
- **Transient**

In this analysis, the matrix equation becomes:

$$\begin{bmatrix} [\mathrm{M}] & [0] & [0] & [0] & [0] \\ [0] & [0] & [0] & [0] & [0] \\ [0] & [0] & [0] & [0] & [0] \\ [0] & [0] & [0] & [0] & [0] \\ [\mathrm{L}]^{\mathrm{T}} & [0] & [0] & [0] & \dfrac{[\mathrm{M}_1]}{p^2c^2} \end{bmatrix} \begin{bmatrix} \underline{\ddot{\mathbf{U}}} \\ \underline{\ddot{\mathbf{\Phi}}} \\ \underline{\ddot{\mathbf{\phi}}} \\ \underline{\ddot{\mathbf{I}}} \\ \underline{\ddot{\mathbf{P}}} \end{bmatrix} + \frac{1}{\omega_0} \begin{bmatrix} [\mathrm{K}''_{uu}] & [\mathrm{K}''_{u\Phi}] & [\mathrm{K}''_{u\phi}] & [\mathrm{K}''_{u\mathrm{I}}] & [0] \\ [\mathrm{K}''_{u\Phi}]^T & [\mathrm{K}''_{\Phi\Phi}] & [0] & [0] & [0] \\ [\mathrm{K}''_{u\phi}]^T & [0] & [\mathrm{K}''_{\phi\phi}] & [\mathrm{K}''_{\phi\mathrm{I}}] & [0] \\ [\mathrm{K}''_{u\mathrm{I}}]^T & [0] & [\mathrm{K}''_{\phi\mathrm{I}}]^T & [\mathrm{K}''_{\mathrm{II}}] & [0] \\ [0] & [0] & [0] & [0] & \left[\dfrac{\mathrm{H}}{p^2c^2}\right]'' \end{bmatrix} \begin{bmatrix} \underline{\dot{\mathbf{U}}} \\ \underline{\dot{\mathbf{\Phi}}} \\ \underline{\dot{\mathbf{\phi}}} \\ \underline{\dot{\mathbf{I}}} \\ \underline{\dot{\mathbf{P}}} \end{bmatrix}$$

$$+ \begin{bmatrix} [\mathrm{K}'_{uu}] & [\mathrm{K}'_{u\Phi}] & [\mathrm{K}'_{u\phi}] & [\mathrm{K}'_{u\mathrm{I}}] & -[\mathrm{L}] \\ [\mathrm{K}'_{u\Phi}]^{\mathrm{T}} & [\mathrm{K}'_{\Phi\Phi}] & [0] & [0] & [0] \\ [\mathrm{K}'_{u\phi}]^{\mathrm{T}} & [0] & [\mathrm{K}'_{\phi\phi}] & [\mathrm{K}'_{\phi\mathrm{I}}] & [0] \\ [\mathrm{K}'_{u\mathrm{I}}]^{\mathrm{T}} & [0] & [\mathrm{K}'_{\phi\mathrm{I}}]^{\mathrm{T}} & [\mathrm{K}'_{\mathrm{II}}] & [0] \\ [0] & [0] & [0] & [0] & \left[\dfrac{\mathrm{H}}{\rho^2c^2}\right]' \end{bmatrix} \begin{bmatrix} \underline{\mathbf{U}} \\ \underline{\mathbf{\Phi}} \\ \underline{\mathbf{\phi}} \\ \underline{\mathbf{I}} \\ \underline{\mathbf{P}} \end{bmatrix} = \begin{bmatrix} \underline{\mathbf{F}} \\ -\underline{\mathbf{q}} \\ -\underline{\mathbf{f}} \\ -\mathbf{f}_{\underline{\mathbf{b}}} \\ \dfrac{1}{P}\underline{\psi} \end{bmatrix} \tag{1}$$

where cap . and .. denote the first and second time derivative, respectively, K' and K'' denote the real and imaginary parts of K respectively and ω_0 is the pulsation at which the material's losses are provided. The preceding system of equations may be rewritten as follows:

$$[A]\ddot{\underline{X}} + [B]\dot{\underline{X}} + [c]\underline{X} = \underline{Y} \tag{2}$$

This differential equation is solved by an iterative method, taking a constant time step At; the successive values of $\underline{\mathrm{X}}$ and $\underline{\mathrm{Y}}$ are noted as $\underline{\mathrm{X}}_n$ and $\underline{\mathrm{Y}}_n$ and correspond to values calculated at the time $n\Delta t$. Three methods are implemented in **ATILA**: the **Central Difference Method,** the **Newmark Method,** and the **Wilson-θ Method.**

All methods are based on the same algorithm to calculate $\underline{\mathrm{Z}}$:

$$([\mathrm{A}] + a_1h[\mathrm{B}] + a_2h^2[\mathrm{C}])\underline{\mathrm{Z}} = \underline{\mathrm{Y}}_{n+1} - [\mathrm{C}](\underline{\mathrm{X}}_n + kh\dot{\underline{\mathrm{X}}}_n + b_1\ddot{\underline{\mathrm{X}}}_n) - [\mathrm{B}](k\dot{\underline{\mathrm{X}}}_n + b_2\ddot{\underline{\mathrm{X}}}_n + \tfrac{(k-1)}{2k}\underline{\mathrm{X}}_n) + \tfrac{(k+1)}{k^2}[\mathrm{A}](2\underline{\mathrm{X}}_n - \underline{\mathrm{X}}_{n-1})$$

Z and the values of a_1, a_2, b_1, b_2, k, l, and h depend on the chosen method and are explained as follows.

A) The Central Difference Method

This method consists in using a second-order (parabolic) interpolation along the time axis and deriving the first- and second-order derivative from it:

$$\dot{\underline{\mathrm{X}}}_n = \frac{1}{2\Delta t}(\underline{\mathrm{X}}_{n+1} - \underline{\mathrm{X}}_{n-1})$$

$$\ddot{\underline{\mathrm{X}}}_n = \frac{1}{\Delta t^2}(\underline{\mathrm{X}}_{n+1} - 2\underline{\mathrm{X}}_n + \underline{\mathrm{X}}_{n-1})$$

Replacing these values in equation (2) written at time step n leads to the following parameter values:

$$\underline{\text{Z}} = \underline{\text{X}}_{n+1} / \Delta t^2, a_1, = \tfrac{1}{2}, a_2, = b_1, = b_2, = k = 1 = 0 \text{ and } h = \Delta t.$$

Assuming that the initial first derivative of $\underline{\text{X}}$ is zero, the algorithm is supplied with an initial value of $\underline{\text{X}}_1$:

$$\underline{\text{X}}_{-1} = \underline{\text{X}}_0 + \frac{\Delta t^2}{2}[\text{A}]^{-1}(\underline{\text{Y}}_0 - [\text{C}]\underline{\text{X}}_0)$$

This method is conditionally stable; that is, knowing the highest eigen frequency f_{max} of the lossless system of equations, the time step must satisfy the following:

$$\pi f_{\text{max}} \Delta t < 1$$

B) The Newmark Method

This method consists of using a truncated Taylor series expansion of Xn_{+1} and its first derivative:

$$\begin{aligned}
\underline{\text{X}}_{n+1} &= \underline{\text{X}}_n + \Delta t \dot{\underline{\text{X}}}_n + \frac{\Delta t^2}{2}\ddot{\underline{\text{X}}}_n + (6\beta)\left(\frac{\underline{\text{X}}_{n+1} - \ddot{\underline{\text{X}}}_n}{\Delta \text{t}}\right) \\
\dot{\underline{\text{X}}}_{n+1} &= \dot{\underline{\text{X}}}_n + \Delta t \ddot{\underline{\text{X}}}_n + (2\gamma)\frac{\Delta t^2}{2}\left(\frac{\ddot{\underline{\text{X}}}_{n+1} - \ddot{\underline{\text{X}}}_n}{\Delta \text{t}}\right)
\end{aligned} \tag{3}$$

The parameters β and γ may be arbitrarily chosen. Introducing these equations in equation (2) written at time step n + 1 leads to the following parameter values:

$$\underline{Z} = \ddot{\underline{\text{X}}}_{n+1}, \; a_1, = \gamma, \; a_2, = \beta, \; \beta_1, = (1 - 2\beta)\Delta t^2 / 2, \; b_2, = (1 - \gamma)\Delta t, \; k = 1 = 1 \text{ and } h = \Delta t.$$

Assuming that the initial first derivative of $\underline{\text{X}}$ is zero, the algorithm is supplied with an initial value of $\ddot{\underline{\text{X}}}_0$:

$$\ddot{\underline{\text{X}}}_0 = [\text{A}]^{-1}(\underline{\text{Y}}_0 - [\text{C}]\underline{\text{X}}_0)$$

This method will be unconditionally stable provided the following inequalities are satisfied:

$$\gamma \geq \frac{1}{2},$$

$$\left(\gamma + \frac{1}{2}\right)^2 - 4\beta < \left(\frac{1}{\pi f_{\text{max}} \Delta t}\right)^2$$

where f_{max} is the highest eigen frequency of the lossless system of equations. Note that a value of $\gamma > \tfrac{1}{2}$ introduces a numerical damping, which is generally avoided. Names are given to usual couples of (γ, β): (1/2, 0) stands for the Central Explicit Difference method, (½, ¼) for the Average Acceleration method, and (1/2, 1/6) for the Linear Acceleration method.

c) The Wilson-Θ Method

This method is very similar to the Newmark method, except that the equations (equation 3) are first written for the time step $n + \theta$, $\theta \geq 1$, instead of $n + 1$ (Δt is replaced with $\theta\Delta t$), and the parameters γ and β are respectively set to ½ and 1/6. Once $X_{n+\theta}$ and its first and second time derivatives are thus calculated, a linear interpolation between X_n and $X_{n+\theta}$ gives the solution X_{n+1}:

$$\underline{X}_{n+1} = \frac{\underline{X}_{n+\theta} + (\theta - 1)\underline{X}_n}{\theta}$$

$$\dot{\underline{X}}_{n+1} = \frac{\dot{\underline{X}}_{n+\theta} + (\theta - 1)\dot{\underline{X}}_n}{\theta}$$

$$\ddot{\underline{X}}_{n+1} = \frac{\ddot{\underline{X}}_{n+\theta} + (\theta - 1)\ddot{\underline{X}}_n}{\theta}$$

This leads to the following parameter values:

$$\underline{Z} = \ddot{\underline{X}}_{n+\theta},\ a_1,= \theta / 2,= a_2,= \theta^2 / 6,\ b_1,= \theta^2\Delta t^2 / 3,\ b_2 = \theta\Delta t / 2,\ k = 1,\ 1 = \theta \text{ and } h = \theta\Delta t.$$

The stability of this algorithm is more difficult to determine but is proved for values of 6 greater than 1.366.

Important Note

Extending these algorithms to piezoelectricity or magnetostriction may lead to inaccurate results because of the quasi-static nature of electrical degrees of freedom: the matrix [A] of equation (2) becomes singular, and f_{max} tends to infinity. In the case of the Central Difference method, the implemented algorithm can deal with this provided the loss matrix [B] contains elastic losses only; that is, $[K''_{u\phi}]$, $[K''_{u\phi}]$, $[K''_{uI}]$, $[K''_{\phi\phi}]$, $[K''_{\phi I}]$, and $[K''_{II}]$, are both null. In the case of the Newmark method, a value of γ slightly greater than 0.5 will help in damping the spurious oscillations related to electric transients.

CALCULATION EXAMPLE

Transident -> Wilson -> Calculation -> Postprocess -> Observe the Current and Phase results.
Note that this pulse is the uniform stress on the bimorph.

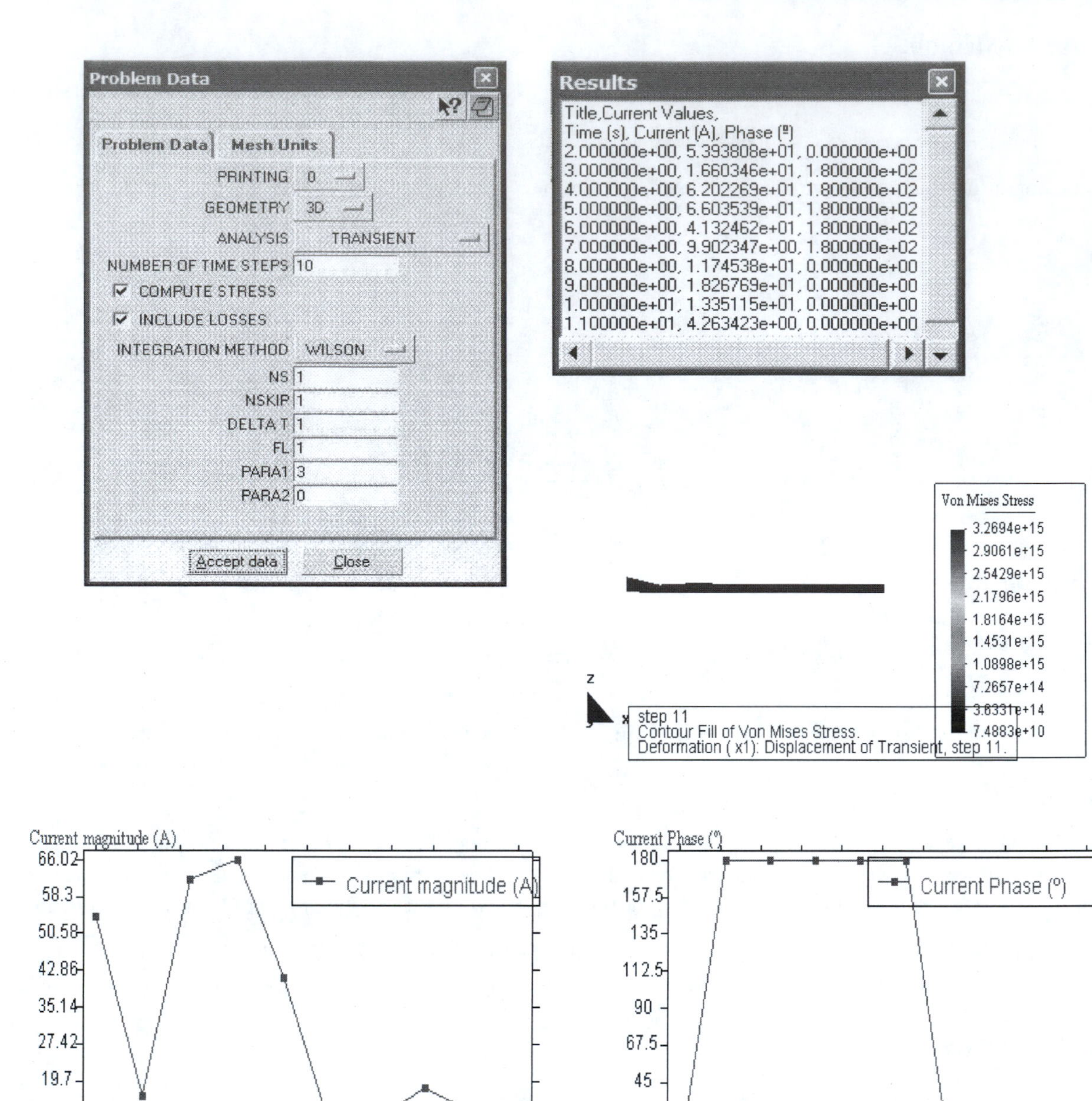

Back to Top

Back to Home

3.2 UNIMORPH (1)

GEOMETRY/DRAWING MATERIAL ASSIGNMENT

PROBLEM—3D RECTANGULAR PLATE, PZT4 AND BRASS = 40 MM × 6 MM × 1 MM, CANTILEVER SUPPORT

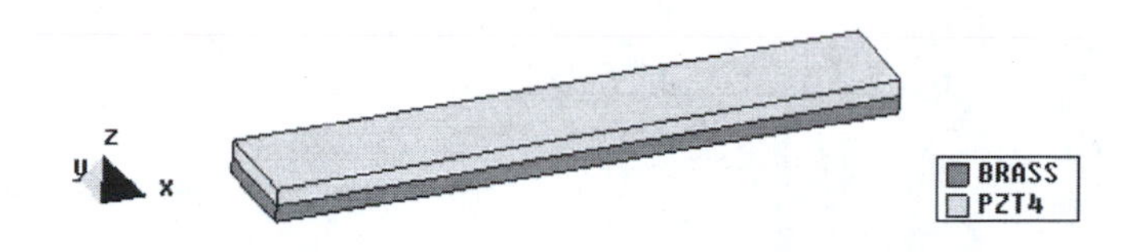

GEOMETRY/DRAWING

CREATE SURFACE SIMPLE STRUCTURE LIST

Geometry -> Create -> Object -> Rectangle -> Enter first corner point -> 0, 0 -> Enter second corner point -> 40, 6 -> Enter.

CREATE VOLUMES COPY MODE

Utilities -> Copy -> Surfaces -> Second point z=1.0 -> Do Extrude Surfaces -> Select -> Multiple copies -> 2. After selecting the rectangular (click on the surface line) -> Finish.

When you select the surface (click on the Magenta line), the surface color Magenta changes to Red.

Geometry -> Create -> Volume -> Automatic 6-sided volumes -> Two new volumes are created.

When you make the volume, you will find Cyan color.

MATERIAL ASSIGNMENT

Piezoelectric Materials

Data -> Materials -> Piezoelectric -> PZT4 -> Assign -> Pick Volumes.

Elastic Materials

Data -> Materials -> Elastic -> Brass -> Assign -> Pick Volumes.

Click Draw -> All materials -> Check the material are assigned as PZT4 and Brass.

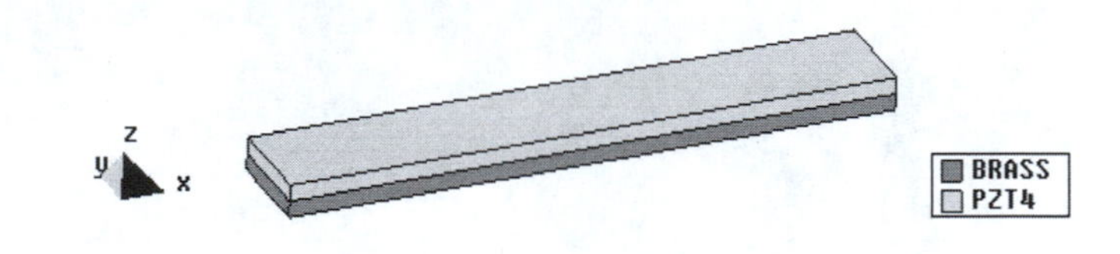

Back to Top

Back to Home

3.2 UNIMORPH (2)

BOUNDARY CONDITIONS

PROBLEM—3D RECTANGULAR PLATE, PZT4 AND BRASS = 40 MM × 6 MM × 1 MM, CANTILEVER SUPPORT

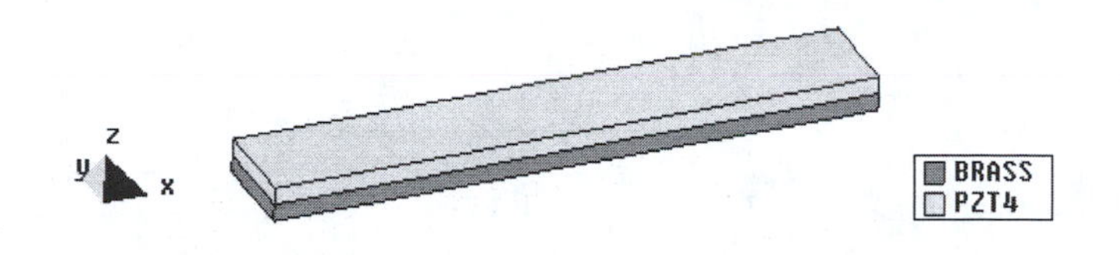

BOUNDARY CONDITIONS

POLARIZATION/LOCAL AXES

Data -> Conditions -> Volumes -> Geometry -> Polarization (Cartesian) -> Define Polarization (P1) (0, 0, −1) direction.

Choose 3PointsXZ -> Define X axis by putting (1, 0, 0) -> Then define Z axis by putting (0, 0, −1) -> Escape.

Assign Polarization (Local-Axes P1) -> Choose the top volume -> Finish.

POTENTIAL

Data -> Conditions -> Surfaces -> Potential 0.0 (or Ground in ATILA 5.2.4 or higher version) -> Assign the bottom surface-> Forced Potential 1.0 -> Assign the top surface.

SURFACE CONSTRAINTS (CANTILEVER: ONE-END CLAMP)

Data -> Conditions -> Surfaces -> Surface Constraints -> X, Y, and Z components -> Clamped -> Assign left two surfaces.

To check the assignment -> Draw -> All conditions -> Include local axes.

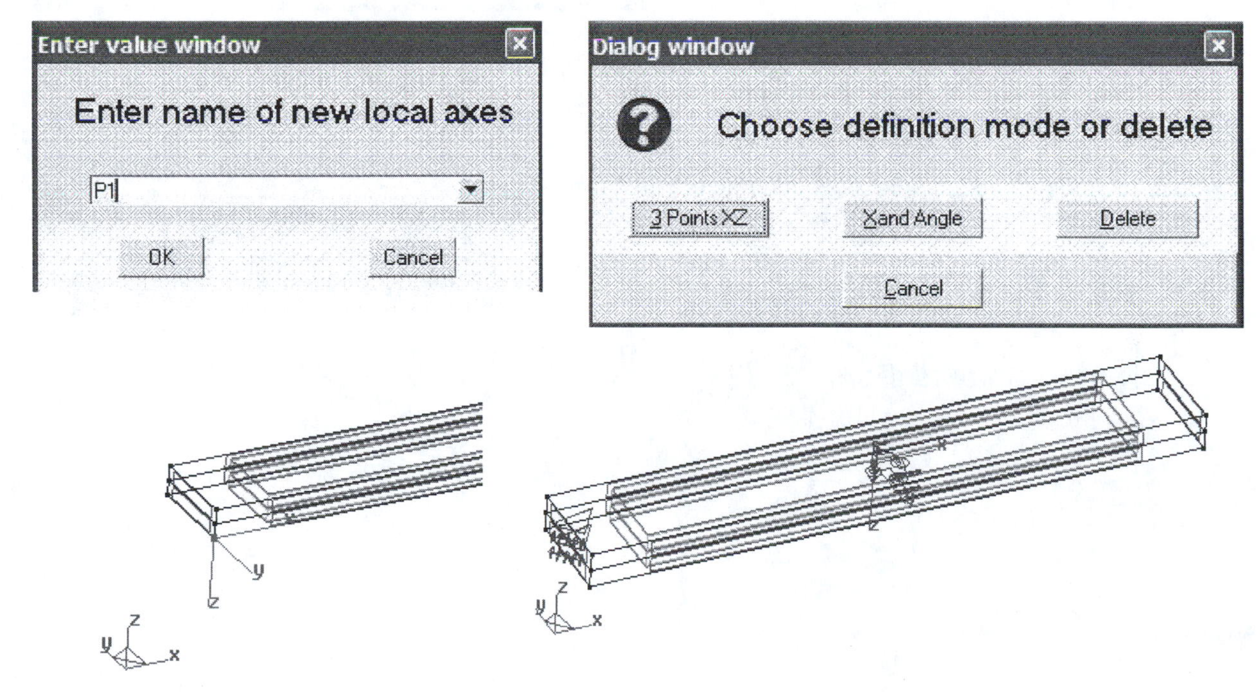

Back to Top

Back to Home

3.2 UNIMORPH (3)

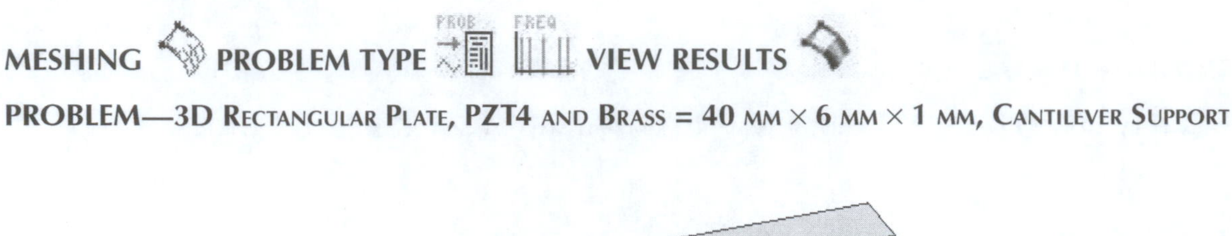

PROBLEM—3D RECTANGULAR PLATE, PZT4 AND BRASS = 40 MM × 6 MM × 1 MM, CANTILEVER SUPPORT

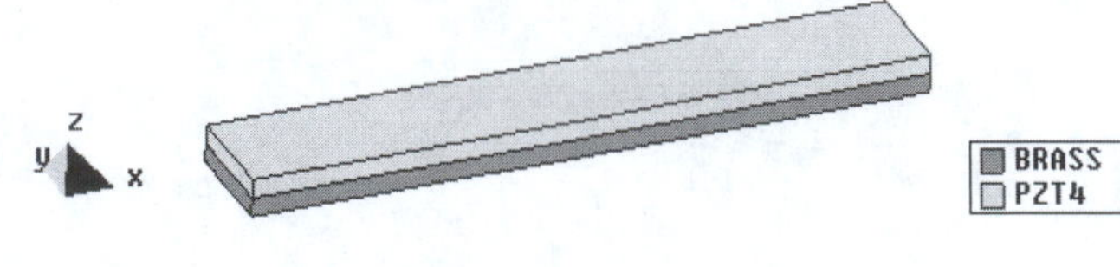

MESHING

Cant view is not very suitable for Meshing -> View -> Rotate -> Plane xy (original).

MAPPED MESH

Meshing -> Structured -> Volumes -> Select all volumes -> Then Cancel the further process.
Meshing -> Structured -> Lines -> Enter number to divide line -> Choose "1" for all lines (which makes the thickness line no division) -> Choose 2 for only the right and left lines -> Choose 4 for only the top and bottom lines -> Generate.

Mapped mesh (good)

PROBLEM TYPE

PROB Data -> Problem data -> 3D, Harmonic, Click both Compute Stress and Include Losses.

FREQUENCY

FREQ Data -> Interval data Interval 1 (100 Hz), Interval 2 (4,000–5,000 Hz).

CALCULATION

Calculate -> Calculate window -> Start.

Problem Data
Problem Data | Mesh Units
PRINTING 0
GEOMETRY 3D
ANALYSIS HARMONIC
COMPUTE STRESS
INCLUDE LOSSES
Accept data | Close

Interval Data
2
TYPE LINEAR DISTRIBUTION
MIN FREQUENCY 4000.
MAX FREQUENCY 5000.
NUMBER OF FREQUENCIES 11
Accept data | Close

POST PROCESS/VIEW RESULTS

Click to Open the Post Calculation (Flavia) Domain.

Note the new Toolbar appearance.

CLICK ADM ADMITTANCE/IMPEDANCE RESULT

Admittance Peak Point: 4,500 Hz -> This peak corresponds to the First Resonance.

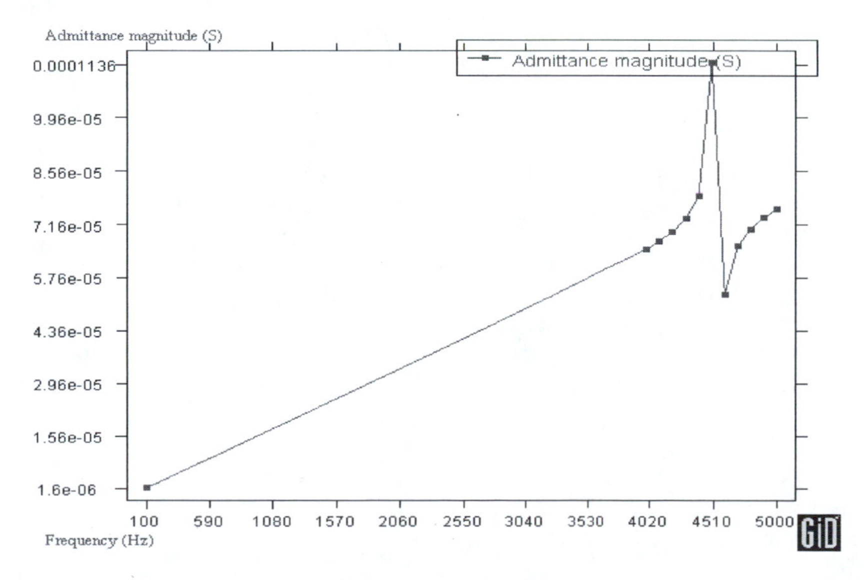

Admittance Curve

To take this graph out -> Go To

View Animation—Vibration Mode at a Particular Frequency

View results -> Default Analysis/Step -> Harmonic-Magnitude -> Set a Frequency
(100 Hz for Off-Resonance Mode: Peak Point: 4,500 Hz for Resonance Mode).

Double Click [ANI icon] Loading frequency: 4,500 Hz Harmonic Case -> Click OK (single-click and wait for a while).

Set Deformation -> Displacement

View results -> Deformation -> Displacement.

Deformation Scale Change

Click Windows -> Deform mesh -> Change the Factor in Mesh Deformation.

Parameter other than Displacement

Set Results View -> View results -> Contour Fill -> Von Mises Stress.

After setting both "Results view" and "Deformation" -> Click [icon] to see the Animation.

To save the GIF Animation Movie, use Save GIF, then Click [icon] after setting the file-saving file.

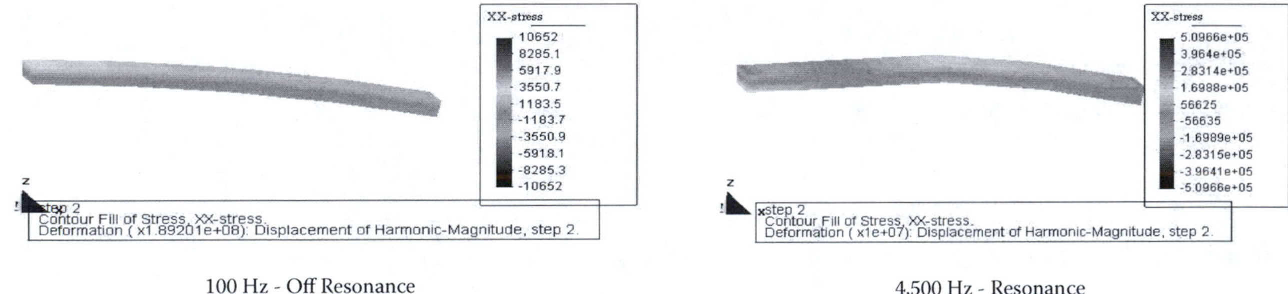

100 Hz - Off Resonance

4,500 Hz - Resonance

> NB: Notice the vibration configuration difference between the off-resonance and the resonance modes. The canti-lever support is suitable to the off-resonance usage, but it does not enhance the tip displacement for the resonance usage.

Back to Top

Back to Home

3.3 MULTILAYER (1)

GEOMETRY/DRAWING MATERIAL ASSIGNMENT

PROBLEM—3D Two-Layer, PZT4, 10 mm × 5 mm × layer thickness 1 mm, Interdigital Electrode

- Composite of two different polarization layers, with an interdigital electrode configuration
- Learn the internal stress distribution

GEOMETRY/DRAWING

Create Surface

Geometry -> Create -> Object -> Rectangle -> Enter first corner point -> 0, 0 -> Enter second corner point -> 10, 5 -> Enter.

Divide Surface

Geometry -> Edit -> Divide -> Surfaces -> Near point -> Select the Surface -> Choose NURBS sense -> Input Near Points 1, 0 and 9, 0.

Create Volumes Copy mode

Utilities -> Copy -> Surfaces -> Translation -> Second point z=1.0 -> Do Extrude Surfaces -> Select -> Multiple copies -> 1 -> After selecting the rectangular (click on the surface line) -> Finish.

Geometry -> Create -> Volume -> Automatic six-sided volumes -> 3 Volumes appear in blue lines.

Utilities -> Copy -> Volumes -> Translation by 1 mm Select (Do not Extrude).

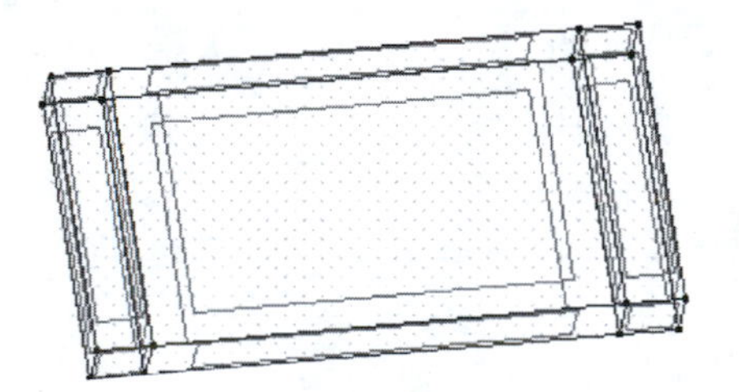

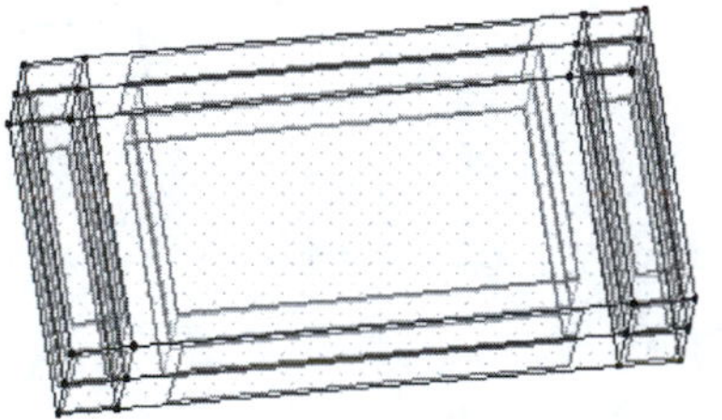

MATERIAL ASSIGNMENT

Piezoelectric Materials

Data -> Materials -> Piezoelectric -> PZT4 -> Assign -> Pick Volumes.
Click Draw -> All materials -> Check the material is assigned as PZT4.

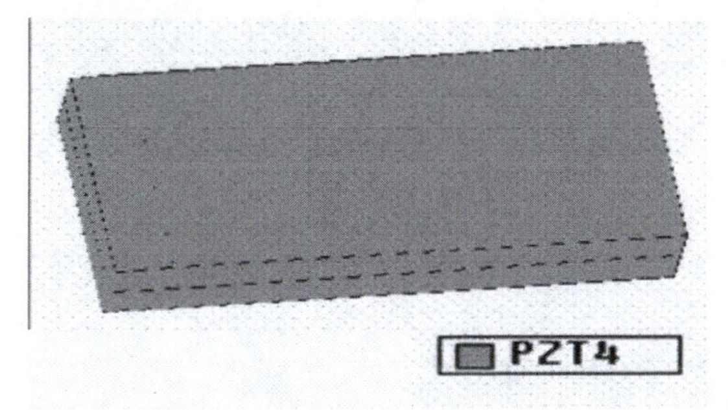

Back to Top

Back to Home

3.3 MULTILAYER (2)

BOUNDARY CONDITIONS MESHING

PROBLEM—3D Two-Layer, PZT4, 10 mm × 5 mm × layer thickness 1 mm, Interdigital Electrode

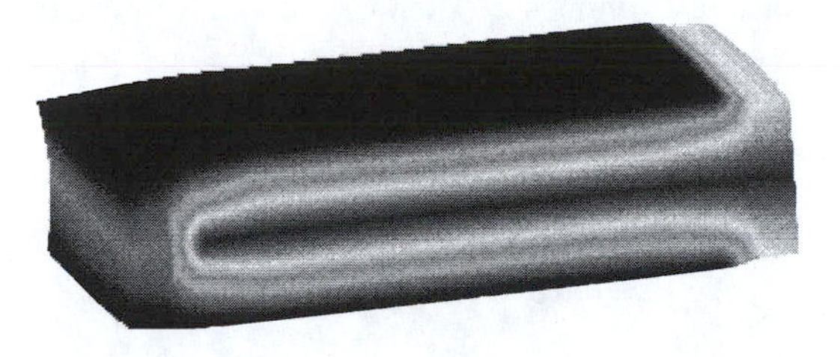

BOUNDARY CONDITIONS

Potential

Data -> Conditions -> Surfaces -> Potential 0.0 (or Ground in ATILA 5.2.4 or higher version) -> Assign the top and bottom surfaces -> Forced Potential 1.0 -> Assign the center surface.

Polarization/Local Axes

Data -> Conditions -> Volumes -> Geometry -> Polarization (Cartesian) -> Define Polarization (P1) (0, 0, 1) direction.

Choose X and Angle -> Define X axis -> Click two points to define X axis -> Choose the suitable Cartesian P1 direction (Z axis) -> Escape.
Or, choose 3PointsXZ -> Define X axis by putting (1, 0, 0) -> Then define Z axis by putting (0, 0, 1) -> Escape.

Data -> Conditions -> Volumes -> Geometry -> Polarization (Cartesian) -> Define Polarization (P2) (0, 0, –1) direction.

Assign Polarization (Local-Axes P1 and P2) -> Choose the volume -> Finish.

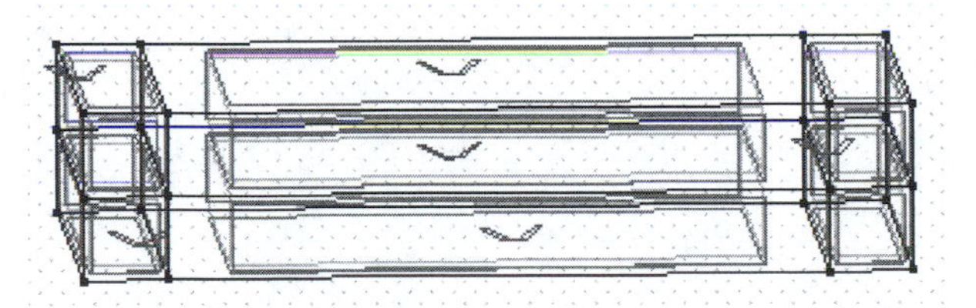

Dialog window

Choose definition mode or delete

3 Points XZ | Xand Angle | Delete

Cancel

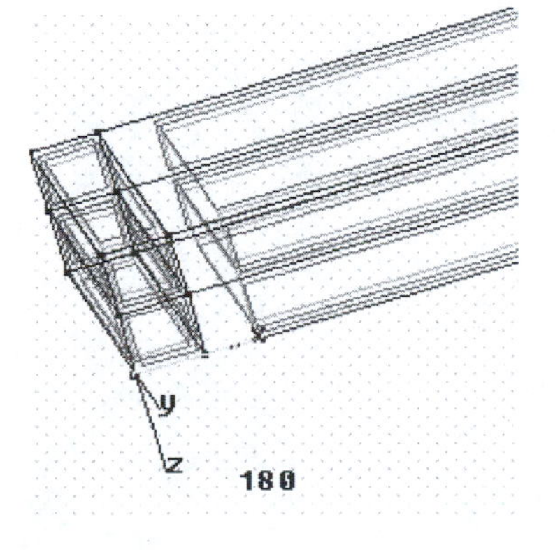

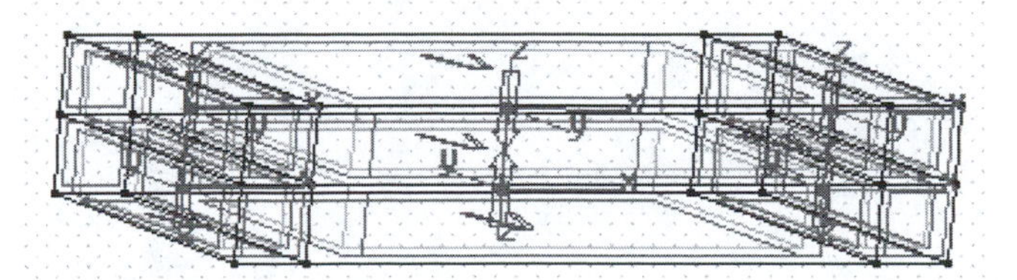

MESHING

Cant view is not very suitable for Meshing -> View -> Rotate -> Plane xy (original).

Mapped Mesh

Meshing -> Structured -> Volumes -> Select all volumes -> Then Cancel the further process.
Meshing -> Structured -> Lines -> Enter number to divide line -> Choose 2 and 8 for the lines -> Generate.

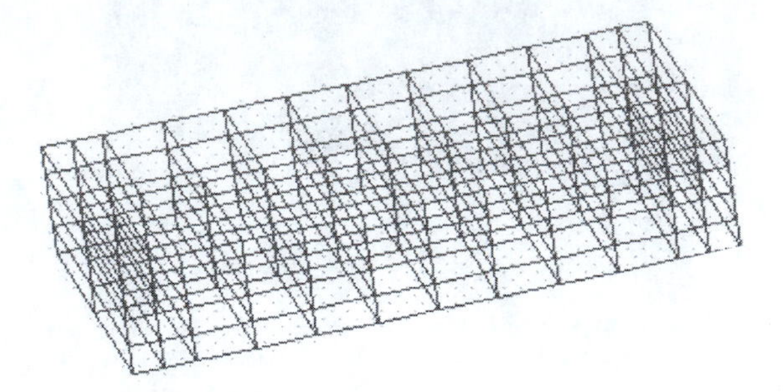

Back to Top

Back to Home

3.3 MULTILAYER (3)

PROBLEM TYPE PROB FREQ VIEW RESULTS

PROBLEM—3D Two-Layer, PZT4, 10 mm × 5 mm × layer thickness 1 mm, Interdigital Electrode

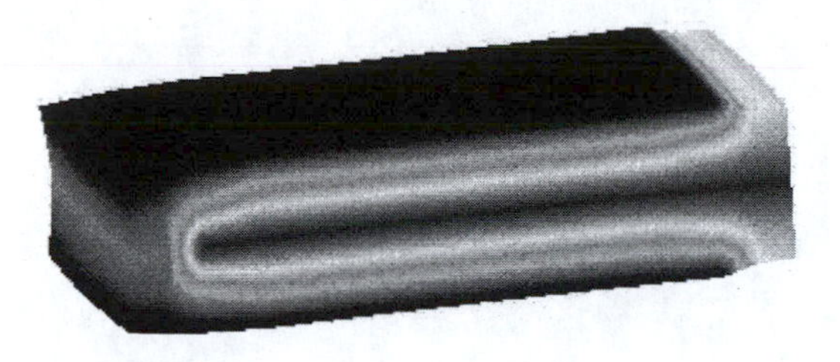

PROBLEM TYPE

PROB Data-> Problem data -> 3D, Static, Click both Compute Stress.

PROB Data -> Problem data -> 3D, Harmonic, Click both Compute Stress and Include Losses.

FREQUENCY

FREQ Data -> Static -> Interval data Interval 1 (1,000 Hz).

FREQ Data -> Harmonic -> Interval 2 (500,000–1,000,000 Hz).

CALCULATION

Calculate -> Calculate window -> Start.

POST PROCESS/VIEW RESULTS

Click to Open the Post Calculation (Flavia) Domain.

Admittance Peak Point: 700,000 Hz -> This peak corresponds to the First Resonance.

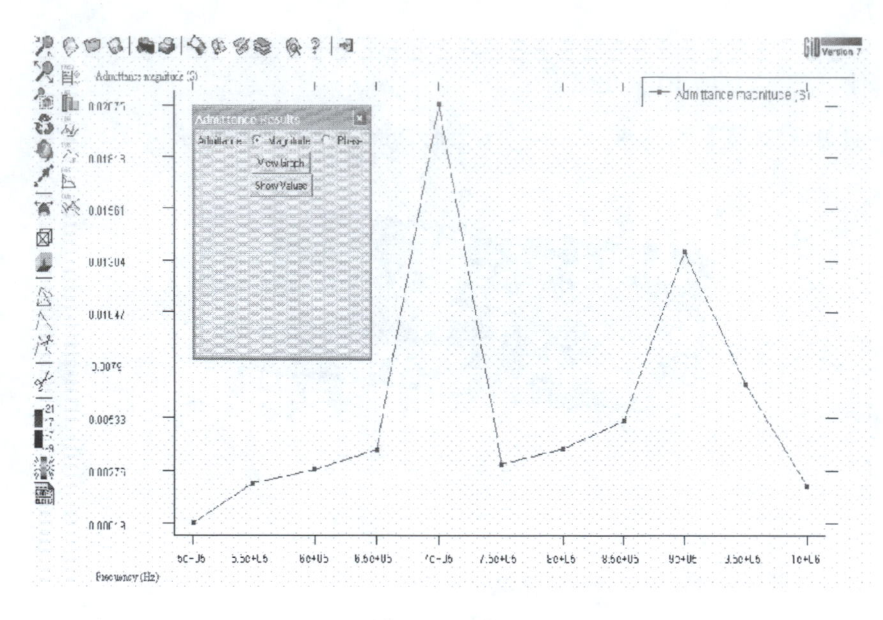

Admittance Curve

View Animation—Vibration Mode at a Particular Frequency

View results -> Default Analysis/Step -> Harmonic-Magnitude -> Set a Frequency
(1,000 Hz for Off-Resonance Mode; Peak Point: 700,000 Hz for Resonance Mode).

Double Click Loading frequency: 1,000 Hz Harmonic Case -> Click OK (Single-click and wait for a while).

Set Deformation -> Displacement

View results -> Deformation -> Displacement.

Parameter other than Displacement

Set Results View -> View results -> Contour Fill -> Electric Potential or Stress XX.

After setting both "Results view" and "Deformation" -> Click to see the Animation.

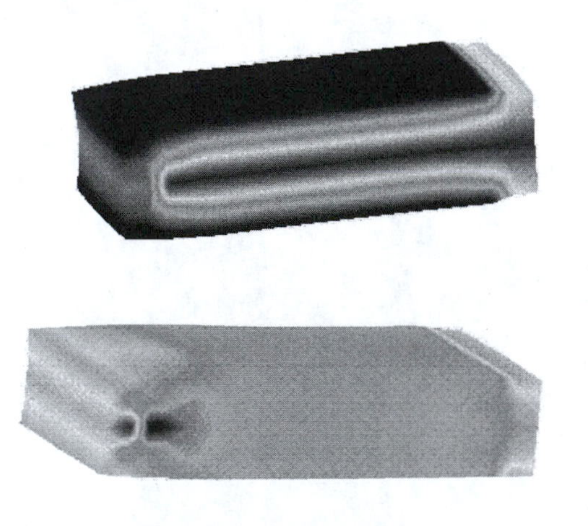

1,000 Hz - Off Resonance

700,000 Hz - Resonance

Back to Top

Back to Home

3.4 CYMBAL (1)

GEOMETRY/DRAWING MATERIAL ASSIGNMENT

PROBLEM—Cymbal 12 mm Φ × 2.4 mm T (Metal thickness 0.2 mm, PZT thickness 0.5 mm × two layer, Cavity height 0.5 mm)

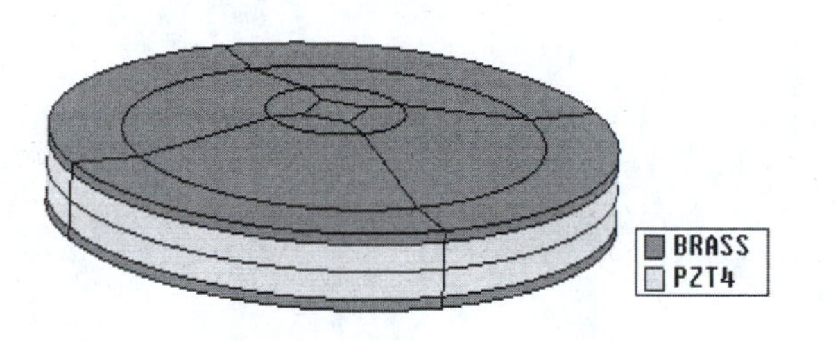

- Composite of two complicated phases
- Learn the layer usage
- Learn the symmetry treatment

GEOMETRY/DRAWING

Layer Definition for "Cap"

Utilities -> Layers -> Name the layer as "Cap."
View Plane
View -> Rotate -> Plane XZ.

Create Line

Surface : Geometry -> Create -> Line -> Enter First and Second Corner Points -> 0, 0, 1.2 -> 1.5, 0, 1.2 -> 4.5, 0, 0.7 -> 6, 0, 0.7.

Create Endcap Surface

Copy/Transform -> Line Rotation by 45° twice -> Select the lines.

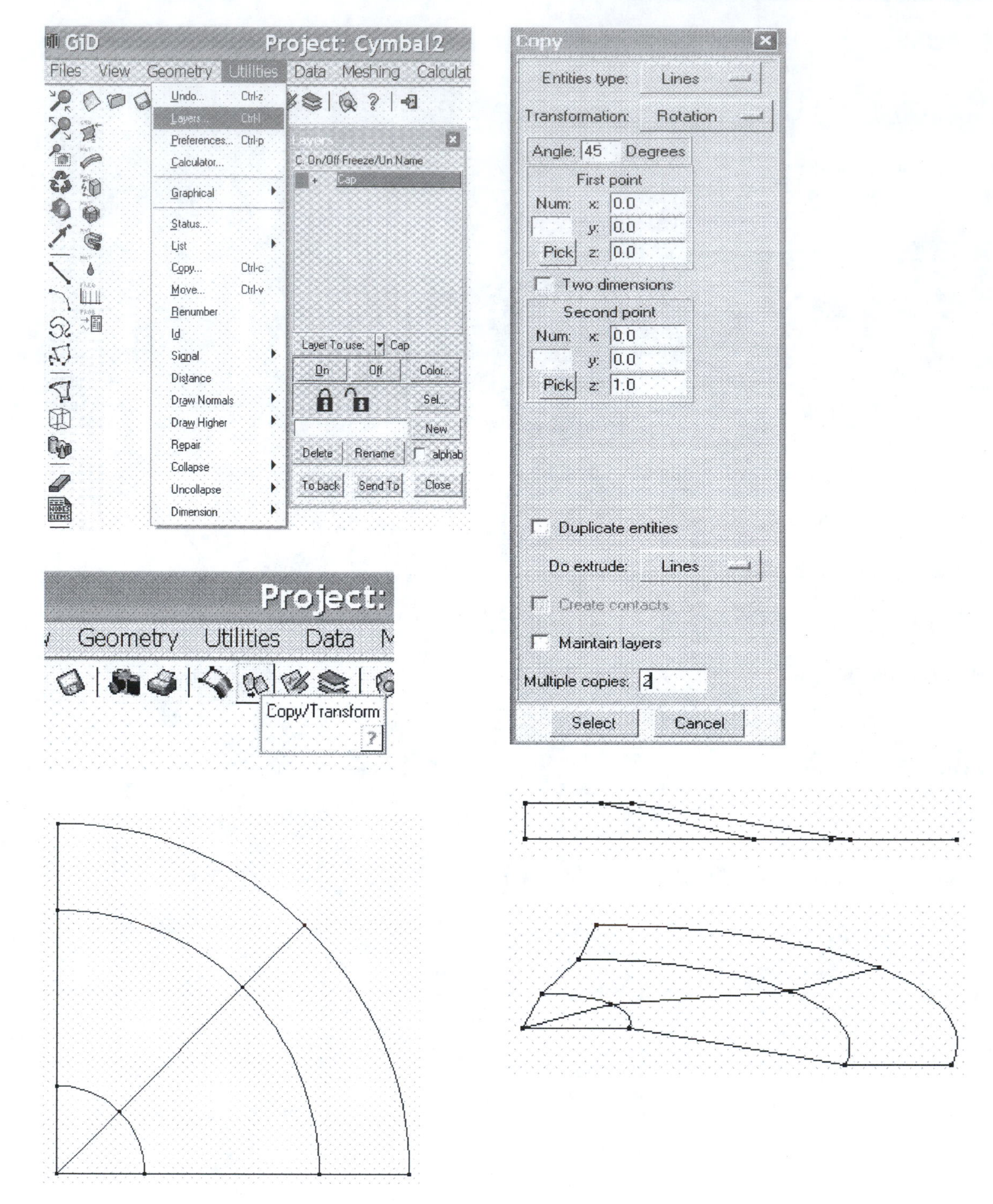

Create Center Surface

Geometry -> Create -> Object -> Rectangle -> Enter First and Second Corner Points -> 0, 0, 1.2, and 0.7, 0.7, 1.2.

Divide Surface and Delete Lines

Line division : Geometry -> Edit -> Divide -> Near point -> Geometry -> Delete Lines -> Create NURBS surface by Automatic and Contours.

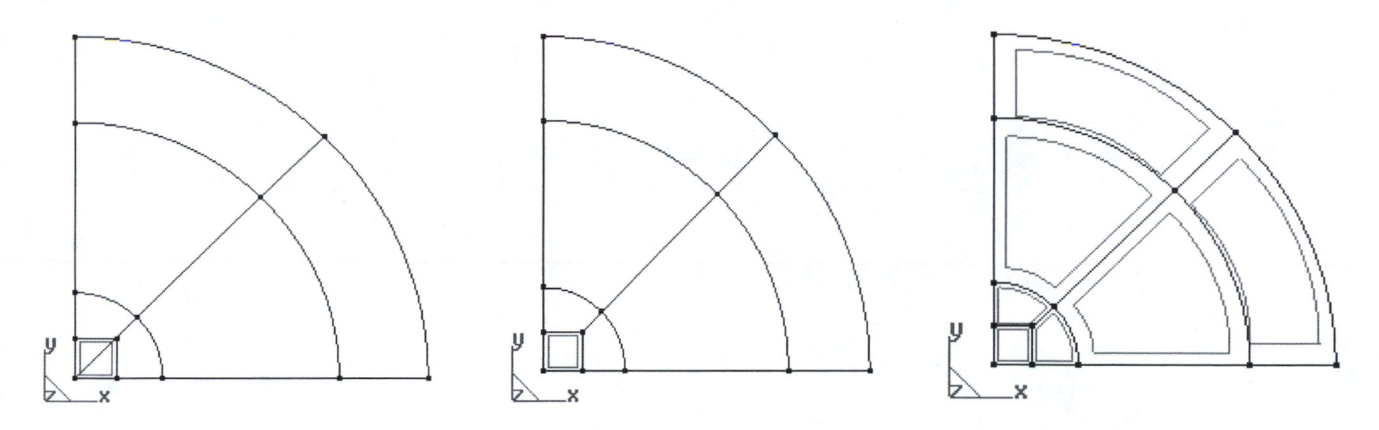

Create Volumes

Utilities -> Copy -> Surfaces -> Extrude Surfaces by −0.2 mm -> Select.
Geometry -> Create -> Volume -> Automatic -> 6-sided volumes -> Seven Volumes appear in blue lines.

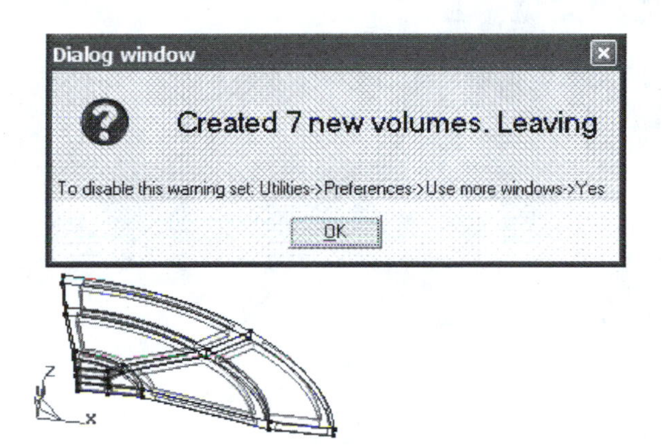

Layer Definition for "PZT Disk"

Utilities -> Layers -> Name the layer as "PZT."

View Plane

View -> Rotate -> Plane XZ.

Create Lines and Divide the Surface

Geometry -> Create -> Lines -> First point 0, 0, 0 -> Second point 4.5, 0, 0 -> Third point -> 6, 0, 0.

Create Endcap Surface

Copy/Transform -> Line Rotation by 45° twice -> Select the lines.

Create Center Surface

Geometry -> Divide Lines -> 0.7, 0, 0 and 0, 0.7, 0.
Geometry -> Create -> Lines -> Point 0.7, 0, 0 -> 0.7 , 0.7, 0 -> 0, 0.7, 0.
Geometry -> Create -> NURBS Surfaces -> Automatic.

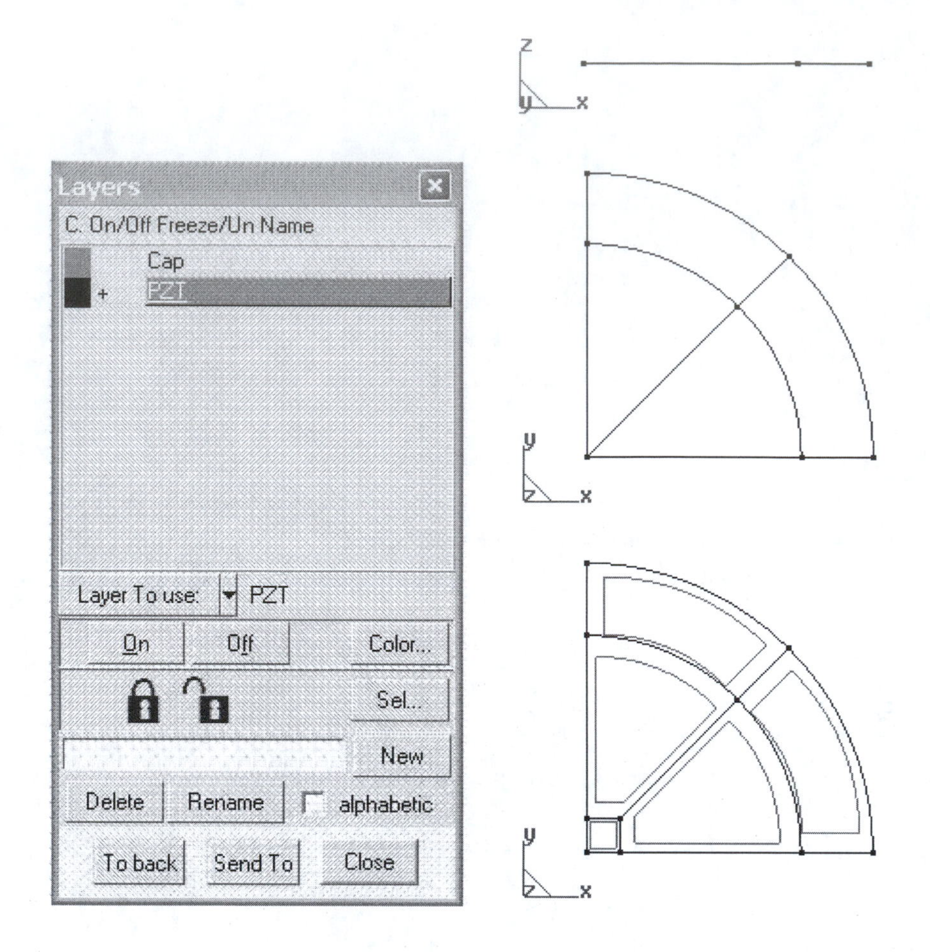

Create Volumes

Utilities -> Copy -> Surfaces -> Extrude Surfaces by 0.5 mm -> Select.
Geometry -> Create -> Volume -> Automatic -> 6-sided volumes -> Five Volumes appear in blue lines.

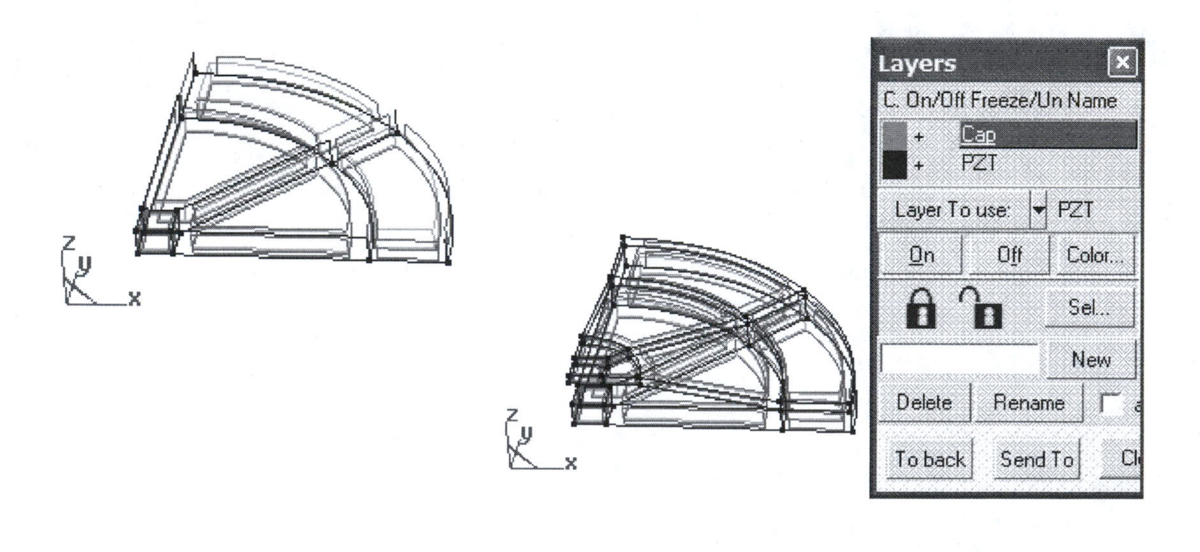

> NB: Two important points: (1) The contact surface between the two-layer structures should have one common surface. (2) The PZT layer does not show this common surface.

MATERIAL ASSIGNMENT

PIEZOELECTRIC MATERIALS 

Data -> Materials -> Piezoelectric -> PZT4 -> Assign -> Pick Volumes.
Click Draw -> All materials -> Check the material is assigned as PZT4.

ELASTIC MATERIALS

Data -> Materials -> Elastic -> Brass -> Assign -> Pick Volumes.

Click Draw -> All materials -> Check the material are assigned as PZT4 and Brass.

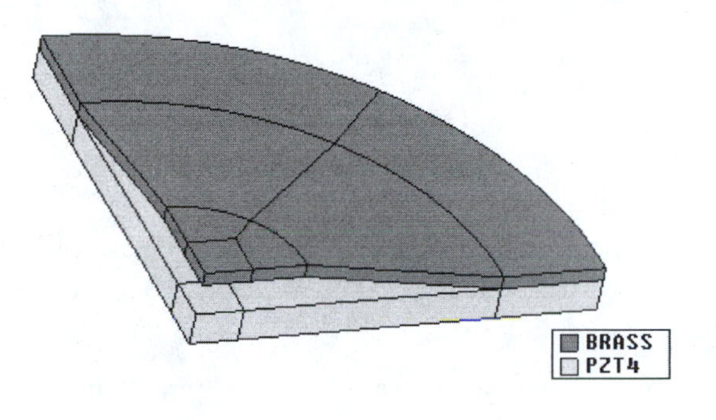

> NB: This 1/8 part is enough for the simulation as long as the mode is considered to be symmetrical.

Back to Top

Back to Home

3.4 CYMBAL (2)

BOUNDARY CONDITIONS MESHING

PROBLEM—CYMBAL 12 MM Φ _ × 2.4 MM T (METAL THICKNESS 0.2 MM, PZT THICKNESS 0.5 MM × TWO LAYER, CAVITY HEIGHT 0.5 MM)

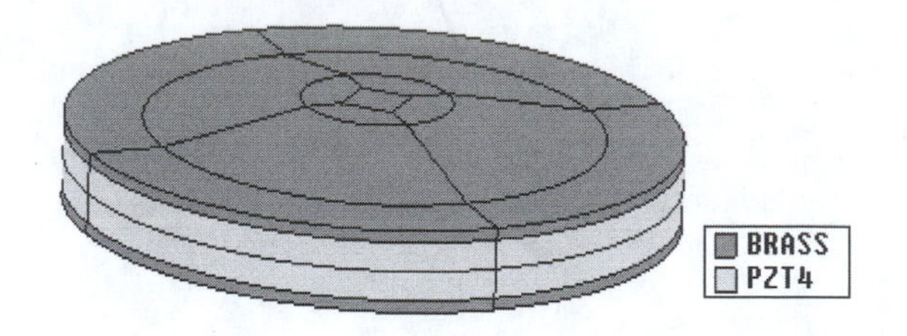

BOUNDARY CONDITIONS

POLARIZATION/LOCAL AXES

High voltage should be applied on the center surface of the PZT two layers, and Ground on the metal end cap. Choose the polarization direction along the electric field direction.

Data -> Conditions -> Volumes -> Geometry -> Polarization (Cartesian) -> Define Polarization (P1) (0,0,1) direction.

Choose 3PointsXZ -> Center (0, 0, 0) -> Define X axis by putting (1, 0, 0) -> Then define Z axis by putting (0,0,1) -> Escape.

Assign Polarization (Local-Axes P1) -> Choose the PZT volume -> Finish.

POTENTIAL

Data -> Conditions -> Surfaces -> Potential 0.0 (or Ground in ATILA 5.2.4 or higher version) -> Assign the top surface of the PZT -> Forced Potential 1.0 -> Assign the bottom surface.

SURFACE CONSTRAINTS (1/8 PART OF THE WHOLE CYMBAL)

We use only 1/8 part of the whole Cymbal, taking into account the symmetry; the bottom surface of the PZT disk - Z component is clamped, whereas X and Y components are free.

Data -> Conditions -> Surfaces -> Surface Constraints -> Z components -> Clamped -> Assign the bottom PZT surfaces.

Similarly, the edge on the X axis -> Surface Constraints -> Y components -> Clamped -> Assign.

The edge on the Y axis -> Surface Constraints -> X components -> Clamped -> Assign.

To check the assignment -> Draw -> All conditions -> Include local axes.

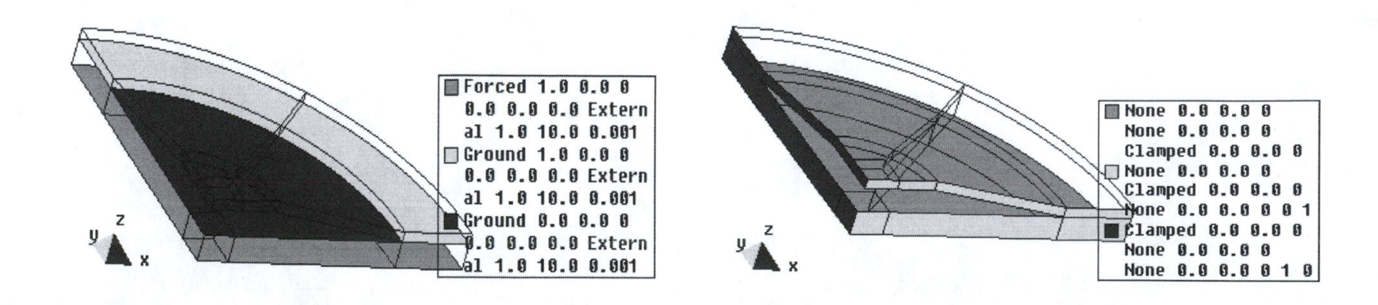

MESHING

Cant view is not very suitable for Meshing -> View -> Rotate -> Plane xy (original).

Mapped Mesh

Meshing -> Structured -> Volumes -> Select all volumes -> Then Cancel the further process.
Meshing -> Structured -> Lines -> Enter number to divide line -> Choose "1" for the lines -> Generate.

Use larger numbers such as 2 or 4 for more accurate calculation.

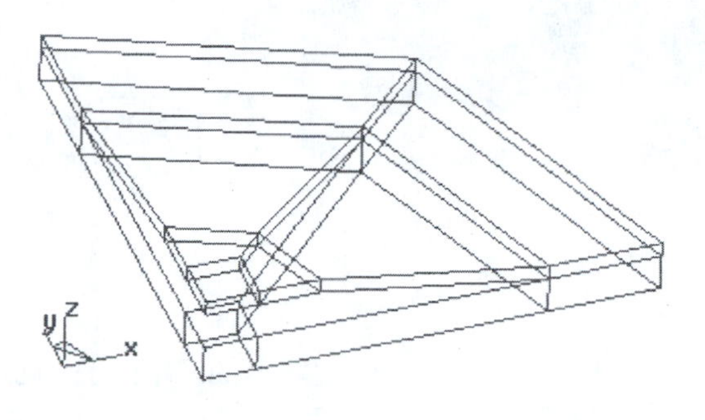

Back to Top

Back to Home

3.4 CYMBAL (3)

PROBLEM TYPE VIEW RESULTS

PROBLEM—CYMBAL 12 MM Φ × 2.4 MM T (METAL THICKNESS 0.2 MM, PZT THICKNESS 0.5 MM × TWO LAYER, CAVITY HEIGHT 0.5 MM)

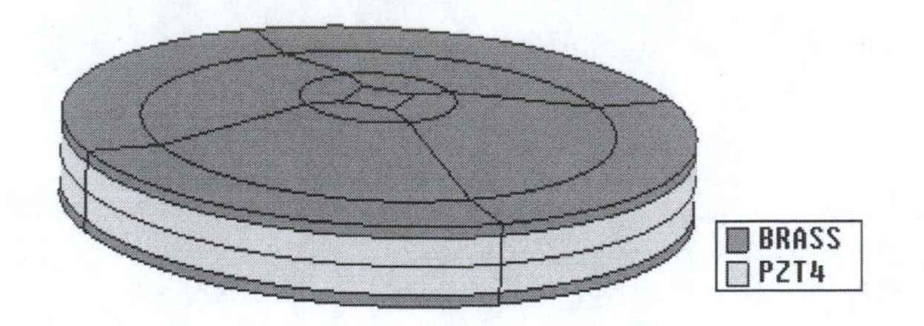

PROBLEM TYPE

Data -> Problem data -> 3D, Harmonic, Click both Compute Stress and Include Losses.

FREQUENCY

Data -> Harmonic -> Interval 2 (35,000–45,000 Hz).

CALCULATION

Calculate -> Calculate window -> Start.

POST PROCESS/VIEW RESULTS

[Click to Open the Post Calculation (Flavia) Domain.

CLICK ADMITTANCE/IMPEDANCE RESULT

Admittance Peak Point: 41,000 Hz -> This peak corresponds to the First Resonance.

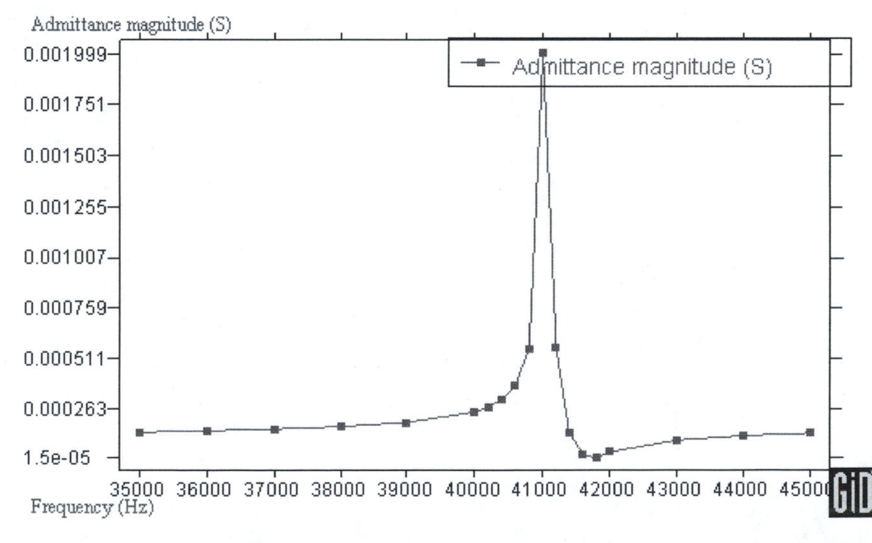

Admittance Curve

View Animation—Vibration Mode at a Particular Frequency

View results -> Default Analysis/Step -> Harmonic-Magnitude -> Set a Frequency
(Peak Point: 41,000 Hz for Resonance Mode).

Double Click Loading frequency: 41,000 Hz Harmonic Case -> Click OK (Single-click and wait for a while).

Set Deformation -> Displacement

View results -> Deformation -> Displacement.

Parameter other than Displacement

Set Results View -> View results -> Contour Fill -> Electric Potential or Von Mises Stress.

After setting both "Results view" and "Deformation"-> Click to see the Animation.

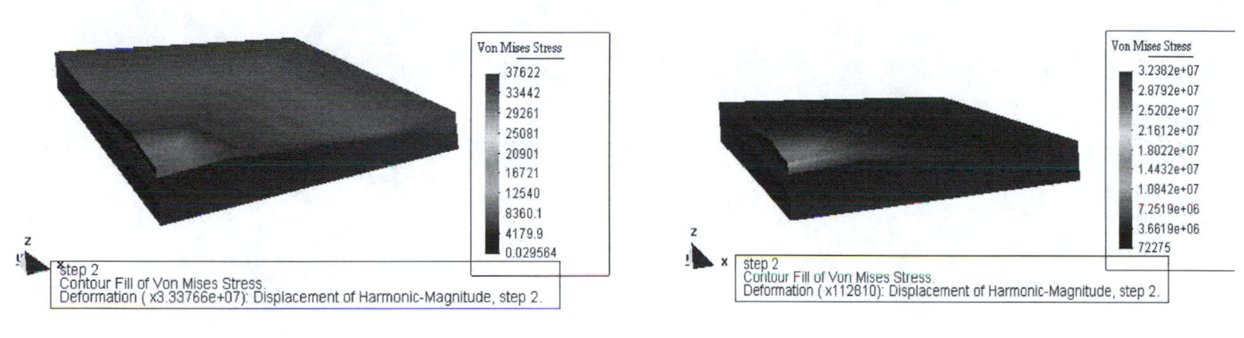

1,000 Hz - Off Resonance

41,000 Hz - Resonance

Back to Top

Back to Home

4 Piezoelectric Transformer

- How to set the "Electrical load"

TEXT CONTENT

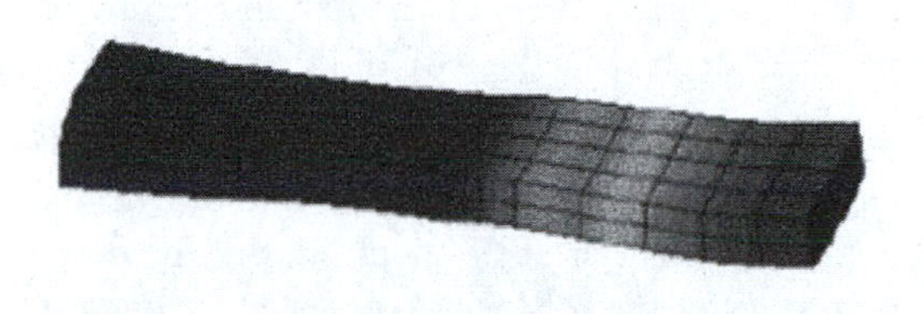

Potential Distribution in a Rosen Transformer

➡	Rosen-Type Transformer
➡	Ring-Dot-Type Transformer
➡	**Homework 6** Step-up ratio versus electric load
How to Use This GiD File	**Example GiD File** Rosen.gid, RingDot.gid

Back to Top

4.1 ROSEN-TYPE TRANSFORMER (1)

GEOMETRY/DRAWING MATERIAL ASSIGNMENT

PROBLEM—ROSEN TRANSFORMER 25 MM × 8 MM × 2.2 MM T

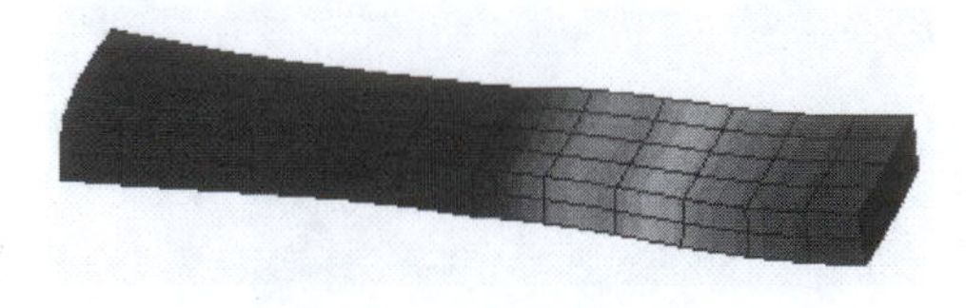

GEOMETRY/DRAWING

Analysis of a piezoelectric transformer. The model is 3D, and two analyses will be used: modal, to determine the resonance frequency of the transformer, and harmonic, to determine the gain of the transformer at resonance.

A piezoelectric transformer is a device that operates at resonance and that makes use of both direct and converse piezoelectric effects. In short, a low-voltage signal is applied at the input section to excite the mechanical resonance of the transformer. The stresses induced by the resonance effect produce charges at the electrodes of the output section. As a result, the input voltage is transformed either into a higher (step-up transformer) or lower voltage (step-down transformer).

The transformer modeled in this tutorial is a so-called Rosen-type transformer (from the name of the inventor), and it is somewhat simplified. It operates in a $\lambda/2$ (half-wavelength) longitudinal vibration mode. The dimensions used for the geometry are shown (in mm) in the following figure.

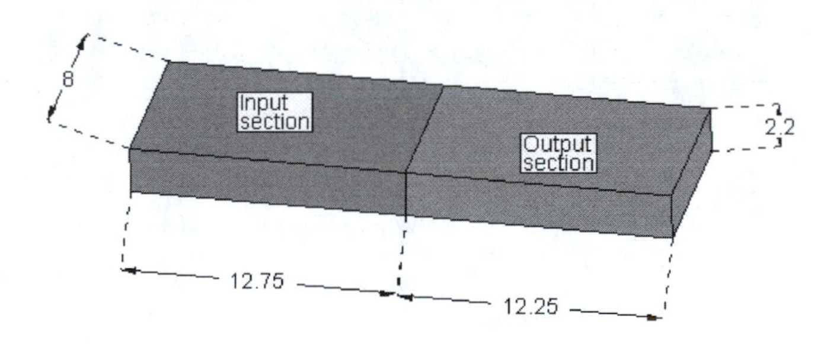

There are many ways to define the geometrical elements (points, lines, surfaces, and volumes) required for the problem. You have already learned enough about constructing the geometry.

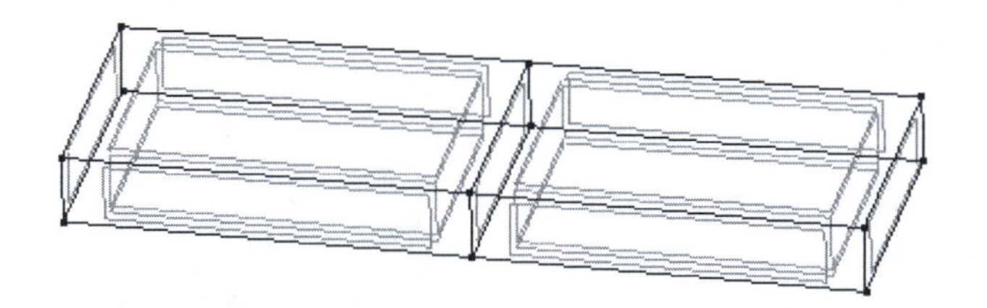

The material selected for this problem is PZT4 for both input and output sections. The electrodes are shown in the next figure.

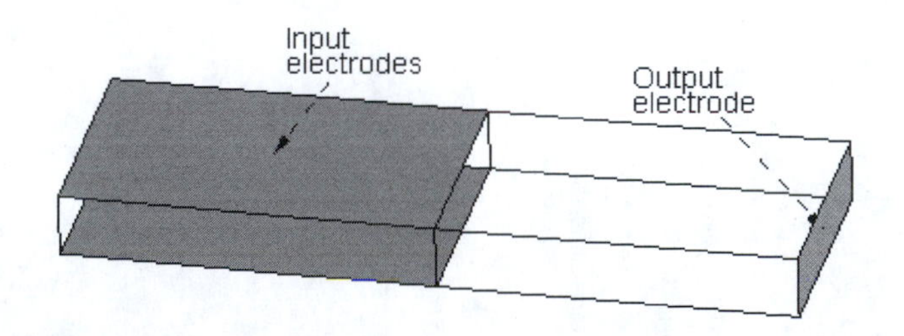

MATERIAL ASSIGNMENT

Piezoelectric Materials

Data -> Materials -> Piezoelectric -> PZT4 -> Assign -> Pick Volumes.

Assign material PZT4 to the two volumes. Note that the definition of the material PZT4 includes losses. You can verify the loss angles by clicking on the Losses tab. The values Delta_m, Delta_p, and Delta_d correspond to the loss angle for the mechanical, piezoelectric, and dielectric tensors, respectively.

When this is done, the material assignment should be similar to the following image (you can obtain this by selecting, on any material window, the option **Draw | All materials**).

Click Draw -> All materials -> Check the material is assigned as PZT4.

Back to Top

Back to Home

4.1 ROSEN-TYPE TRANSFORMER (2)

BOUNDARY CONDITIONS

PROBLEM—Rosen Transformer 25 mm × 8 mm × 2.2 mm t

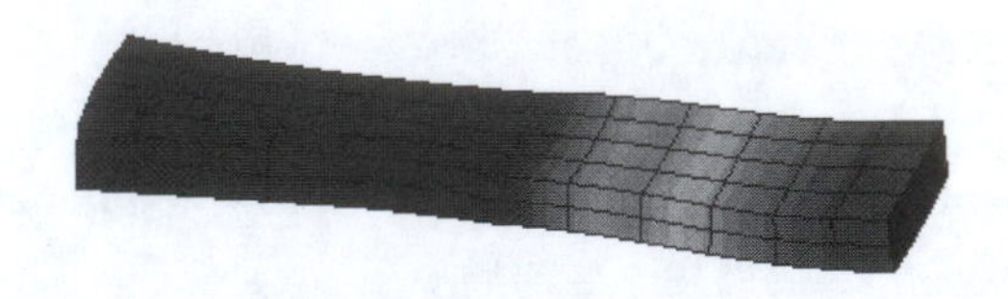

BOUNDARY CONDITIONS

The boundary conditions needed for the problem are the electrical conditions on the piezoelectric ceramics and the definition of the polarization direction of the ceramics. There are no mechanical conditions as the system is considered to vibrate freely in space.

Potential

Data -> Conditions -> Surfaces -> Potential 0.0 (or Ground in ATILA 5.2.4 or higher version) -> Assign the top surface of the PZT -> Forced Potential 1.0 -> Assign the bottom surface.

To apply the electrical boundary conditions on the ceramics, select the **Surfaces** tab, and then **Electric Potential** in the drop-down list. First, apply the excitation signal by selecting the **Forced** potential option. Define an excitation signal of amplitude 1V and zero phase, and apply it to one of the input electrodes.

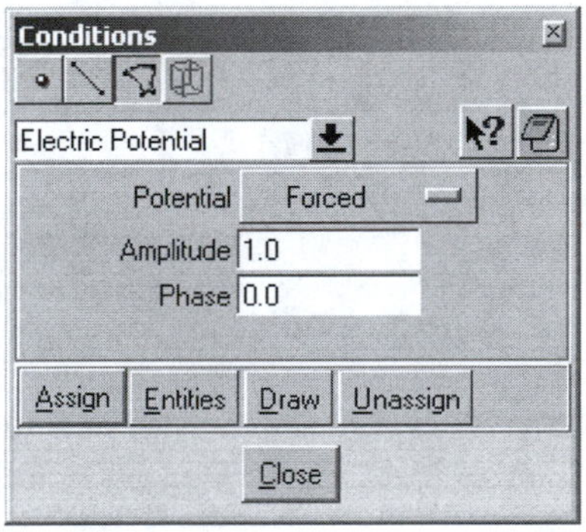

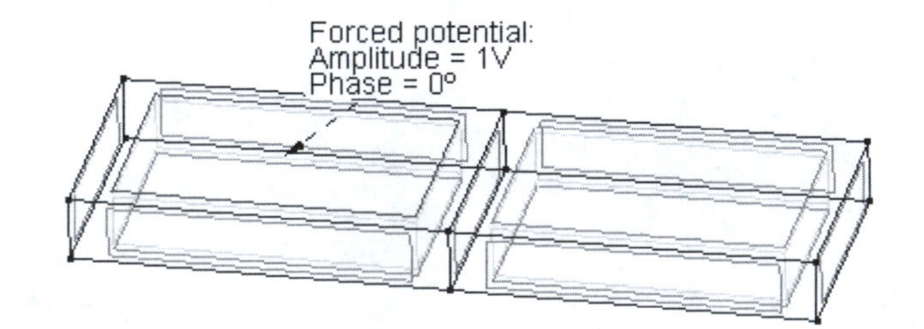

Define the reference electrode by selecting the **Ground** potential condition and applying it to the other input electrode.

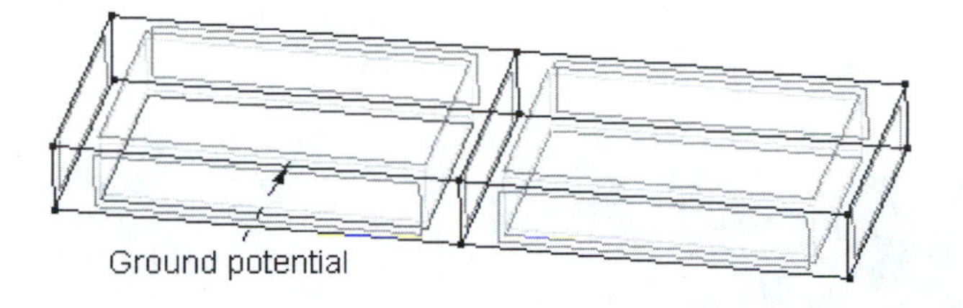

The condition on the output electrode is a little special and is called **Floating**. This condition indicates that the surface must display the same potential value at all nodes. However, this value is not imposed. It is computed.

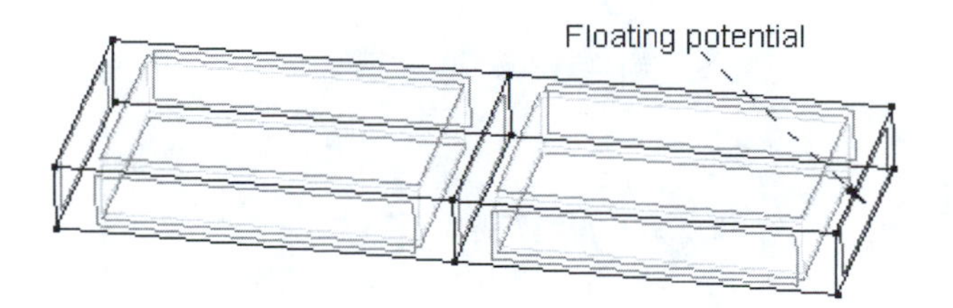

You can verify that all potentials are correctly applied by selecting **Draw | Colors**.

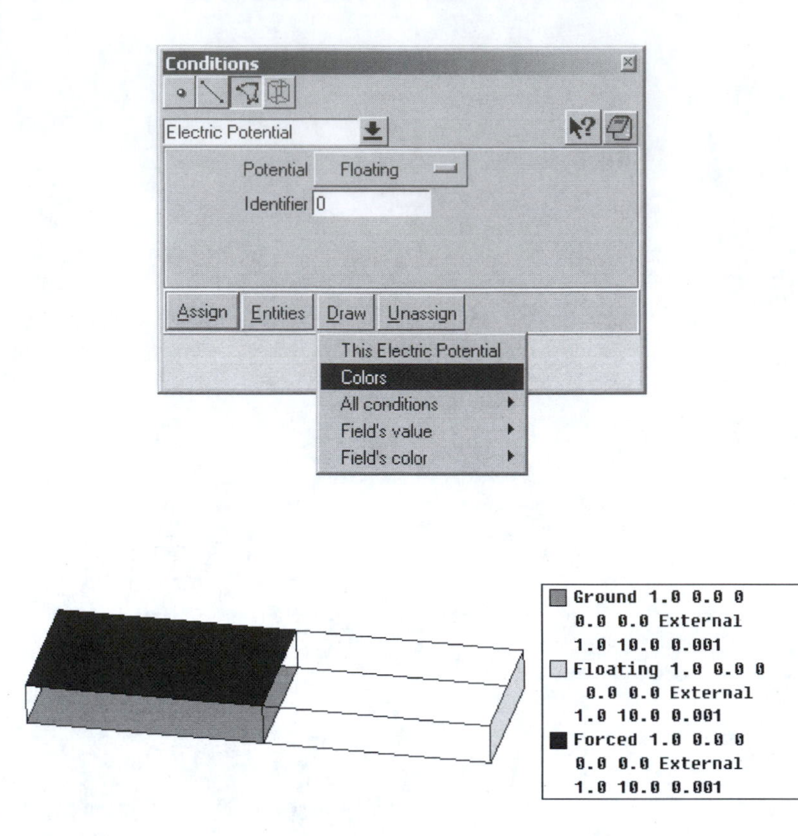

In this last picture, it is easy to see where the ground, floating, and forced potentials are applied. The other values displayed here are unimportant, except for the two immediately following the word "Forced." These values, 1.0 and 0.0, represent the amplitude and phase of the excitation signal, respectively.

Polarization/Local Axes

The next step consists of defining the required polarizations. Here, we will consider the polarization to be uniform in each section of the transformer. Therefore, we will apply the polarization named **Cartesian**. The polarization in the input section is oriented in the thickness direction, whereas that of the output section is oriented along the length of the transformer such that each section of the transformer operates in the so-called d_{33} mode.

The definition of the polarizations is accomplished by using local coordinate systems, also called **Local Axes**. To understand how the polarization is defined with respect to local axes, please read the Help files about the ATILA–GiD interface (under the **ATILA | Help** menu).

Select the **Volume** tab of the conditions window, then **Polarization** from the drop-down list. Select **Local-Axes | Define** to create a polarization.

Enter *P_input* for the name of the local axes, and select, for instance, the **3 Points XZ** definition mode.

Here, it is convenient to select three points of the geometry to define the local coordinate system, but remember that point coordinates can also be typed in the command line.
Next, define the polarization for the output section, *P_output* by following the same procedure.

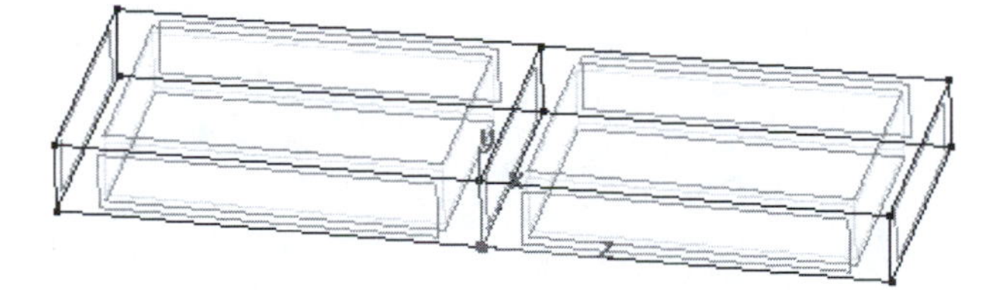

Next, assign the polarizations that you have just created. Select the **Local Axes** name and then **Assign** to the appropriate volume. For the *P_input* polarization, first,

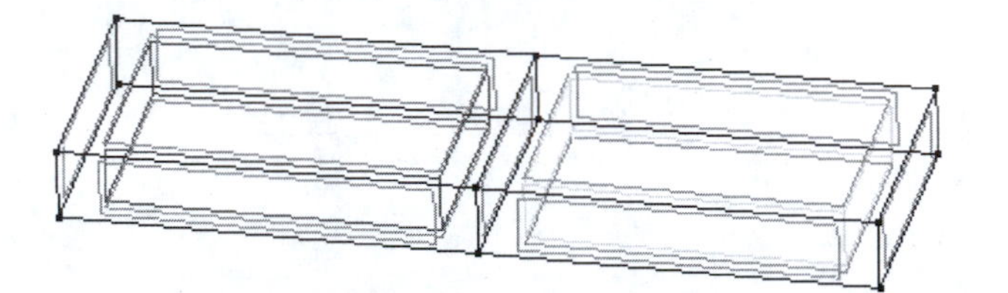

and the *P_output* polarization,

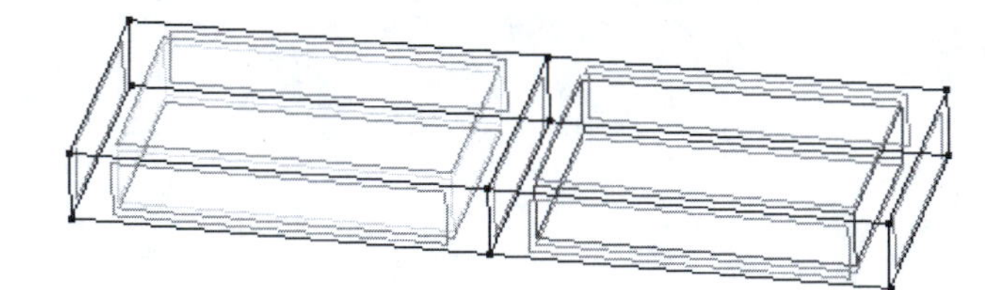

To review the polarizations, select **Draw | This Polarization | Exclude Local Axes**.

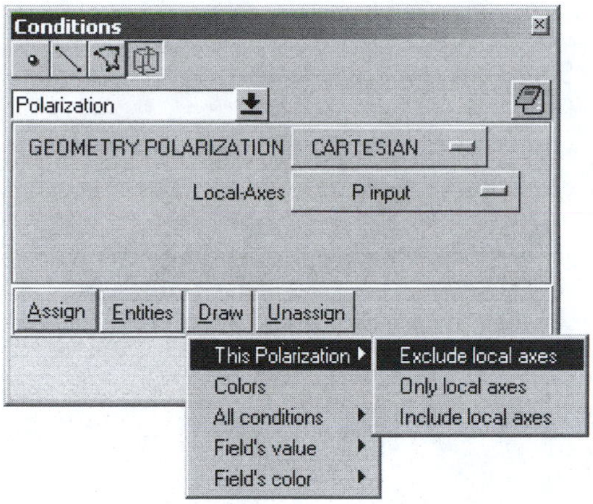

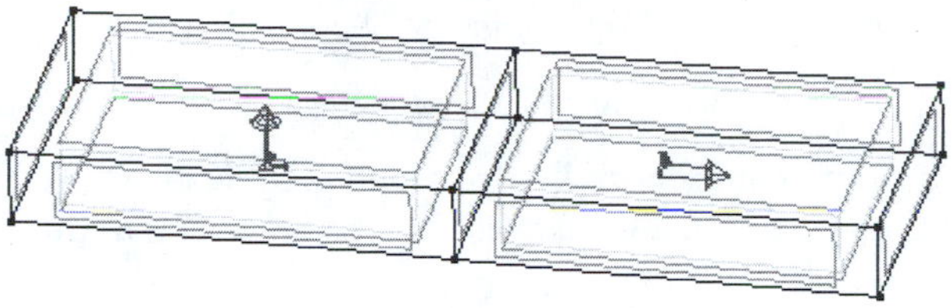

Finally, it is always a good idea to verify all the boundary conditions. Select **Draw | All Conditions | Exclude Local Axes**.

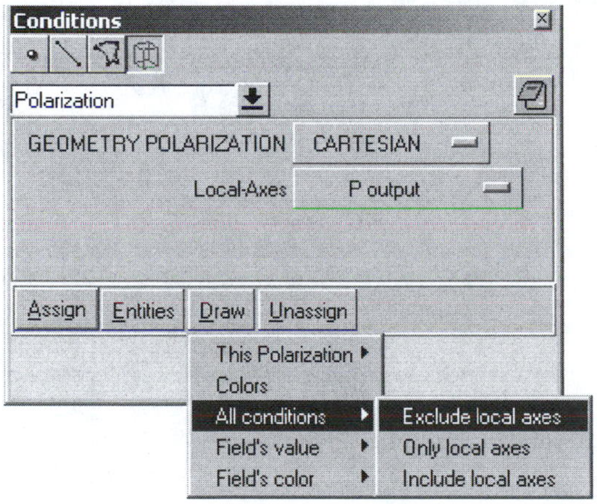

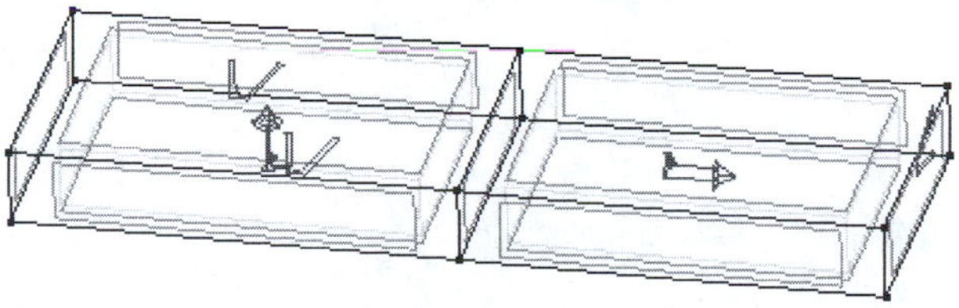

In this last drawing, the **V** indicates that an electrical potential condition is applied. The **arrows** show the polarizations.

Back to Top

4.1 ROSEN-TYPE TRANSFORMER (3)

MESHING

PROBLEM—Rosen Transformer 25 mm × 8 mm × 2.2 mm t

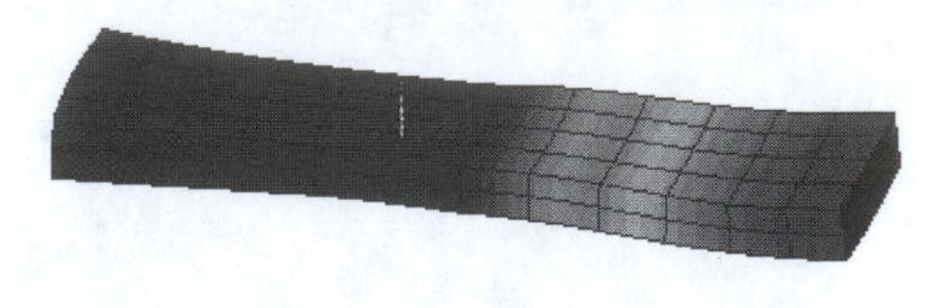

MESHING

Mapped Mesh

Meshing -> Structured -> Volumes -> Select all volumes -> Then Cancel the further process.
Meshing -> Structured -> Lines -> Enter number to divide line -> Choose "2, 4, or 6" for the lines -> Generate.

Use a greater number such as 4 or 6 for more accurate calculation.

ATILA requires that the mesh be composed exclusively of second-order elements. The corresponding setting can be found under the **Meshing** menu. Select **Quadratic Elements**, then **Quadratic**. Note that this option is automatically set when you start the ATILA–GiD interface.
For the transformer, we will define a structured mesh. Select **Structured | Volumes** from the **Meshing** menu.

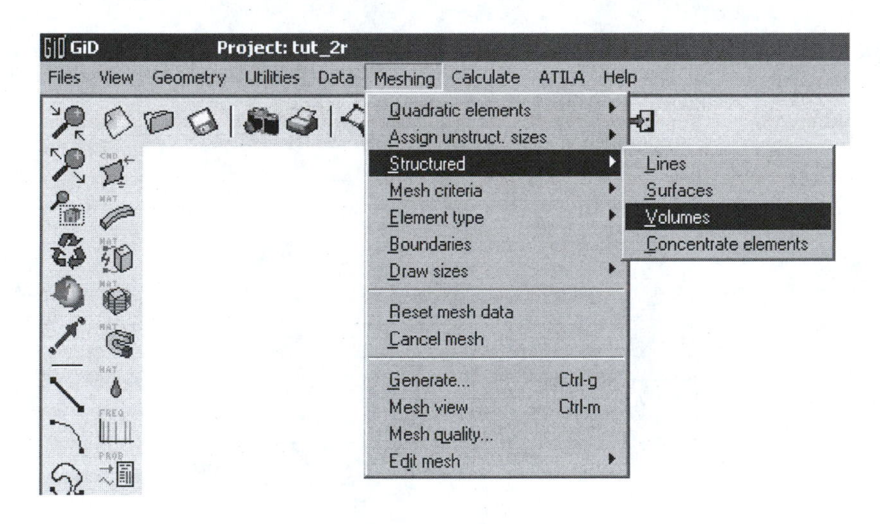

Apply this meshing condition to all the volumes in the geometry.

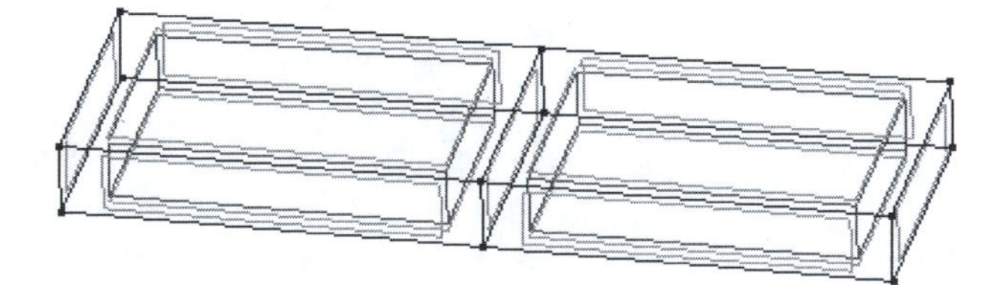

Then, define the number of subdivisions on all lines. Define two subdivisions along the thickness of the transformer.

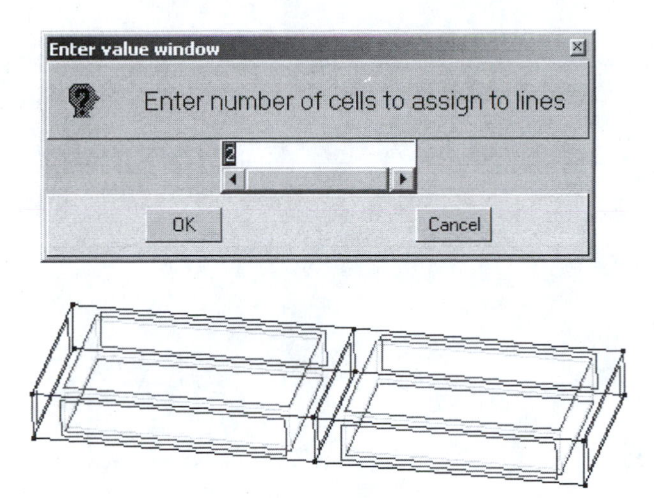

Next, impose four subdivisions in the width of the transformer.

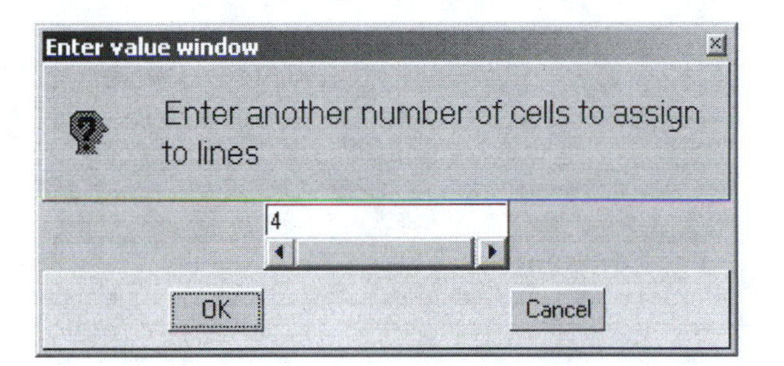

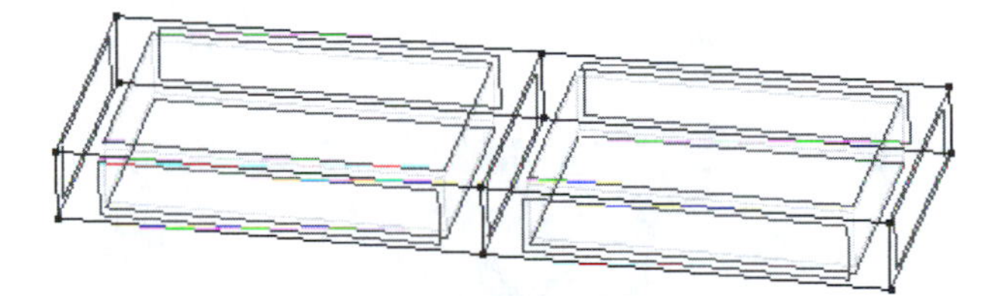

And finally, define six subdivisions along the length of each section of the transformer.

Enter value window

Enter another number of cells to assign to lines

6

OK Cancel

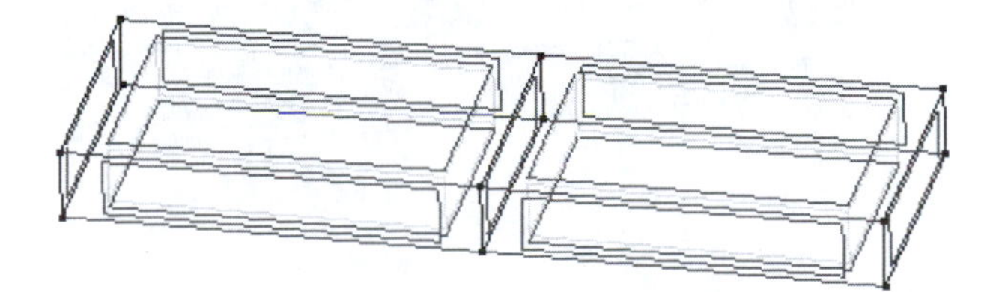

To construct the mesh, select the **Meshing** command from the **Meshing** menu.

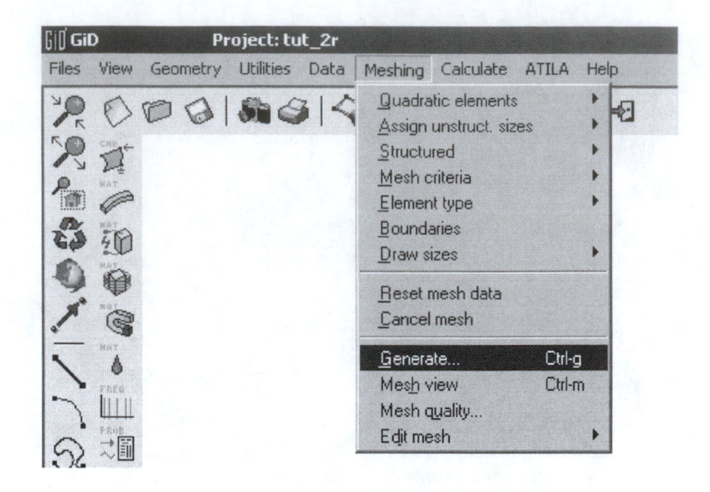

GiD asks for a default element size, which is here of no importance, because the whole geometry will receive a structured mesh. You can therefore use any arbitrary value—the mesh will be the same.

Enter value window

Enter size of elements to be generated

2.6

OK Cancel

When the meshing is complete, GiD displays information about the number of nodes and elements that were generated.

Dialog window

Mesh Generated. Press 'OK' to see it

Num. of Hexahedra elements=96
Num. of nodes=661

OK Cancel

The resulting mesh is shown in the next figure.

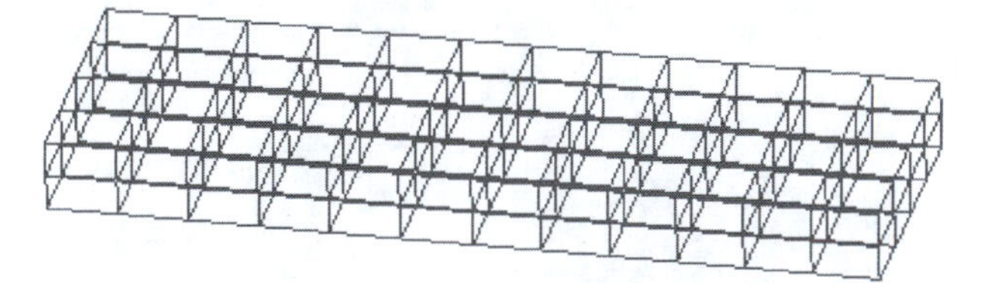

Back to Top

Back to Home

4.1 ROSEN-TYPE TRANSFORMER (4)

PROBLEM TYPE [VIEW RESULTS

PROBLEM—Rosen Transformer 25 mm × 8 mm × 2.2 mm t

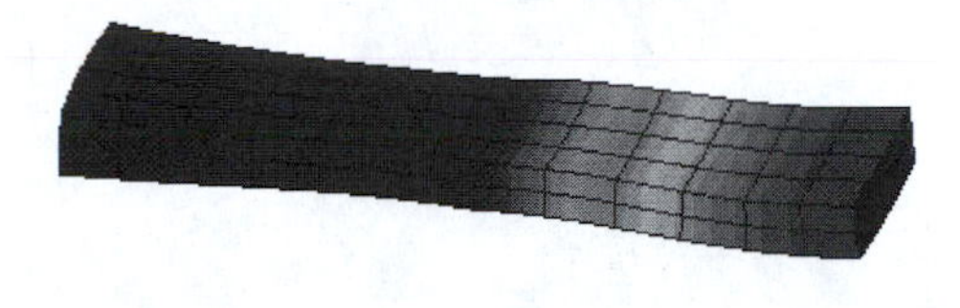

PROBLEM TYPE—Modal Analysis

Data -> Problem data -> 3D, Modal Resantires.

We wish to perform a modal analysis of the transformer to identify its resonance frequency. With the boundary conditions that we have defined, the resonance mode will correspond to a low-signal excitation at the input side of the transformer and open-circuit conditions at the output side.

You can use any **Printing** value. This will not affect the results, but only the amount of information printed in the LST file generated by ATILA. The **Geometry** should clearly be set to **3D**. The **Analysis** that we wish to perform here is a **Modal Resantires** because this particular type of modal analysis takes into account the piezoelectric properties of the material. In particular, it computes the resonance modes as well as the antiresonance modes of the structure. Finally, the **Number of Modes** is set arbitrarily. However, because the model consists of a 3D structure that is free of mechanical conditions, the first six modes computed by the modal analysis will correspond to the six rigid-body modes of the structure. It is therefore recommended to use a number greater than six. Here, we pick 20.

Problem Data
Problem Data | Mesh Units
PRINTING 0
GEOMETRY 3D
ANALYSIS MODAL RESANTIRES
NUMBER OF MODES 20
Accept data | Close

Note that the mesh density limits the number of modes that can be accurately computed. With the mesh we have defined, we cannot expect to accurately compute modes that display more than half a wavelength in the thickness direction, or more than one wavelength in the width direction, or more than three wavelengths in the length direction.
Note also that computation time increases geometrically with mesh density and with the number of modes.

Finally, make sure the **Mesh Units** are set to millimeters.

Do not forget to validate your changes by clicking on **Accept Data**. Then **File | Save** the problem.

CALCULATION—Modal Analysis

Calculate -> Calculate window -> Start.

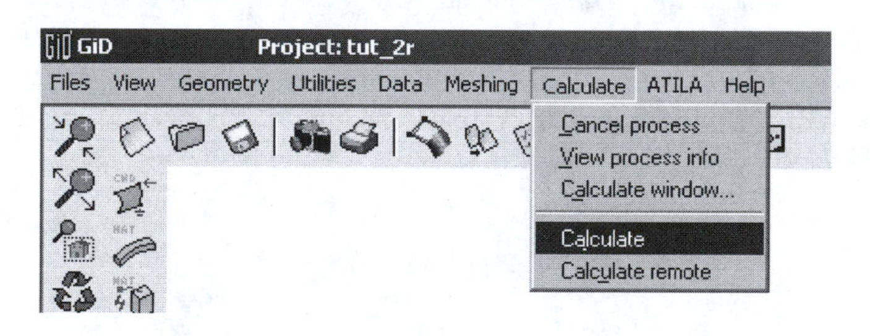

You can use the **Calculate Window** as well as **View Process Info** commands to monitor the progression of the computation.

Note: Because we have set **Printing = 0** in this tutorial (**Problem Data** section), very little information will be displayed in the **View Process Info** window. To obtain more details about the progression of the computation, set **Printing** to 1.

POST PROCESS/VIEW RESULTS—Modal Analysis

Click to Open the Post Calculation (Flavia) Domain

Once the computation is completed, go to the postprocessing environment in GiD.

To see the list of frequencies, click on

The list of frequencies for the resonance and antiresonance modes, as well as electromechanical coupling factors, is shown as follows:

Results

```
Title, Resonance & antiresonance frequencies,,
Mode, Resonance Frequency (Hz), Antiresonance Frequency (Hz),Coupling coefficient (%)
1, -1.601846e-02, -1.631154e-02, 1.887120e+01
2, -8.627961e-03, -8.639847e-03, 5.243509e+00
3, -2.871154e-03, -1.820654e-03, 0.000000e+00
4, 4.988846e-03, 5.490692e-03, 4.176663e+01
5, 7.740423e-03, 7.715846e-03, 0.000000e+00
6, 1.413346e-02, 1.360346e-02, 0.000000e+00
7, 1.127962e+04, 1.128769e+04, 3.782139e+00
8, 1.882577e+04, 1.882577e+04, 0.000000e+00
9, 3.021462e+04, 3.021615e+04, 1.009295e+00
10, 3.201423e+04, 3.201423e+04, 0.000000e+00
11, 3.879230e+04, 3.879230e+04, 0.000000e+00
12, 5.596077e+04, 5.598346e+04, 2.846897e+00
13, 6.110116e+04, 6.110116e+04, 0.000000e+00
14, 6.638962e+04, 6.638962e+04, 0.000000e+00
15, 7.115500e+04, 7.299769e+04, 2.232688e+01
16, 8.627654e+04, 8.627731e+04, 4.234268e-01
17, 8.661308e+04, 8.662538e+04, 1.685436e+00
18, 1.029192e+05, 1.029231e+05, 8.642355e-01
19, 1.036500e+05, 1.036500e+05, 0.000000e+00
20, 1.126885e+05, 1.126962e+05, 1.168601e+00
```

Note that the first six modes clearly correspond to rigid body modes. Then, in the preceding figure we have highlighted one mode (at 71.15 kHz) that displays a high electromechanical coupling coefficient (22.3%). This mode is most likely the $\lambda/2$ longitudinal mode.

Note that the frequency list is available in a comma-separated value (CSV) format in the problem directory (use **ATILA | Open Shell** to directly access it).

This is easy to verify by displaying the deformation. For instance, from the GiD **Deformation** and **View Results** windows, select the resonance mode at 71.15 kHz, and display the mode shape.

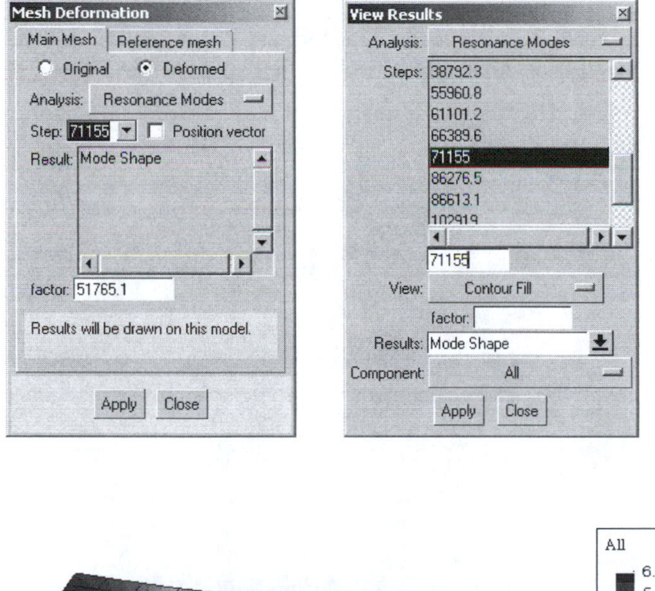

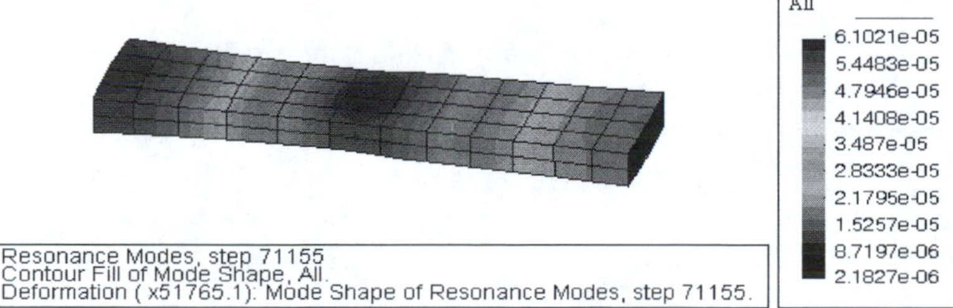

Note that the values obtained for the displacement are only relative values. They do not represent absolute displacement values. To obtain displacement values at a specific frequency, it is necessary to perform a harmonic computation.

PROBLEM TYPE—Harmonic Analysis

Data -> Problem data -> 3D, Harmonic, Click "Include Losses."

We now want to compute the deformation and voltage gain at resonance.
Return to the preprocessing environment and save the problem under a new name, using **File | Save_As**. Change the **Problem Data** so that the **Analysis is** now **Harmonic**. Make sure **Losses** is checked so that losses are taken into account for this computation. Select **Accept Data** to validate your changes.

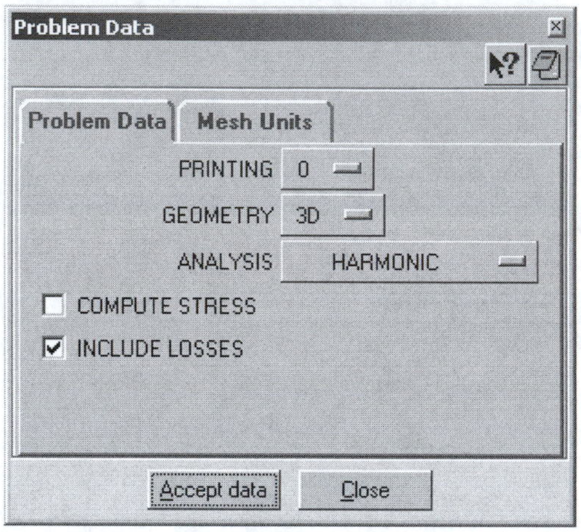

Click on to define the frequency intervals for the harmonic computation.

Here, we will use only one interval centered on the resonance frequency that was previously computed at about 71 kHz. Therefore, we will define a frequency sweep between 70 and 72 kHz, with increments of 40 Hz.
Click **Accept Data** to validate your changes.

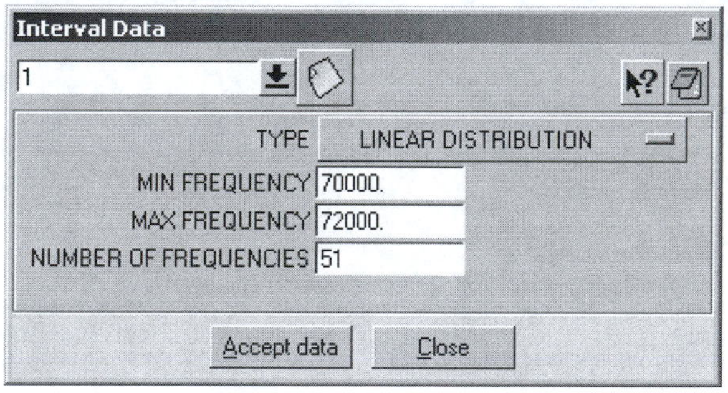

CALCULATION—Harmonic Analysis

In this problem, we are not changing any of the boundary conditions. We are still applying a 1 V amplitude signal to the input section and letting the output in open-circuit conditions. Therefore, we do not need to regenerate the mesh.
Make sure you save your problem (**File | Save**), and start the computation (**Calculate | Calculate**).

POST PROCESS/VIEW RESULTS

Click to Open the Post Calculation (Flavia) Domain.

CLICK ADMITTANCE/IMPEDANCE RESULT

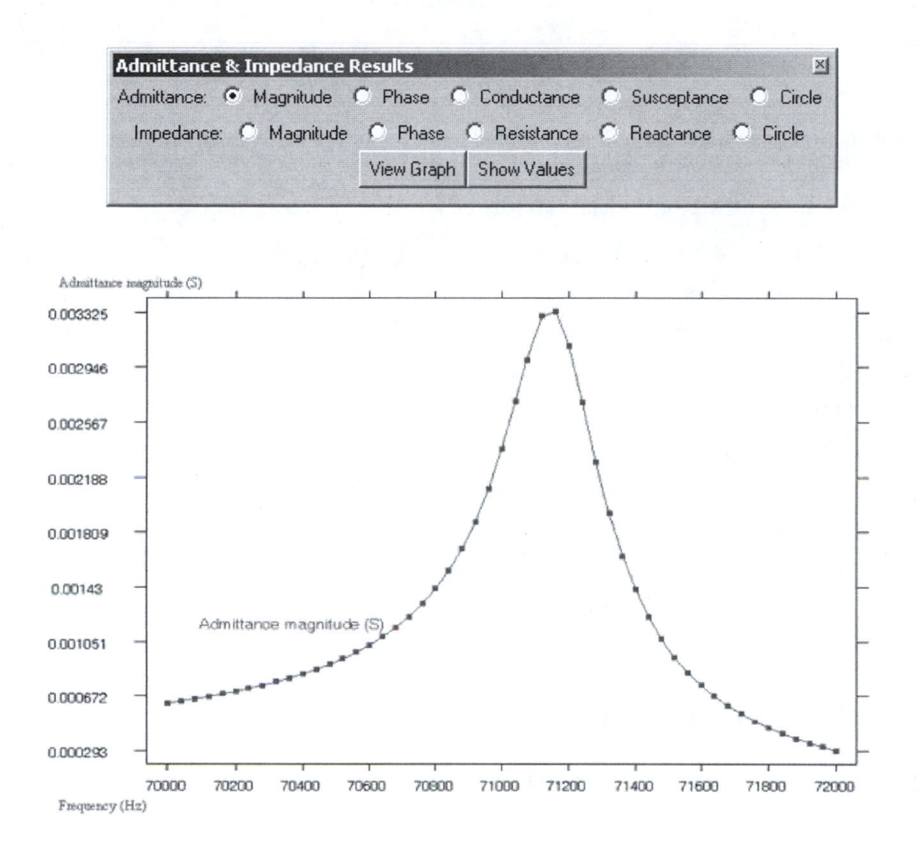

Then, because we have used an excitation of amplitude 1 V, displaying the electrical potential shows directly the transformation gain.

Mesh Deformation
Main Mesh | Reference mesh
Original | Deformed
Analysis: Harmonic-Real Part
Step: 71160 | Position vector
Result: Displacement
factor: 1.07032e+009
Results will be drawn on this model.
Apply | Close

View Results
Analysis: Harmonic-Magnitude
Steps: 71120, 71160, 71200, 71240, 71280, 71320, 71360, 71400
71160
View: Contour Fill
factor:
Results: Potential
Component: Potential
Apply | Close

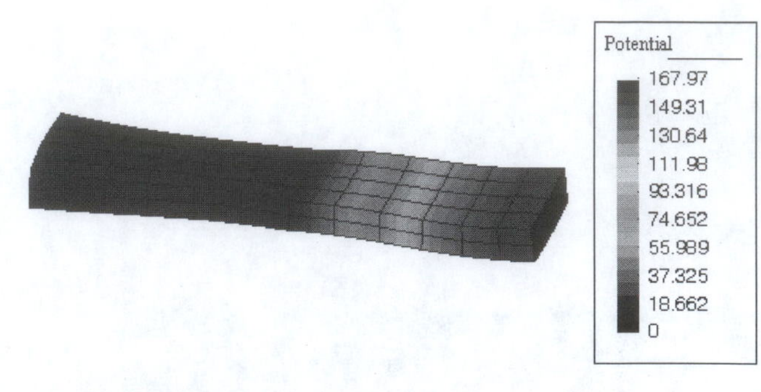

Note that we have displayed the magnitude of the potential on the drawing of the real part of the structure deformation.

View Animation—Vibration Mode at a Particular Frequency

View results -> Default Analysis/Step -> Harmonic-Magnitude -> Set a Frequency
(Peak Point: 71,160 Hz for Resonance Mode).

Double Click Loading frequency: 71,160 Hz Harmonic Case -> Click OK (single-click and wait for a while).

Set Deformation -> Displacement

View results -> Deformation -> Displacement.

Parameter other than Displacement

Set Results View -> View results -> Contour Fill -> Electric Potential.
After setting both "Results view" and "Deformation" -> Click to see the Animation.
Change the page.

4.2 RING-DOT-TYPE TRANSFORMER (1)

GEOMETRY/DRAWING MATERIAL ASSIGNMENT [12.81]

PROBLEM—Ring-Dot-Type Transformer with PZT8, INPUT: Ring = 20 mm outer Φ-12 mm inner Φ × 1 mm t, OUTPUT: Dot = 10 mm × _ 1 mm t

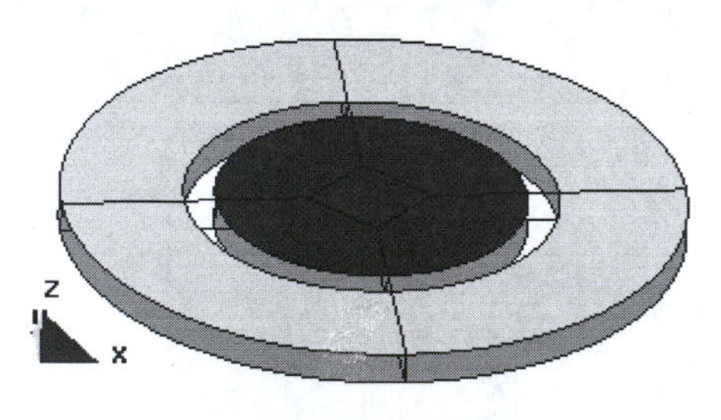

Ground 1.0 0.0 0 0.0 0.0 0.0 External 1.0 10.0 0.001
Forced 1.0 0.0 0 0.0 0.0 0.0 External 1.0 10.0 0.001
Floating 1.0 0.0 0 0.0 0.0 0.0 External 1.0 10.0 0.001

- Transformer with higher *k* and step-up voltage ratio
- Review how the center portion of the circle is treated in meshing

GEOMETRY/DRAWING

Create Circles

Surface : Geometry -> Create -> Object -> Circle -> Enter Center -> 0, 0 -> Normal to Z -> Radius -> 10, 6, and 5, respectively. Then, Geometry -> Delete -> Surfaces.

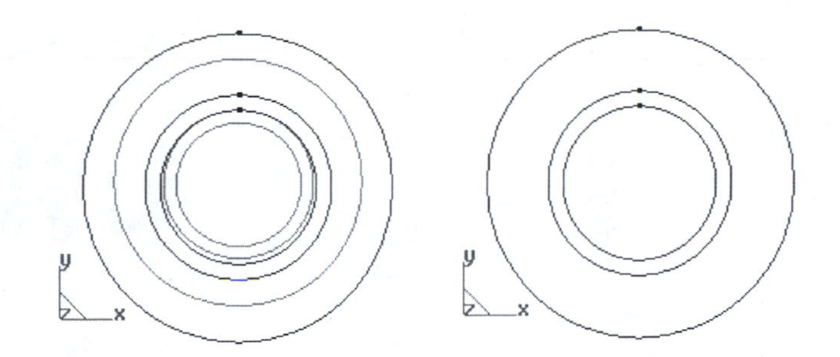

Create Electrode Surface

Geometry -> Edit -> Divide -> Num Divisions -> Select three circle lines -> Number of Division -> 4.

Create Center Portion Area

Geometry -> Create -> Lines -> Enter Point-> 2,0 -> 0,2 -> –2,0 -> 0,–2 -> 2,0 -> Joint -> Escape.

Segment the Surfaces

Geometry -> Create -> Lines -> Choose the Points -> Finally, you obtain the 4-sided rectangles for all areas.

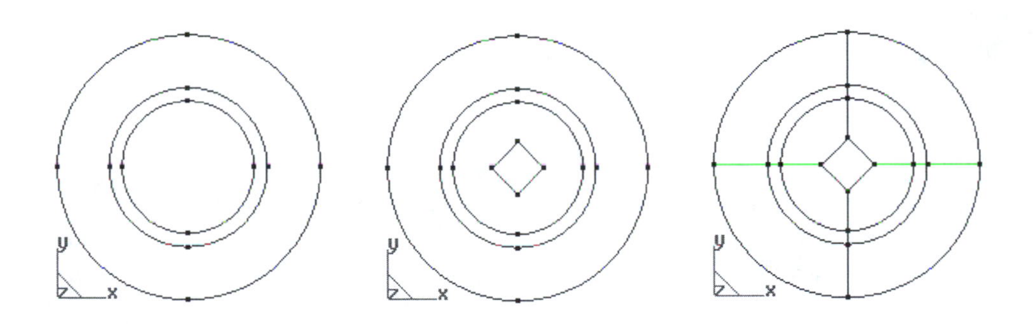

Create the Surfaces

Geometry -> Create -> NURBS Surfaces by Automatic -> Enter Number of Lines -> 4.

Create Volumes

Utilities -> Copy -> Surfaces -> Extrude Surfaces by 1 mm -> Select.
Geometry -> Create -> Volume -> Automatic 6-sided volumes -> 13 Volumes appear in blue lines.

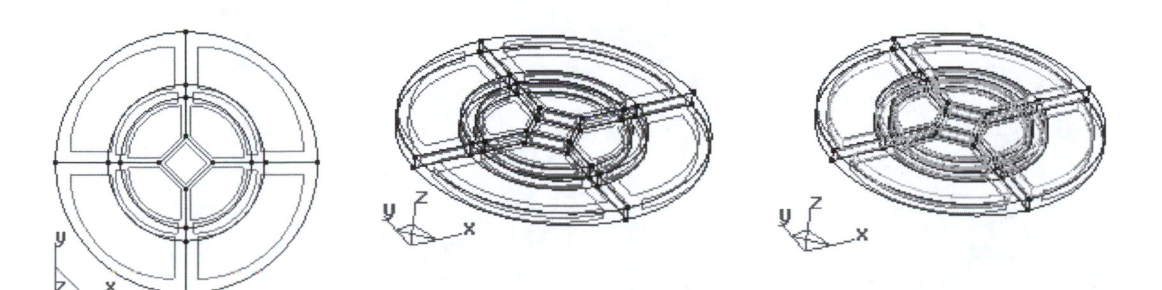

MATERIAL ASSIGNMENT

Piezoelectric Materials

Data -> Materials -> Piezoelectric -> PZT8 -> Assign -> Pick all Volumes.
Click Draw -> All materials -> Check the material is assigned as PZT4.

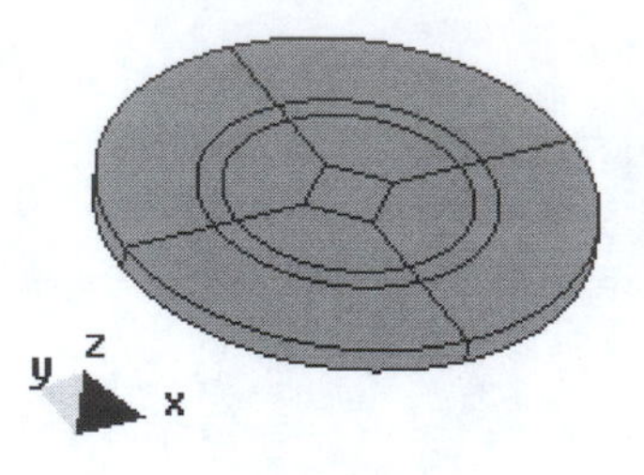

NB: PZT8 is a *hard* PZT; PZT5A and PZT5AH are *soft* PZTs; and PZT4 exhibits an intermediate performance. Soft and hard PZTs are used for off-resonance- and resonance-type actuators, while *intermediate/semi-soft* PZTs are used for transformers and transducers in general.

Back to Top

Back to Home

4.2 RING-DOT-TYPE TRANSFORMER (2)

BOUNDARY CONDITIONS MESHING

PROBLEM—Ring-Dot-Type Transformer with PZT8, INPUT: Ring = 20 mm outer Φ -12 mm inner Φ × 1 mm t, OUTPUT: Dot = 10 mm × 1 mm t

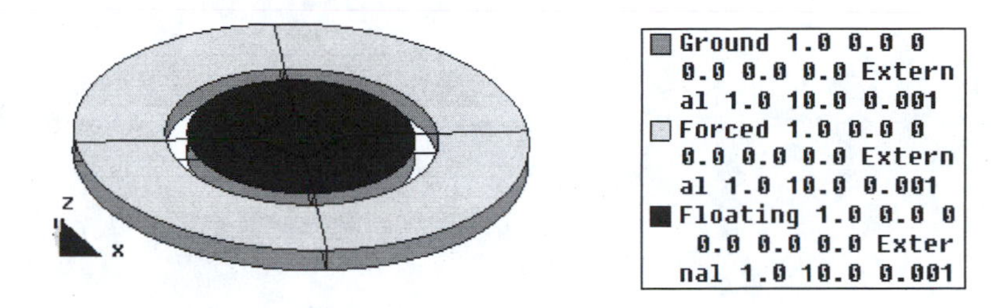

BOUNDARY CONDITIONS

Polarization/Local Axes

Data -> Conditions -> Volumes -> Geometry -> Polarization (Cartesian) -> Define Polarization (P1) (0, 0, 1) direction.

Choose 3PointsXZ -> Center (0,0,0) -> Define X axis by putting (1,0,0) -> Then, define Z axis by putting (0, 0, 1)-> Escape.

Assign Polarization (Local-Axes P1) -> Choose all the PZT volumes -> Finish.

Potential

Data -> Conditions -> Surfaces -> Potential 0.0 (or Ground in ATILA 5.2.4 or higher version) -> Assign the bottom (Ring and Dot) 9 surfaces of the PZT.

Forced Potential 1.0 -> Assign the top Ring 4 surfaces.

Finally, Floating -> Assign the top Dot 5 surfaces.

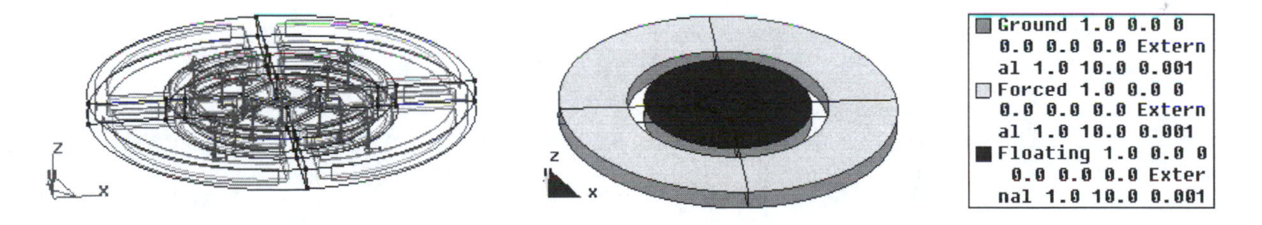

NB: Instead of "Floating," "Load," and "Resistance," either "Capcitance or "Inductance" can be chosen in a higher version. Step-up voltage ratio versus resistance (from 1GΩ down to 1KΩ) can be obtained.

MESHING

Mapped Mesh

Meshing -> Structured -> Volumes -> Select all volumes -> Then Cancel the further process.
Meshing -> Structured -> Lines -> Enter number to divide line -> Choose “1” for the lines -> Generate.
Use a larger number such as 2 for more accurate calculation.

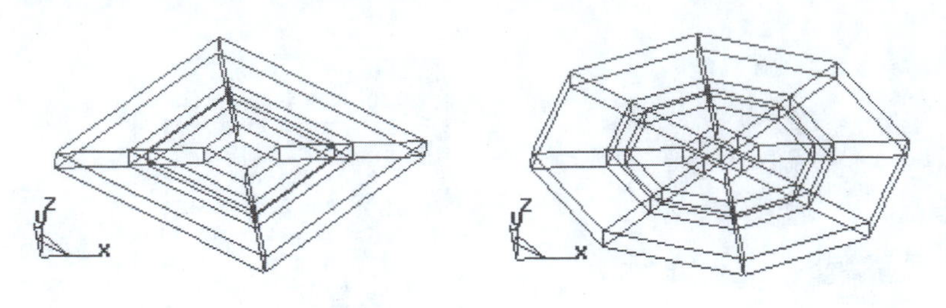

Back to Top

Back to Home

4.2 RING-DOT-TYPE TRANSFORMER (3)

PROBLEM TYPE VIEW RESULTS

PROBLEM—Ring-Dot-Type Transformer with PZT8, INPUT: Ring = 20 mm outer Φ -12 mm inner Φ × 1 mm t, OUTPUT: Dot = 10 mm × 1 mm t

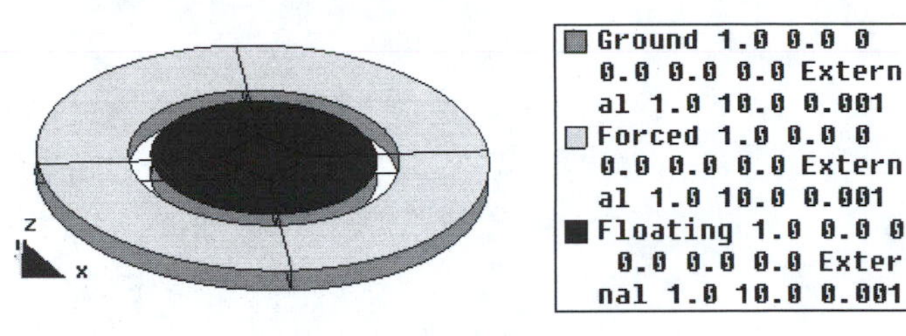

- We use Modal Analysis to find the first resonance frequency, and then Harmonic Analysis to obtain the Impedance curve.

PROBLEM TYPE

Data -> Problem data -> 3D, Modal Resantires, Number of Modes 30.

CALCULATION

Calculate -> Calculate window -> Start.

POST PROCESS/VIEW RESULTS

Click to Open the Post Calculation (Flavia) Domain.

Click to obtain all the Resonance and Antiresonance frequencies. The first six modes are trivial translation modes (very low resonance frequencies). The first high coupling coefficient, 31%, can be found at 24. 131 kHz, which corresponds to the first meaningful electromechanical resonance.

Problem Data

Problem Data | Mesh Units

PRINTING 0
GEOMETRY 3D
ANALYSIS MODAL RESANTIRES
NUMBER OF MODES 30

Accept data Close

Results

Title, Resonance & antiresonance frequencies,,
Mode, Resonance Frequency (Hz), Antiresonance Frequency (Hz),Coupling coeffi
1, -1.885154e-02, -1.902000e-02, 1.328001e+01
2, -1.251808e-02, -8.289577e-03, 0.000000e+00
3, -6.994308e-03, -6.671269e-03, 0.000000e+00
4, -4.973115e-03, -5.491615e-03, 4.241683e+01
5, 5.906385e-03, 6.011461e-03, 1.861537e+01
6, 8.505962e-03, 1.227808e-02, 7.211525e+01
7, 1.120654e+04, 1.120654e+04, 0.000000e+00
8, 1.757346e+04, 1.757346e+04, 0.000000e+00
9, 2.751769e+04, 2.751769e+04, 0.000000e+00
10, 3.982384e+04, 3.982384e+04, 0.000000e+00
11, 3.982384e+04, 3.982384e+04, 0.000000e+00
12, 6.758000e+04, 6.758000e+04, 0.000000e+00
13, 6.758000e+04, 6.758000e+04, 0.000000e+00
14, 7.441462e+04, 7.441770e+04, 9.095269e-01
15, 8.195384e+04, 8.195384e+04, 0.000000e+00
16, 8.202308e+04, 8.202308e+04, 0.000000e+00
17, 8.259346e+04, 8.259346e+04, 0.000000e+00
18, 8.337500e+04, 8.337500e+04, 0.000000e+00
19, 9.711000e+04, 9.711000e+04, 0.000000e+00
20, 9.711000e+04, 9.711000e+04, 0.000000e+00
21, 1.050308e+05, 1.050308e+05, 0.000000e+00
22, 1.193731e+05, 1.193731e+05, 0.000000e+00
23, 1.193731e+05, 1.193731e+05, 0.000000e+00
24, 1.311731e+05, 1.381308e+05, 3.133743e+01
25, 1.431000e+05, 1.431000e+05, 0.000000e+00
26, 1.431038e+05, 1.431038e+05, 0.000000e+00
27, 1.617308e+05, 1.617308e+05, 0.000000e+00
28, 1.619346e+05, 1.619346e+05, 0.000000e+00
29, 1.703846e+05, 1.703385e+05, 0.000000e+00
30, 1.756308e+05, 1.756538e+05, 1.620411e+00

PROBLEM TYPE

Data -> Problem data -> 3D, Harmonic, Click both Compute Stress and Include Losses.

FREQUENCY

Data -> Harmonic -> Interval 2 (120,000–140,000 Hz).

CALCULATION

Calculate -> Calculate window -> Start.

POST PROCESS/VIEW RESULTS

Click to Open the Post Calculation (Flavia) Domain.

Click Admittance/Impedance Result

Admittance Peak Point: 130,000 Hz -> This peak corresponds to the first resonance.

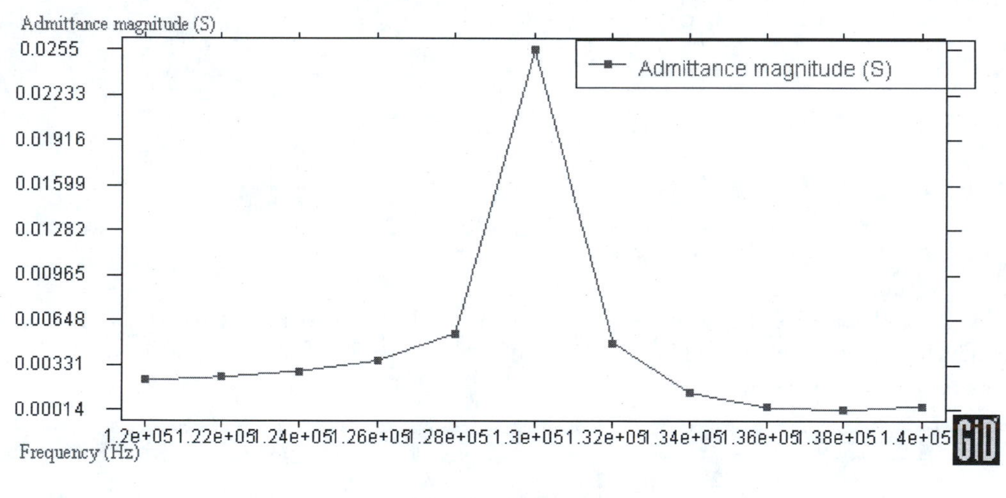

Admittance Curve

View Animation—Vibration Mode at a Particular Frequency

View results -> Default Analysis/Step -> Harmonic-Magnitude -> Set a Frequency (Peak Point: 130,000 Hz for Resonance Mode).

Double Click Loading frequency: 130,000 Hz Harmonic Case -> Click OK (single-click and wait for a while).

Set Deformation -> Displacement

View results -> Deformation -> Displacement.

Parameter other than Displacement

Set Results View -> View results -> Contour Fill -> Electric Potential or Von Mises Stress.

After setting both "Results view" and "Deformation" -> Click [button] to see the Animation.

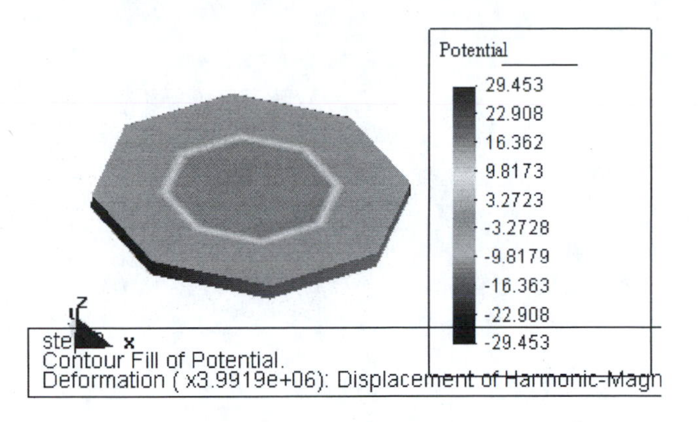

130,000 Hz - First Resonance

Under 1 V applied on the Ring input portion, we obtain 30 V on the Dot output portion -> Step-up voltage ratio under no load (open) is 30.

Back to Top

5 Ultrasonic Motor

- How to apply sine and cosine voltages

TEXT CONTENT

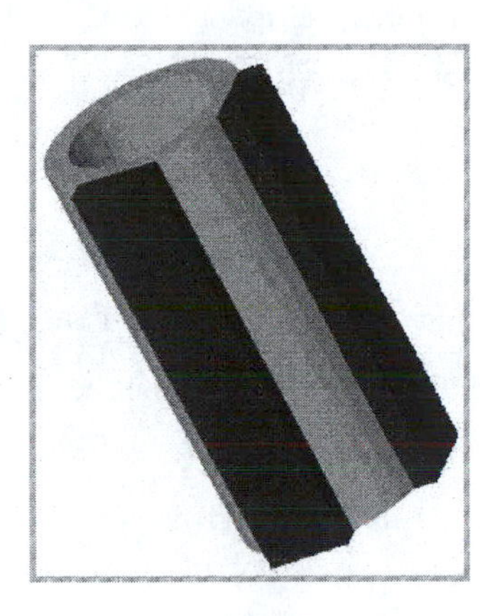

Metal-Tube Motor

⇨	**5.1 L-Shape Motor**
⇨	**5.2 π-Shape Motor**
⇨	**5.3 Metal-Tube Motor**
⇨	**Homework 7** Surface Wave Motor
⇨	**Homework 8** Π-Shape Motor
How To Use this GiD file	**Example GiD File** L-Shape.gid, LBMotor, MTube.gid

Back to Top

5.1 L-SHAPED USM (1)

GEOMETRY/DRAWING MATERIAL ASSIGNMENT

PROBLEM - L-Shaped USM

2 Legs: 8 mm L × 2 mm W × 1 mm t
Corner: 2 mm × 2 mm × 1 mm t (PZT4)
Clamped at two leg ends

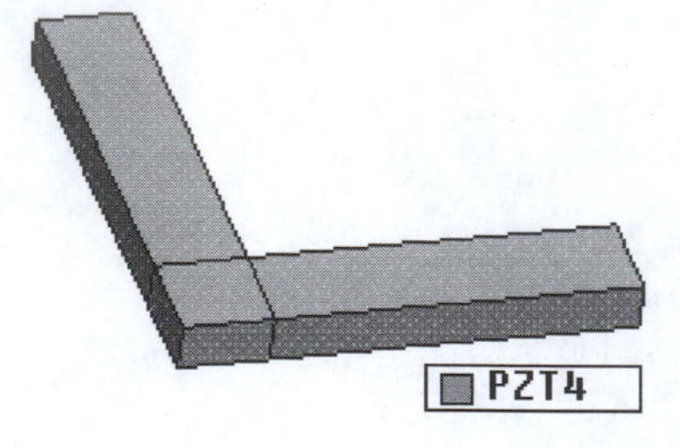

- Learn the ultrasonic motor principle—generate an elliptical locus.
- Learn how to apply sine and cosine voltages simultaneously.

GEOMETRY/DRAWING

Create two "Legs"

Surfaces: Geometry -> Create -> Object -> Rectangle -> Enter First and Second Corner Points -> 2, 2 and 10, 0: 2, 2 and 0, 10.

Make sure you click "Join" for 2, 2. Otherwise, these two legs will not be correctly connected.

Create Line

Surface: Geometry -> Create -> Line -> 2, 0 -> 0, 0 -> 0, 2.

Geometry -> Create -> NURB Surfaces -> Automatic -> Number of sides 4.

Create Volume

Copy/Transform -> Translation -> 1 mm along the *Z* axis with Surface Extrusion.

Geometry -> Create -> Volume -> Automatic 6-sided volumes -> Create three Volumes.

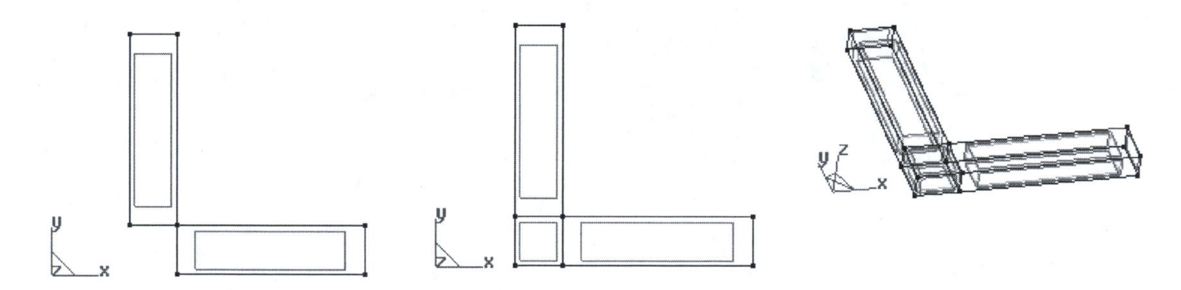

MATERIAL ASSIGNMENT

PIEZOELECTRIC MATERIALS 

Data -> Materials -> Piezoelectric -> PZT4 -> Assign -> Pick all Volumes.
Click Draw -> All materials -> Check the material is assigned as PZT4.

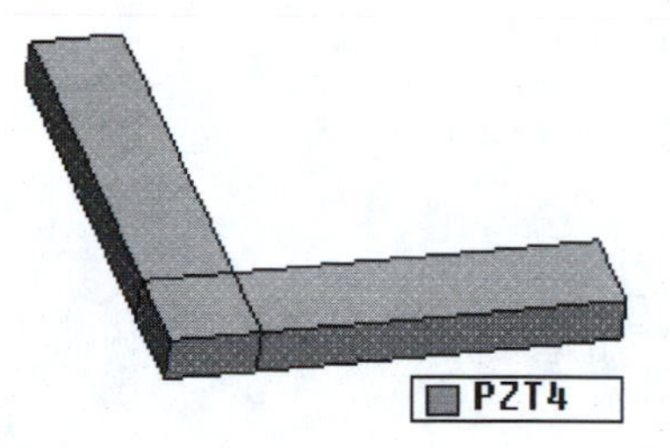

Back to Top

Back to Home

5.1 L-SHAPED USM (2)

BOUNDARY CONDITIONS MESHING

PROBLEM - L-Shaped USM

2 Legs: 8 mm L × 2 mm W × 1 mm t
Corner: 2 mm × 2 mm × 1 mm t (PZT4)
Clamped at two leg ends

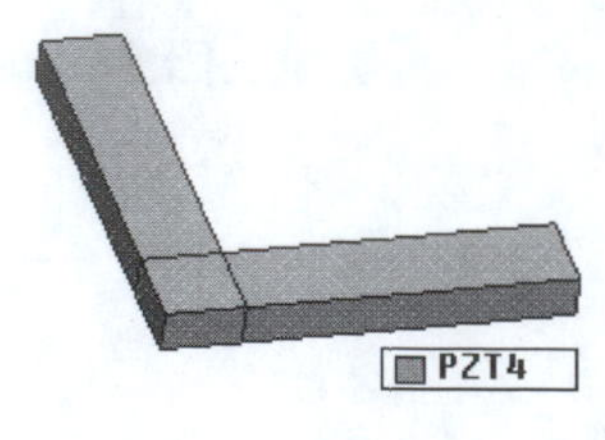

- Learn the ultrasonic motor principle—generate an elliptical locus.
- Learn how to apply sine and cosine voltages simultaneously.

BOUNDARY CONDITIONS

Polarization/Local Axes

Data -> Conditions -> Volumes -> Geometry -> Polarization (Cartesian) -> Define Polarization (P1) (0,0,-1) direction.

Choose 3PointsXZ -> Center (0, 0, 0) -> Define X axis by putting (1, 0, 0) -> Then define Z axis by putting (0, 0, 1) -> Escape.

Assign Polarization (Local-Axes P1) -> Choose the PZT all three volumes -> Finish.

Potential

Data -> Conditions -> Surfaces -> Electric Potential -> Ground -> Assign the bottom two surfaces of the PZT.

Data -> Conditions -> Forced Potential 1.0 with phase 0 -> Assign the top right-side surface.
Data -> Conditions -> Forced Potential 1.0 with phase 90 -> Assign the top left-side surface.

This is how sine and cosine are applied on PZT plates.

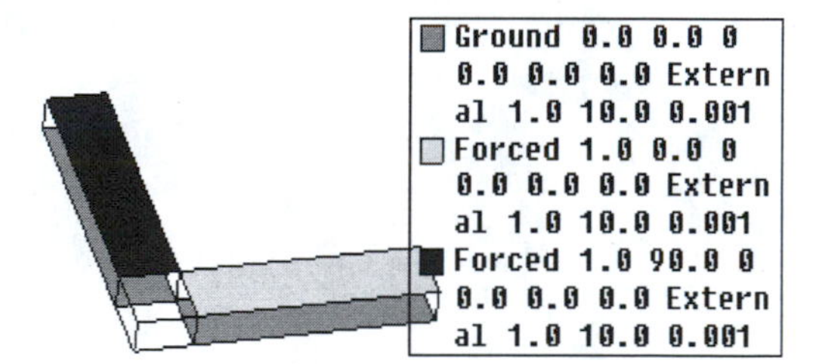

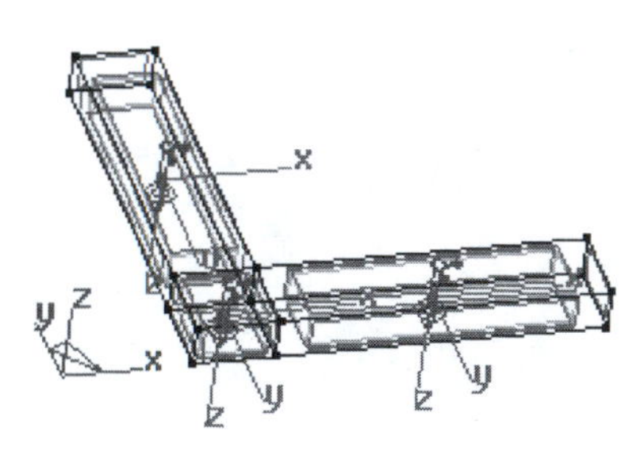

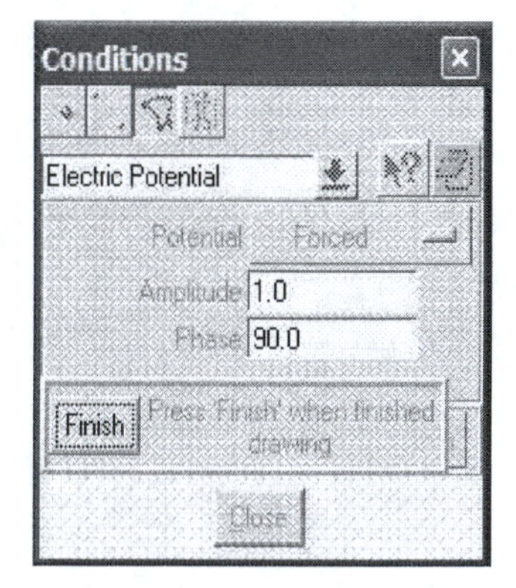

Surface Constraints (Two "Leg" ends are clamped)

Data -> Conditions -> Surfaces -> Surface Constraints -> X, Y, and Z components -> Clamped -> Assign the PZT Leg End surfaces.

To check the assignment -> Draw -> All conditions -> Include local axes.

MESHING

Cant view is not very suitable for Meshing -> View -> Rotate -> Plane xy (original).

Mapped Mesh

Meshing -> Structured -> Volumes -> Select all volumes -> Then Cancel the further process.
Meshing -> Structured -> Lines -> Enter number to divide line -> Choose "1, 2, and 4" for the lines -> Generate.

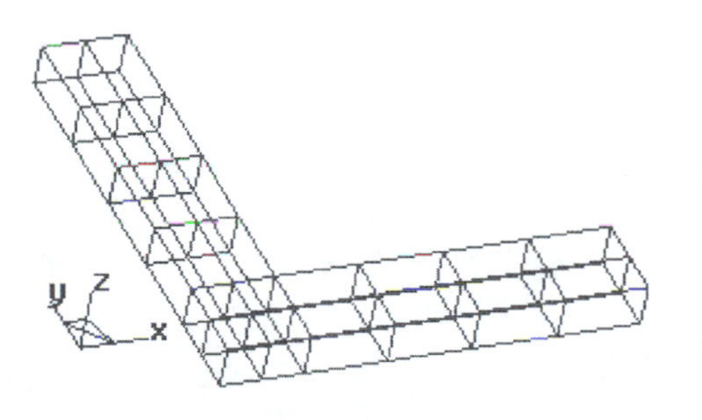

Back to Top

Back to Home

5.1 L-SHAPED USM (3)

PROBLEM TYPE VIEW RESULTS

PROBLEM - L-Shaped USM

2 Legs: 8 mm L × 2 mm W × 1 mm t
Corner: 2 mm × 2 mm × 1 mm t (PZT4)
Clamped at two leg ends

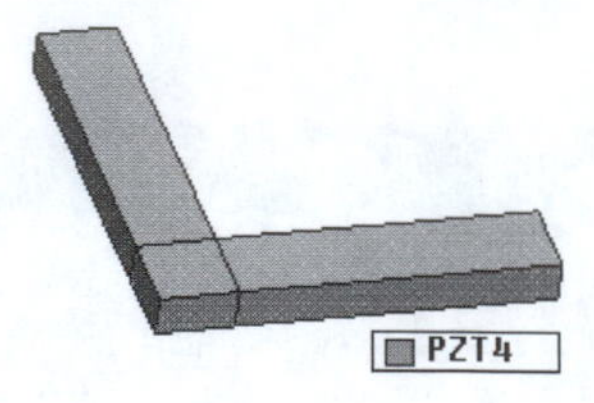

- Learn the ultrasonic motor principle—generate an elliptical locus.
- Learn how to apply sine and cosine voltages simultaneously.

PROBLEM TYPE

Data -> Problem data -> 3D, Harmonic, Click both Compute Stress and Include Losses.

FREQUENCY

Data -> Harmonic -> Interval 2 (35,000–100,000 Hz).

CALCULATION

Calculate -> Calculate window -> Start.

POST PROCESS/VIEW RESULTS

Click to Open the Post Calculation (Flavia) Domain.

CLICK ADMITTANCE/IMPEDANCE RESULT

Admittance Peak Points: 61,000 Hz and 825 Hz -> This peak corresponds to the First and Second Resonance Modes.

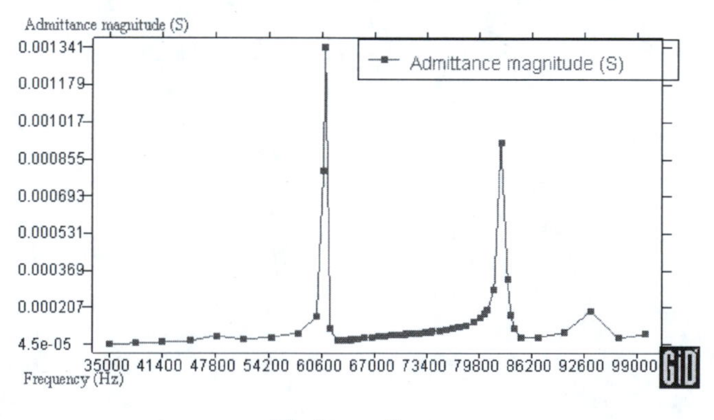

Admittance Curve

VIEW ANIMATION—VIBRATION MODE AT A PARTICULAR FREQUENCY

View results -> Default Analysis/Step -> Harmonic-Magnitude -> Set a Frequency
(Peak Point: 41,000 Hz for Resonance Mode).

Double Click [ANI icon] Loading frequency: 61,000 Hz Harmonic Case -> Click OK (single-click and wait for a while).

Set Deformation -> Displacement

View results -> Deformation -> Displacement.

Parameter other than Displacement

Set Results View -> View results -> Contour Fill -> Electric Potential or Von Mises Stress.

After setting both "Results view" and "Deformation" -> Click to see the Animation.

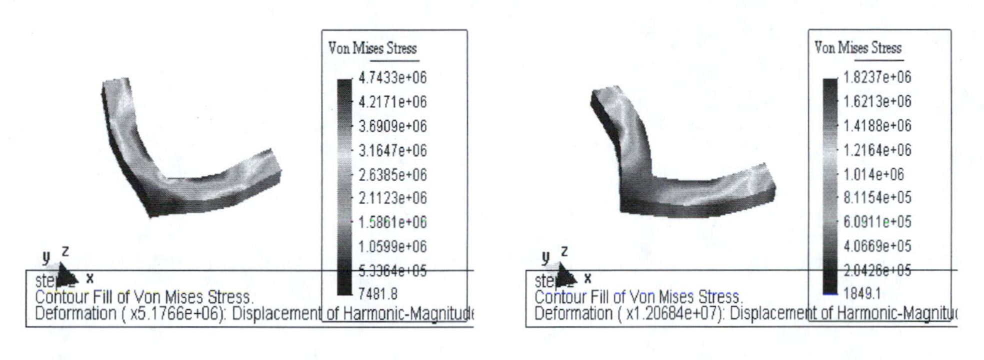

61,000 Hz - Resonance　　82,500 Hz - Resonance

When we drive this actuator at the intermediate frequency of these two resonances (70 kHz), we obtain the superposed elliptical locus at the tip. This is used as a linear ultrasonic motor.

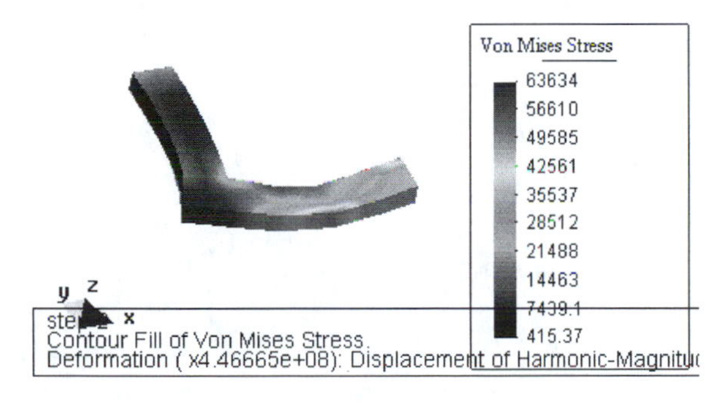

70,000 Hz - Resonance

Back to Top

Back to Home

5.2 Π-SHAPED USM (1)

GEOMETRY/DRAWING MATERIAL ASSIGNMENT

PROBLEM - L1B4 Type USM 2D Model :

Brass Bar L = 46 mm, t = 2.5 mm
Leg L = 10 mm, t = 2 mm
PZT 4 L = 11.5 mm, t = 1 mm

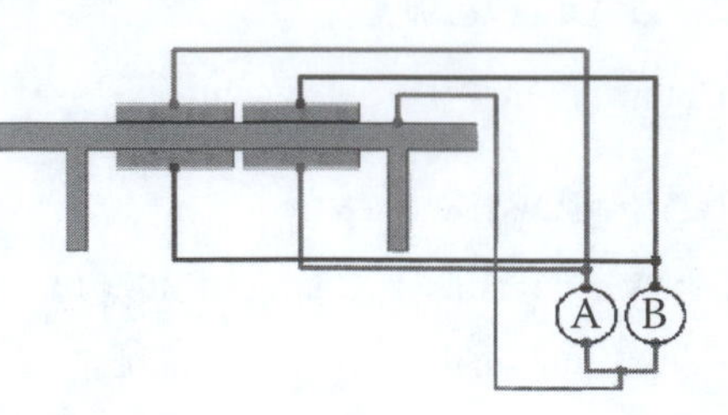

- One of the most famous linear motors.
- Learn how to apply sine and cosine voltages.

GEOMETRY/DRAWING

Create Surface

Create the coordinate map as illustrated below, and draw the motor in a 2D map.

Surface: Geometry -> Create -> Object -> Rectangle -> Enter First Corner Point -> Second Corner Point Geometry -> Create -> Lines.

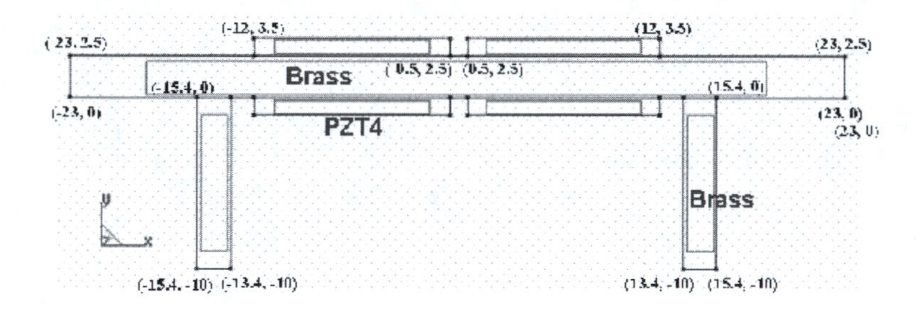

MATERIAL ASSIGNMENT

Piezoelectric Materials

Data -> Materials -> Piezoelectric -> PZT4 -> Assign -> Pick Surfaces.

Elastic Materials

Data -> Materials -> Elastic -> Brass -> Assign -> Pick Surfaces.

Click Draw -> All materials -> Check the materials are assigned as PZT4 and Brass.

Back to Top

Back to Home

5.2 Π-SHAPED USM (2)

BOUNDARY CONDITIONS MESHING

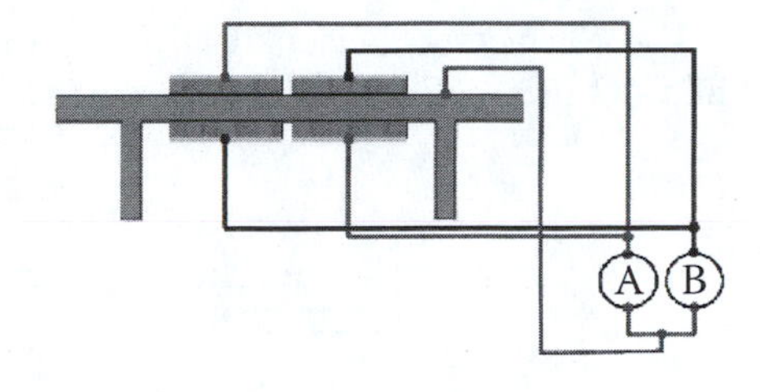

PROBLEM - L1B4 Type USM 2D Model :

Brass Bar L = 46 mm, t = 2.5 mm
Leg L = 10 mm, t = 2 mm
PZT 4 L = 11.5 mm, t = 1 mm

- One of the most famous linear motors.
- Learn how to apply sine and cosine voltages.

BOUNDARY CONDITIONS

Polarization/Local Axes

Data -> Conditions -> Surfaces -> Geometry -> Polarization (Cartesian) -> Define Polarization (P1) (0, 0, −1) direction.

Choose 3PointsXZ -> Center (0, 0, 0) -> Define X axis by putting (1, 0, 0) -> Then define Z axis by putting (0, 0, −1) -> Escape.

Assign Polarization (Local-Axes P1) -> Choose the PZT all four surfaces -> Finish.

Potential

Data -> Conditions -> Lines -> Electric Potential -> Ground -> Assign the four lines of the PZT bonded on the Brass.

Data -> Conditions -> Forced Potential 1.0 with phase 0 -> Assign the top left-side and bottom right-side surfaces.

Data -> Conditions -> Forced Potential 1.0 with phase 90 -> Assign the top right-side and bottom left-side surfaces.

This is how sine and cosine are applied on PZT plates.

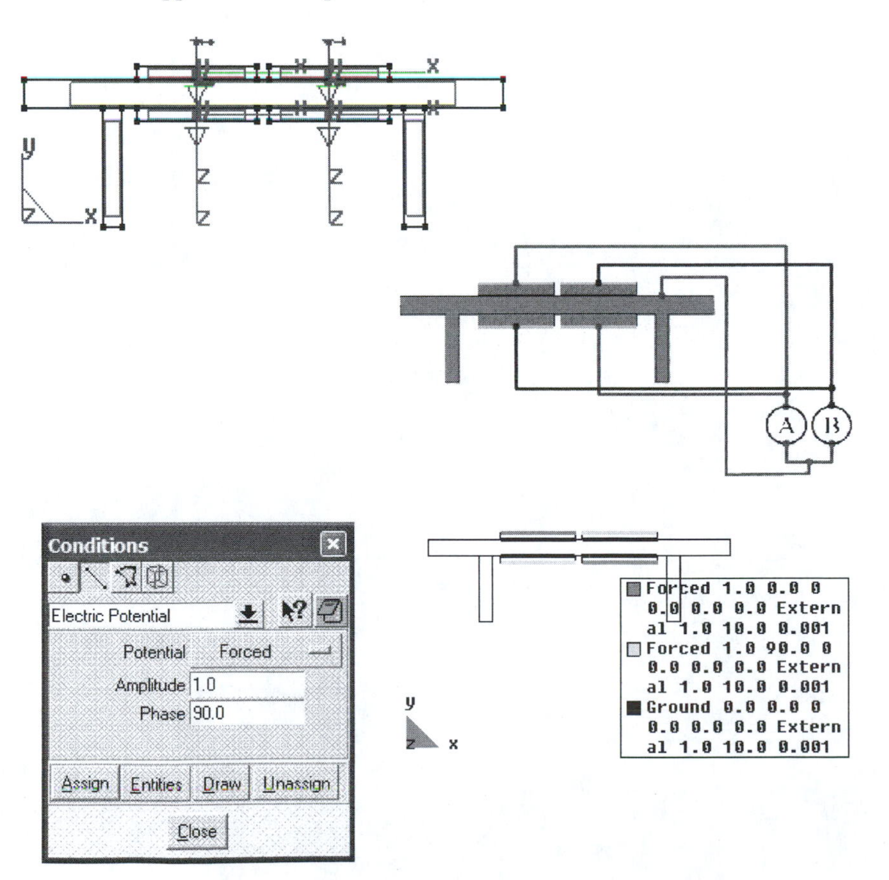

MESHING

Mapped Mesh

Meshing -> Structured -> Surfaces -> Select all surfaces -> Then Cancel the further process.
Meshing -> Structured -> Lines -> Enter number to divide line -> Choose "2, 4, 6" for the lines -> Generate.

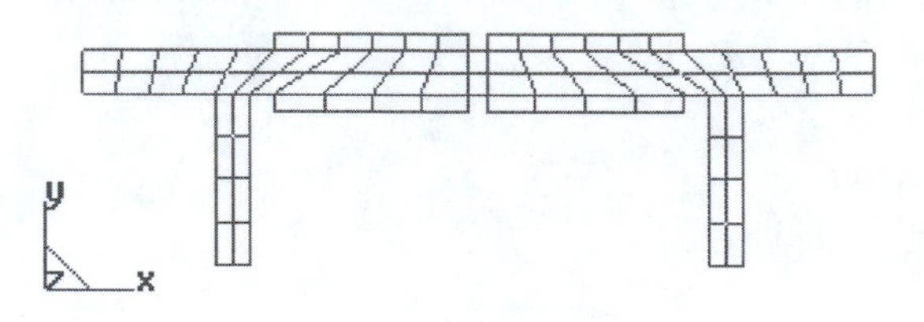

Back to Top

Back to Home

5.2 Π-SHAPED USM (3)

PROBLEM TYPE **VIEW RESULTS**

PROBLEM - L1B4 Type USM 2D Model :

Brass Bar L = 46 mm, t = 2.5 mm
Leg L = 10 mm, t = 2 mm
PZT 4 L = 11.5 mm, t = 1 mm

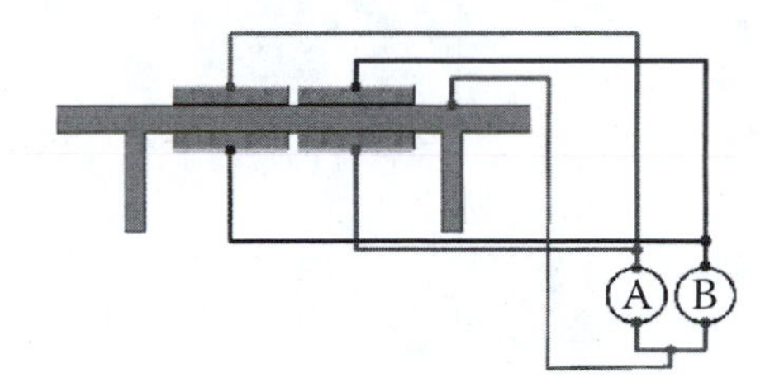

- One of the most famous linear motors.
- Learn how to apply sine and cosine voltages.

PROBLEM TYPE

Data -> Problem data -> 2D, Harmonic, Click both Compute Stress and Include Losses.

Choose Class = PLSTRAIN in this case.

Problem Data
Problem Data | Mesh Units
PRINTING 0
GEOMETRY 2D
CLASS PLSTRAIN
ANALYSIS HARMONIC
☑ COMPUTE STRESS
☑ INCLUDE LOSSES
Accept data | Close

FREQUENCY

Data -> Harmonic -> Interval 1 (10,000–110,000 Hz).

CALCULATION

Calculate -> Calculate window -> Start.

POST PROCESS/VIEW RESULTS

Click to Open the Post Calculation (Flavia) Domain.

CLICK ADMITTANCE/IMPEDANCE RESULT

Admittance Peak Points: 64,000 Hz and 76,000 Hz -> These peaks correspond to the Fourth Bending and the First Longitudinal Resonance Modes.

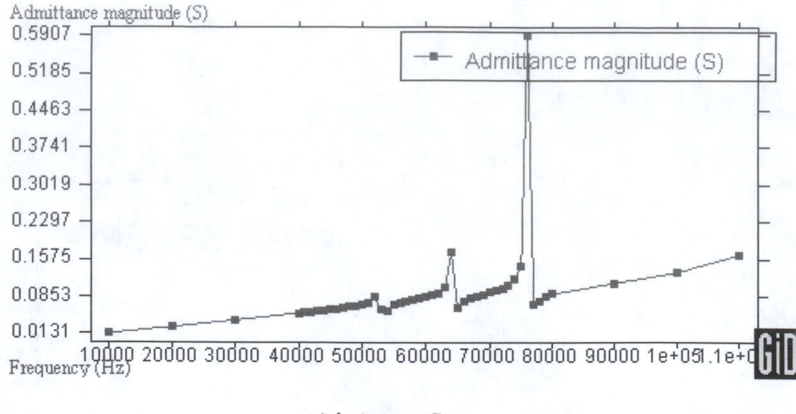

Admittance Curve

View Animation— Vibration Mode at a Particular Frequency

View results -> Default Analysis/Step -> Harmonic-Magnitude -> Set a Frequency
(Peak Point: 64,000 Hz for Resonance Mode).

Double Click Loading frequency: 64,000 Hz Harmonic Case -> Click OK (single-click and wait for a while).

Set Deformation -> Displacement

View results -> Deformation -> Displacement.

Parameter other than Displacement

Set Results View -> View results -> Contour Fill -> Electric Potential or Von Mises Stress.

After setting both "Results view" and "Deformation" -> Click to see the Animation.

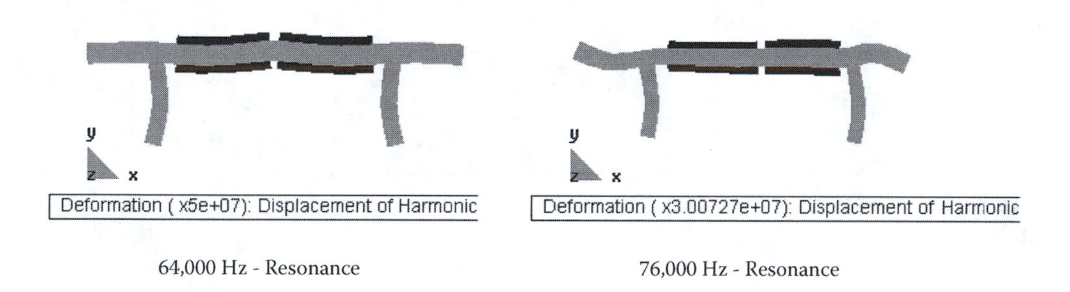

64,000 Hz - Resonance   76,000 Hz - Resonance

When we drive this actuator at the intermediate frequency of these two resonances (72 kHz), we obtain the superposed elliptical locus at the tip. This is used as a linear ultrasonic motor.

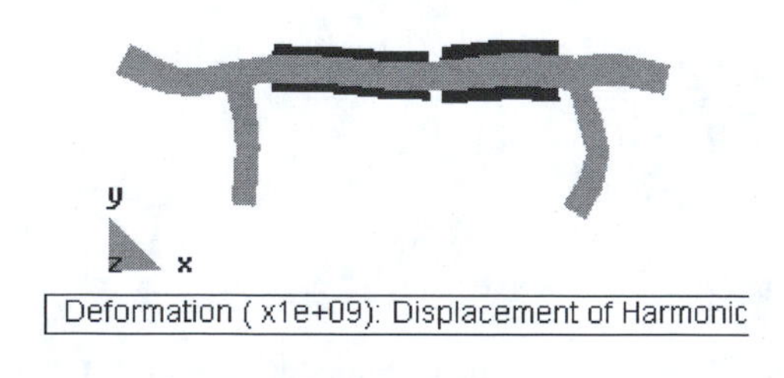

72,000 Hz - Resonance

Back to Top

Back to Home

5.2 Π-SHAPED USM (4)

PROBLEM TYPE: 2D AND 3D

PROBLEM - L1B4 Type USM 2D and 3D Models :

Brass Bar L = 46 mm, t = 2.5 mm
Leg L = 10 mm, t = 2 mm
PZT 4 L = 11.5 mm, t = 1 mm

- Learn the difference between the 2D and 3D models.

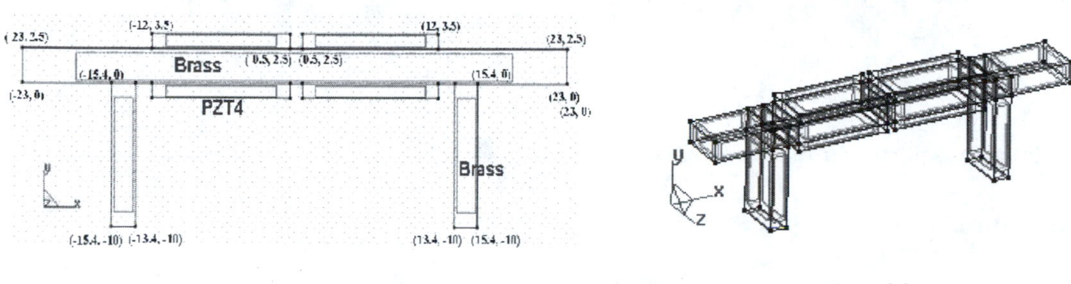

2D model 3D model

PROBLEM TYPE

Data -> Problem data -> 2D or 3D, Harmonic, Click both Compute Stress and Include Losses.

Choose Class = PLSTRAIN in the 2D case.

CALCULATION

Calculate -> Calculate window -> Start.

POST PROCESS/VIEW RESULTS

Click to Open the Post Calculation (Flavia) Domain.

CLICK ADMITTANCE/IMPEDANCE RESULT

2D Result: Admittance Peak Points: 64,000 Hz and 76,000 Hz -> These peaks correspond to the Fourth Bending and the First Longitudinal Resonance Modes.

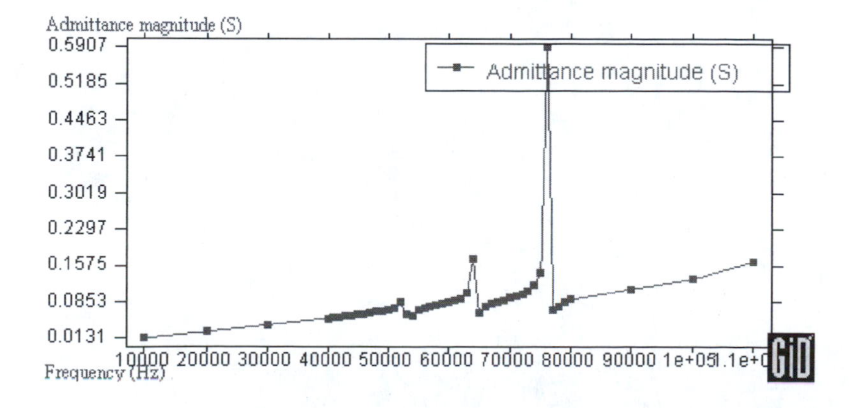

3D Result: Admittance Peak Points: 64,800 Hz and 72,800 Hz -> These peaks correspond to the Fourth Bending and the First Longitudinal Resonance Modes.

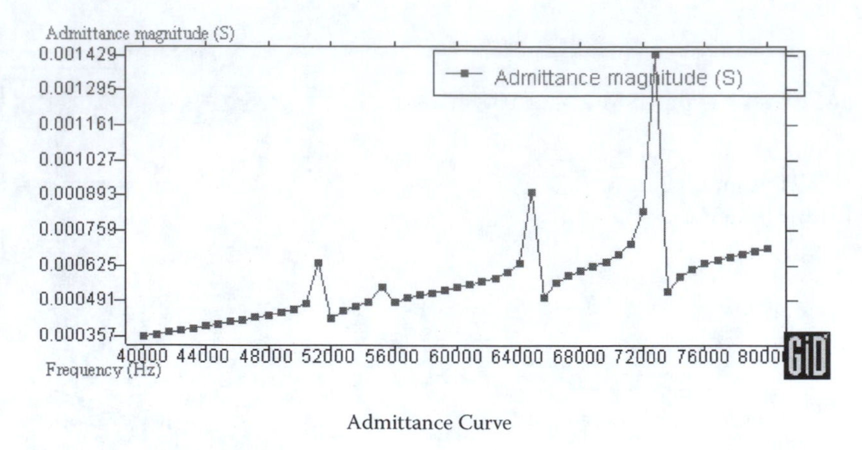

Admittance Curve

Note the almost identical spectra for the 2D and 3D simulations, except for a slight deviation in the resonance frequencies.

View Animation—Vibration Mode at a Particular Frequency

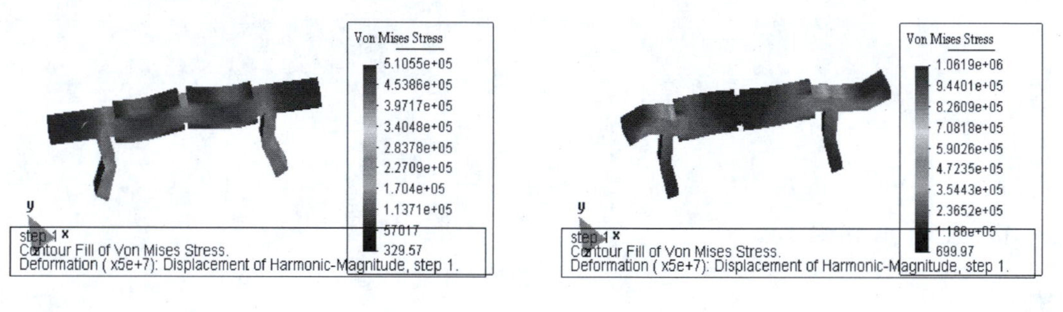

64,800 Hz - Resonance

72,800 Hz - Resonance

Back to Top

5.3 METAL-TUBE USM (1)

GEOMETRY/DRAWING MATERIAL ASSIGNMENT

PROBLEM - Metal tube: Brass OD = 2.4 mm ϕ,
ID = 1.8 mm ϕ × 5 mm L; PZT 4: 0.3 mm t × 1 mm × 5 mm

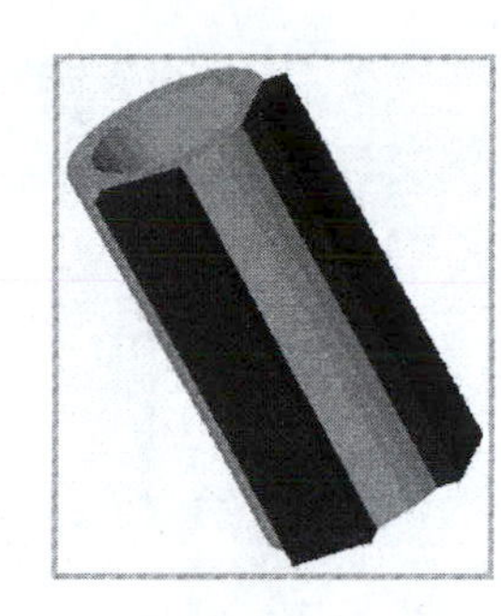

- Asymmetric motor with split modes.
- Learn how to cut volume.

GEOMETRY/DRAWING

CREATE SURFACE

Surface: Geometry -> Create -> Object -> Circle -> Enter Points for Center -> 0,0 -> Radius -> 1.2 and 0.9.
Surface subtraction: Geometry -> Create -> Surface Boolean op. -> Subtraction.

DIVIDE SURFACE

Geometry -> Edit -> Divide -> Surfaces -> Near point.

- U Sense Enter point (0,*y*).
- V Sense Enter point (*x*,0).

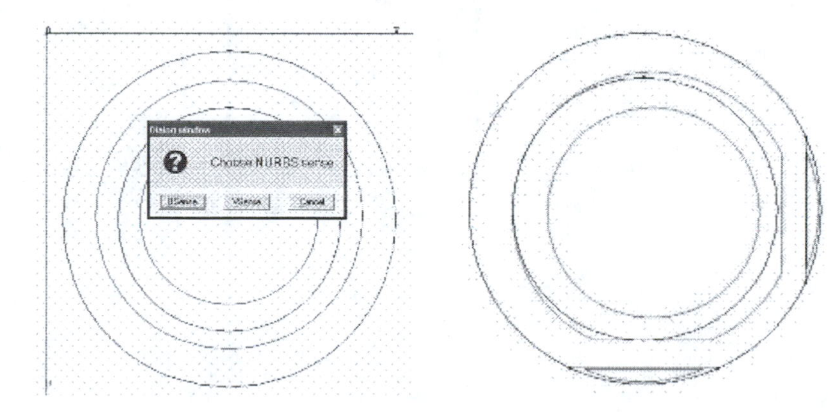

Delete Surface: Geometry -> Delete -> Surfaces -> Pick Surfaces.
Copy Points: Utilities -> Copy -> Select Points.
Create Surfaces: Geometry -> Create -> NURBS Surface -> By Points.

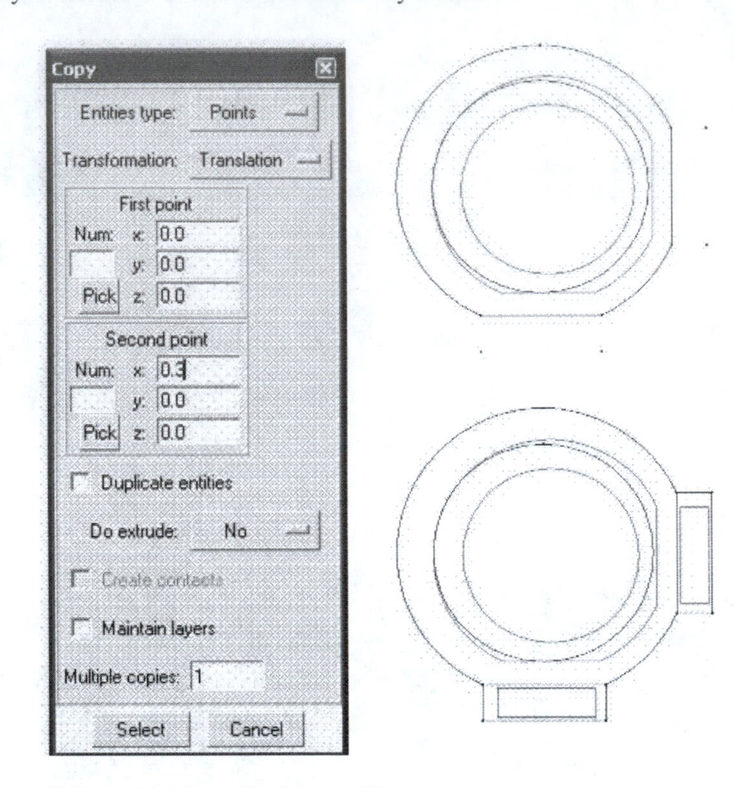

Divide Surface: Geometry -> Edit -> Divide -> Surfaces -> Near point.

- U Sense Enter point $(0,y)$.
- V Sense Enter point $(x,0)$.

CREATE VOLUMES

Utilities Copy Surfaces Extrude Volumes Select all Surfaces.

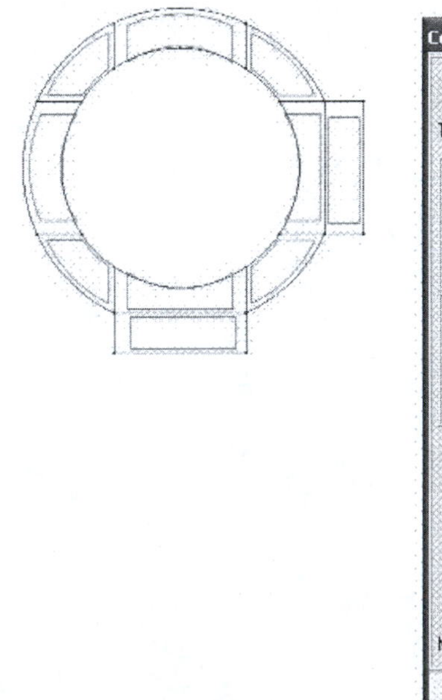

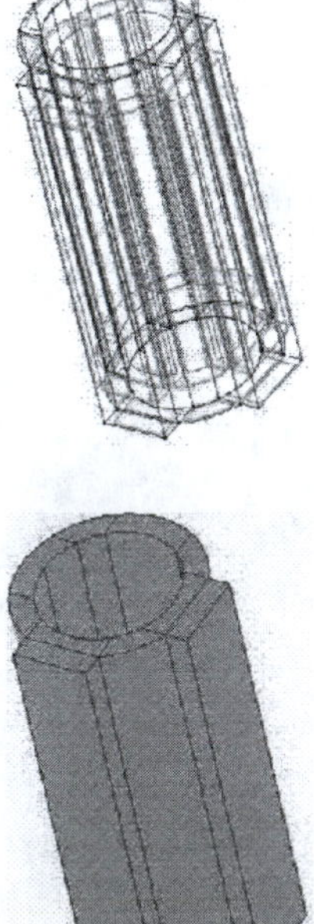

MATERIAL ASSIGNMENT

Piezoelectric Materials

Data -> Materials -> Piezoelectric -> PZT4 -> Assign -> Pick Volumes.

Elastic Materials

Data -> Materials -> Elastic -> Brass -> Assign -> Pick Volumes.
Click Draw -> All materials -> Check the materials are assigned as PZT4 and Brass.

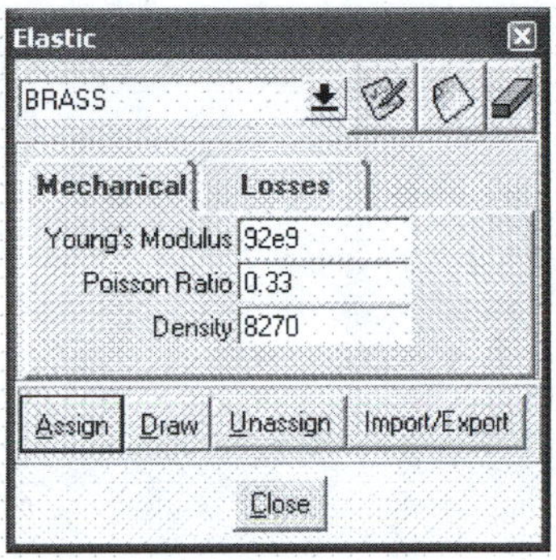

Elastic

BRASS

Mechanical	Losses
Young's Modulus	92e9
Poisson Ratio	0.33
Density	8270

Assign Draw Unassign Import/Export

Close

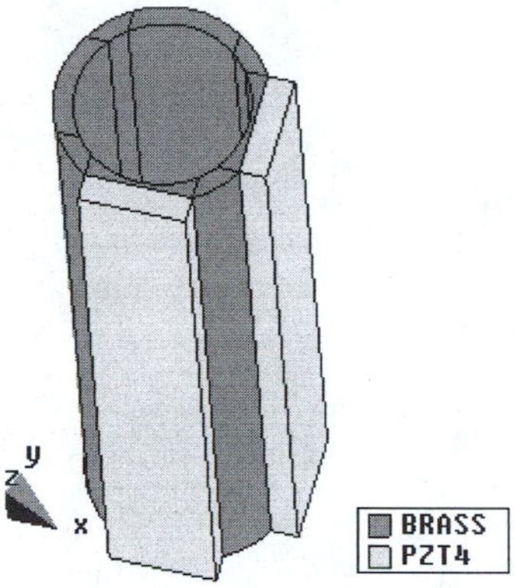

BRASS
PZT4

Back to Top

Back to Home

5.3 METAL-TUBE USM (2)

BOUNDARY CONDITIONS MESHING

PROBLEM - Metal tube: Brass OD = 2.4 mm ϕ, ID = 1.8 mmϕ × 5 mm L; PZT 4: 0.3 mm t × 1 mm × 5 mm

BOUNDARY CONDITIONS

POLARIZATION/LOCAL AXES

Data -> Conditions -> Volumes -> Geometry -> Polarization (Cartesian) -> Define Polarization (P1) (–1,0,0) direction.
Choose 3PointsXZ -> Center (0,0,0) -> Define X axis by putting (0,1,0) -> Then define Z axis by putting (–1,0,0) -> Escape.

Define Polarization (P2) (0,1,0) direction.
Choose 3PointsXZ -> Center (0,0,0) -> Define X axis by putting (–1,0,0) -> Then define Z axis by putting (0,1,0)-> Escape.

Assign Polarization (Local-Axes P1) -> Choose the PZT volume -> Assign Polarization (Local-Axes P2) -> Choose the PZT volume - > Finish.

You may choose:
X and Angle: Pick 2 Points for Center and X, and Y with angle.

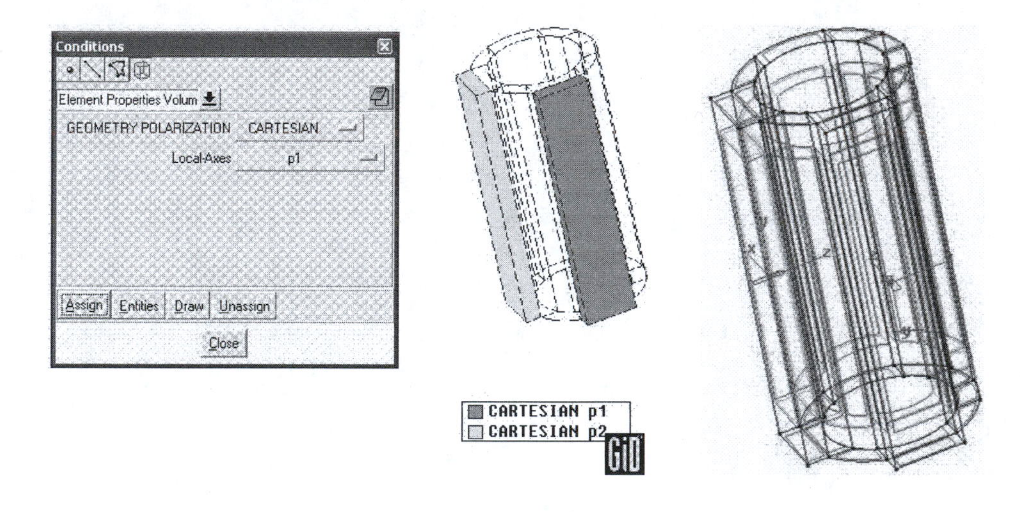

POTENTIAL

Data -> Conditions -> Surfaces -> Potential 0.0 (or Ground in ATILA 5.2.4 or higher version) -> Assign the metal side surface of the PZT -> Forced Potential 1.0 -> Assign the top surface.

Only one PZT plate is driven, keeping the other PZT plate short-circuited.

MESHING

Mapped Mesh

Meshing -> Structured -> Volumes -> Select all volumes -> Then Cancel the further process.
Meshing -> Structured -> Lines -> Enter number to divide line -> Generate.

Use the larger numbers such as 2 or 4 for more accurate calculation.

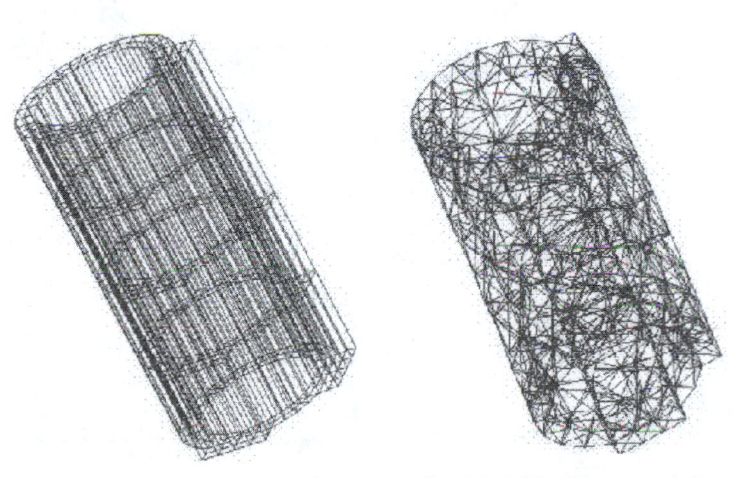

Structured Mesh (Recommended) Free Mesh (Not Recommended)

Back to Top

Back to Home

5.3 METAL-TUBE USM (3)

PROBLEM TYPE **VIEW RESULTS**

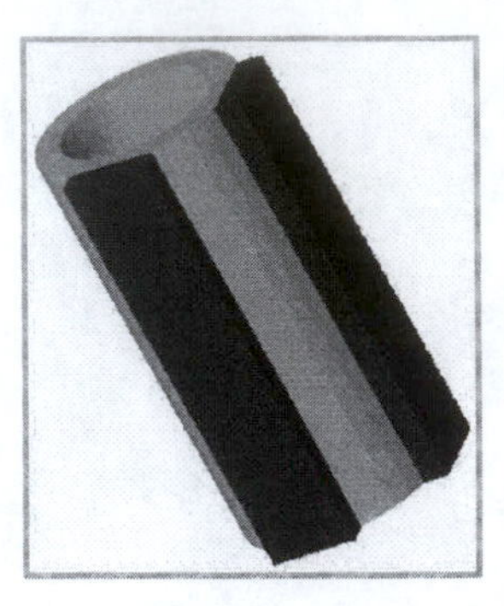

PROBLEM - Metal tube: Brass OD = 2.4 mm ϕ,
ID = 1.8 mm ϕ × 5 mm L; PZT 4: 0.3 mm t × 1 mm × 5 mm

PROBLEM TYPE

Data -> Problem data -> 3D, Harmonic, Click both Compute Stress and Include Losses.

FREQUENCY

Data -> Harmonic -> Interval 2 (170,000–220,000 Hz).

Problem Data
Problem Data | Mesh Units
GEOMETRY 3D
ANALYSIS HARMONIC
☑ COMPUTE STRESS
☑ INCLUDE LOSSES
Accept data | Close

Interval Data
1
TYPE LINEAR DISTRIBUTION
MIN FREQUENCY 170000
MAX FREQUENCY 220000
NUMBER OF FREQUENCIES 51
Accept data | Close

CALCULATION

Calculate -> Calculate window -> Start.

POST PROCESS/VIEW RESULTS

Click to Open the Post Calculation (Flavia) Domain.

Click Admittance/Impedance Result

Admittance Peak Points: 207.4 kHz and 208.4 kHz -> These peaks correspond to the First split Resonances

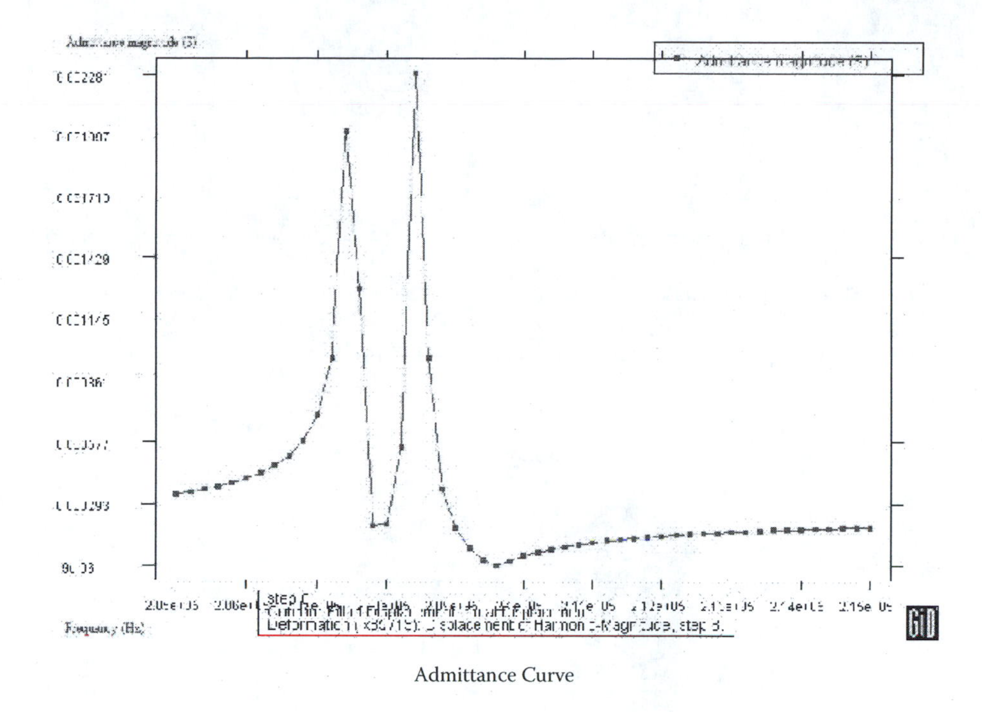

Admittance Curve

View Animation - Vibration Mode at a Particular Frequency

View results -> Default Analysis/Step -> Harmonic-Magnitude -> Set a Frequency
(Peak Point: 207,400 Hz or 208,400 Hz for the Resonance Mode).

Double Click Loading frequency: 207,400 or 208,400 Hz Harmonic Case -> Click OK (single-click and wait for a while).

Set Deformation -> Displacement

View results -> Deformation -> Displacement.

Parameter other than Displacement

Set Results View -> View results -> Contour Fill -> Von Mises Stress.

After setting both "Results view" and "Deformation" -> Click to see the Animation.

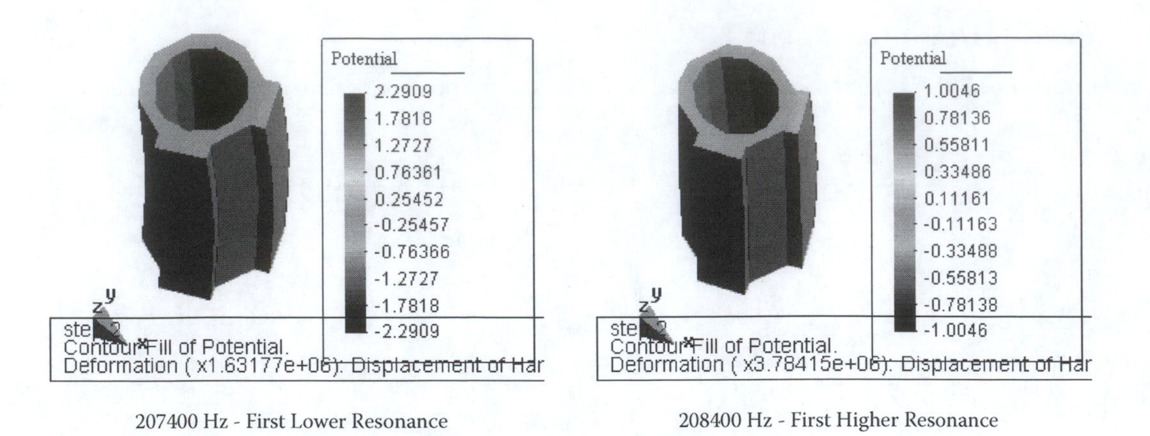

207400 Hz - First Lower Resonance

208400 Hz - First Higher Resonance

When we drive at the intermediate frequency, we obtain the superposed "hula-hoop" mode, which can be used as a rotary ultrasonic motor.

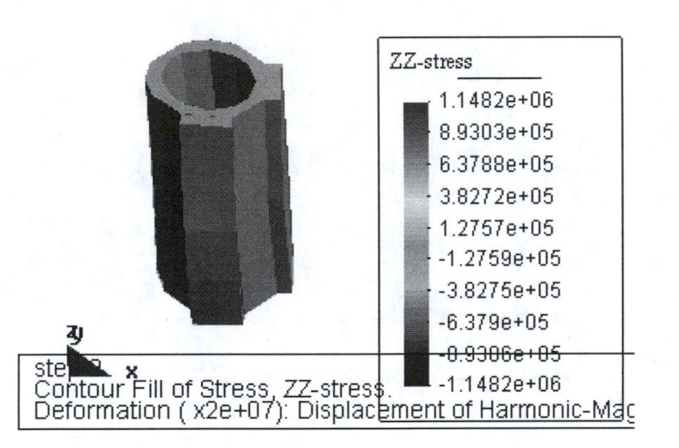

Back to Top

Back to Home

6 Underwater Transducer

- How to assign "water"
- How to set the boundary condition
- How to calculate the medium pressure

TEXT CONTENT

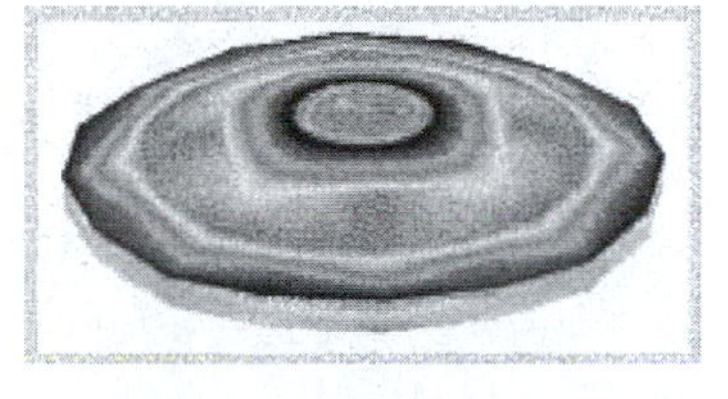

Metal-PZT Composite, Cymbal

Sound Pressure Calcuation in a Tonpilz Transduce

Back to Top

6.1 LANGEVIN-TYPE TRANSDUCER (1)

GEOMETRY/DRAWING MATERIAL ASSIGNMENT

PROBLEM - Langevin Transducer 50 mm ϕ × 40 mm t (Two PZT 8 disks with 50 mm ϕ × 1 mm t are sandwiched by two Steel 1 blocks with 50 mm ϕ × 19 mm thickness); 2D Axisymmetric with 1/2 portion

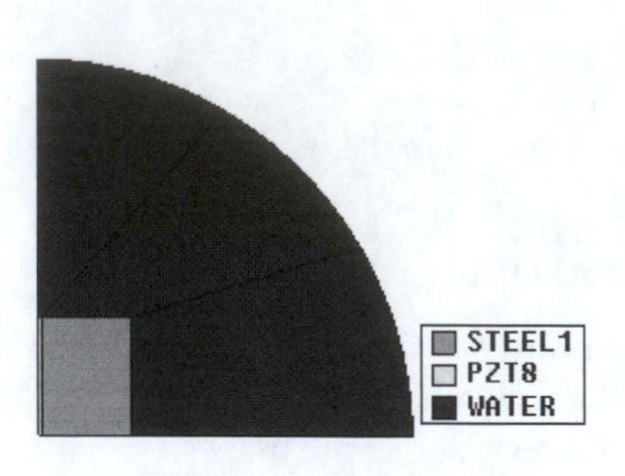

- Simplest underwater transducer
- Learn how to use a small sample in a low-frequency application
- Learn about underwater sound propagation

GEOMETRY/DRAWING

Create Surface

Geometry -> Create -> Object -> Rectangle -> Enter First and Second Corner Points -> 0, 0 ; 20, 25.

Divide Surface

Geometry -> Edit -> Divide -> Surface -> Near point -> Select Surface to Divide -> Choose NURBS sense -> VSense -> Near point 1, 0.

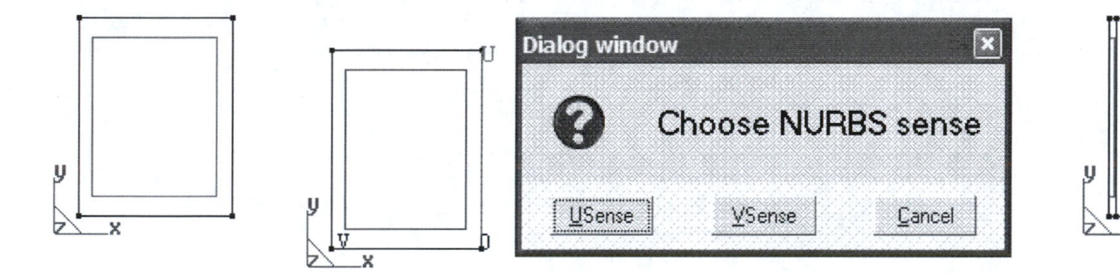

Create Water Area

Geometry -> Create -> Arc -> First, Second and Third points -> 80, 0; 0, 80; −80, 0.
Geometry -> Edit -> Divide -> Num Divisions -> Select line -> Num Divisions 2.
Delete -> Line and point.
Then, Geometry -> Create -> Line.

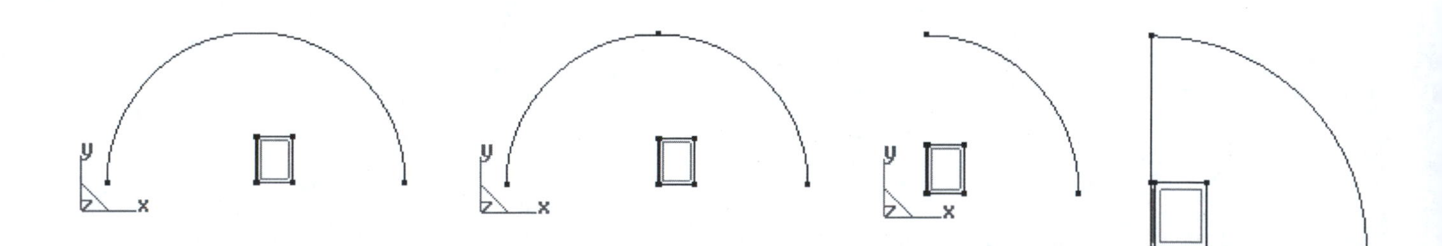

Divide the Water Surface

In order to simplify the meshing in the Water area (small number of meshes), we will use the water surface area division technique.

Geometry -> Edit -> Divide -> Num Divisions -> Select line -> Num Divisions 3.
Then, Geometry -> Create -> Line.
Geometry -> Create -> NUEBS surface -> Automatic -> 4-sided lines.

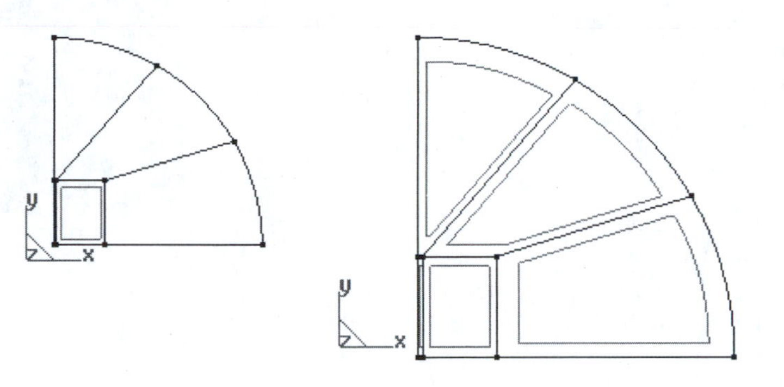

MATERIAL ASSIGNMENT

Piezoelectric Materials

Data -> Materials -> Piezoelectric -> PZT8 -> Assign -> Pick Surface.

Elastic Materials

Data -> Materials -> Elastic -> Steel 1-> Assign -> Pick Surface.

Fluid Materials

Data -> Materials -> Fluid -> Water-> Assign -> Pick Surface.
Click Draw -> All materials -> Check the materials are assigned as PZT8, Steel 1, and Water.

- Note that this half is enough for the simulation, as long as we consider the symmetrical mode.

Back to Top

Back to Home

6.1 LANGEVIN-TYPE TRANSDUCER (2)

BOUNDARY CONDITIONS MESHING

PROBLEM - Langevin Transducer 50 mm φ × 40 mm t (Two PZT 8 disks with 50 mm φ × 1 mm t are sandwiched by two Steel 1 blocks with 50 mm φ × 19 mm thickness); 2D Axisymmetric with 1/2 portion

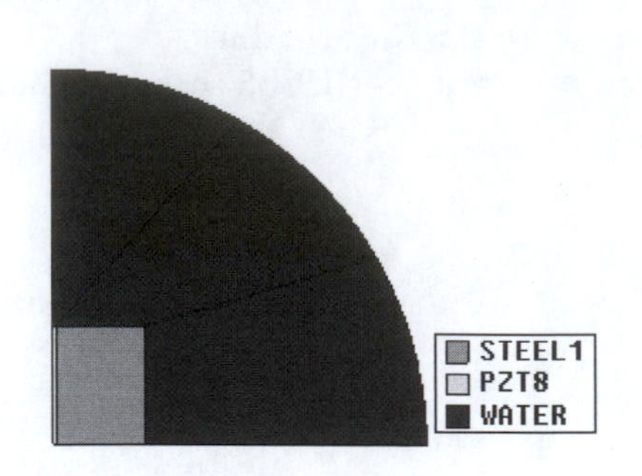

- Simplest underwater transducer
- Learn how to use a small sample in a low-frequency application
- Learn about underwater sound propagation

BOUNDARY CONDITIONS

POLARIZATION/LOCAL AXES

High voltage should be applied on the center surface of the two layers of the PZT, and Ground on the metal side surface. Choose the polarization direction along the electric field direction.

Data -> Conditions -> Surfaces -> Geometry -> Polarization (Cartesian) -> Define Polarization (P1) (1, 0) direction.

Choose 3PointsXZ -> Center (0, 0, 0) -> Define X axis by putting (0, 1, 0) -> Then define Z axis by putting (1, 0, 0) -> Escape.

Assign Polarization (Local-Axes P1) -> Choose the PZT surface -> Finish.

POTENTIAL

Data -> Conditions -> Lines -> Potential 0.0 (or Ground in ATILA 5.2.4 or higher version) -> Assign the right-side line of the PZT -> Forced Potential 1.0 -> Assign the left-side line of the PZT (which is the center surface of the two PZT plates).

SURFACE CONSTRAINTS (HALF OF THE WHOLE CYMBAL)

We use only half of the whole Langevin transducer, taking into account the symmetry; the left-side line of the PZT and water should be clamped (X component is clamped).

The bottom axisymmetric line of the whole system (including water) - Y component is clamped. The axisymmetric axis is taken along X axis (this is the rule in ATILA).

Data -> Conditions -> Lines -> Line Constraints -> X components -> Clamped -> Assign the left-side PZT and Water lines.

Similarly, on the Axisymmetric X axis -> Line Constraints -> Y & Z components -> Clamped -> Assign.

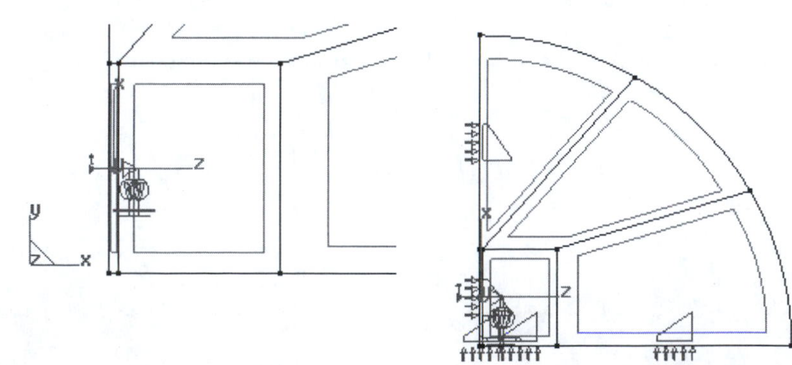

Radiation Boundary

Data -> Conditions -> Lines -> Line Constraints -> X components -> Clamped -> Assign the left-side PZT and Water lines.

Radiation boundary is applied not to reflect the water-propagating sound at this boundary (like an anechoic wall). In this sense, the radius of this water boundary is not essential.

To check the assignment -> Draw -> All conditions -> Include local axes.

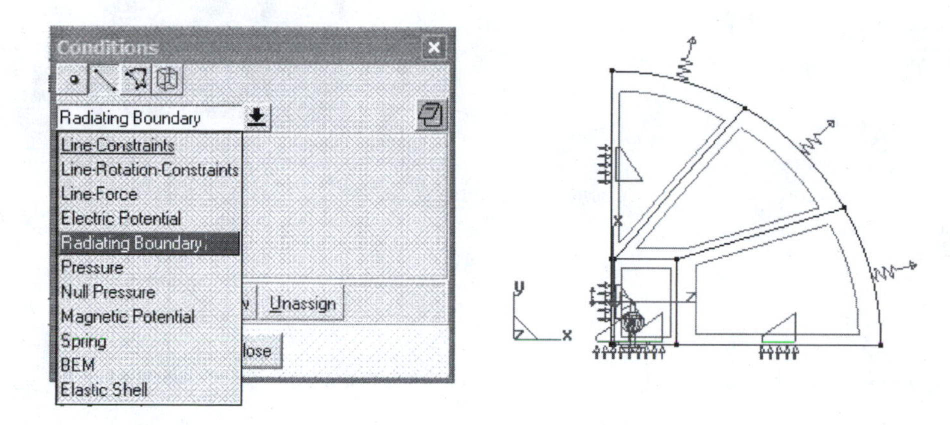

MESHING

Mapped Mesh

Meshing -> Structured -> Surfaces -> Select all Surfaces -> Then Cancel the further process.
Meshing -> Structured -> Lines -> Enter number to divide line -> Choose "1, 2, 4" for the lines -> Generate.

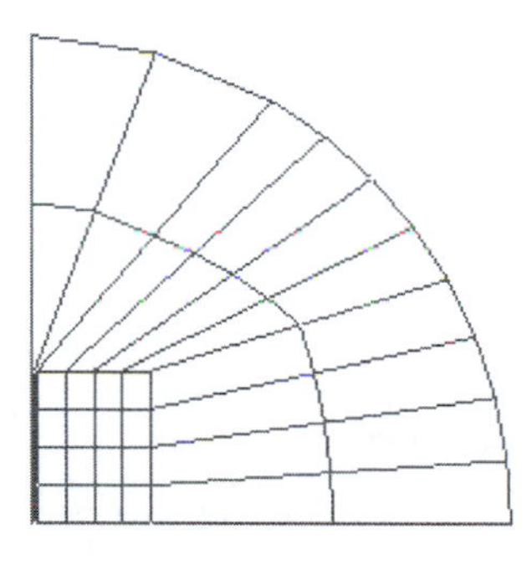

Back to Top

Back to Home

6.1 LANGEVIN-TYPE TRANSDUCER (3)

PROBLEM TYPE PROB FREQ VIEW RESULTS

PROBLEM - Langevin Transducer 50 mm ϕ × 40 mm t (Two PZT 8 disks with 50 mm ϕ × 1 mm t are sandwiched by two Steel 1 blocks with 50 mm ϕ × 19 mm thickness); 2D Axisymmetric with 1/2 portion

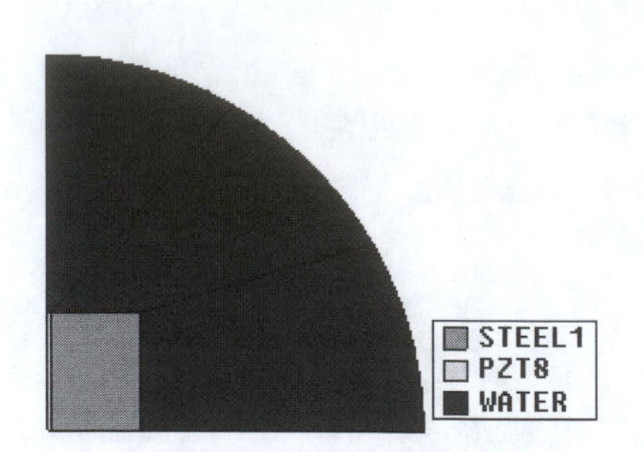

- Simplest underwater transducer
- Learn how to use a small sample in a low-frequency application
- Learn about underwater sound propagation

PROBLEM TYPE

PROB Data -> Problem data -> 2D, Axisymmetric, Harmonic, Click both Compute Stress and Include Losses.

FREQUENCY

FREQ Data -> Harmonic -> Interval 1 (20,000–100,000 Hz), Interval 2 (45–55 kHz).

CALCULATION

Calculate -> Calculate window -> Start.

POST PROCESS/VIEW RESULTS

Click to Open the Post Calculation (Flavia) Domain.

CLICK ADM ADMITTANCE/IMPEDANCE RESULT

To find the exact resonance frequency, Click "Show Values."

Admittance Peak Point: 52,000 Hz -> This peak corresponds to the First Resonance.

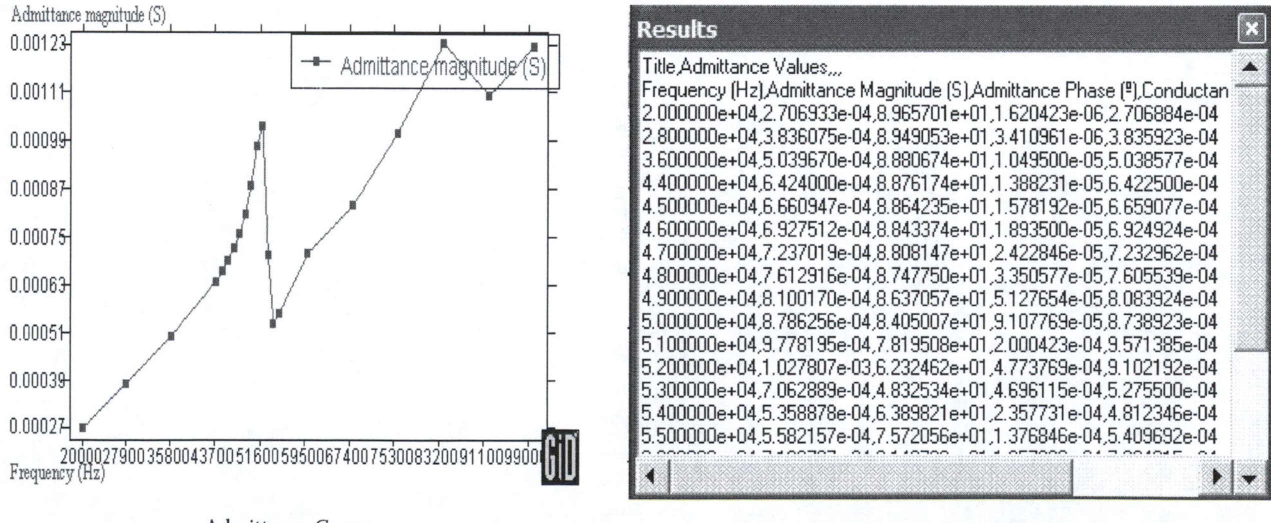

Admittance Curve

Admittance Value Results

View Animation—Vibration Mode at a Particular Frequency

View results -> Default Analysis/Step -> Harmonic-Magnitude -> Set a Frequency (Peak Point: 52,000 Hz for Resonance Mode).

Double Click Loading frequency: 52,000 Hz Harmonic Case -> Click OK (single-click and wait for a while).

Set Deformation -> Displacement

View results -> Deformation -> Displacement.

Parameter other than Displacement

Set Results View -> View results -> Contour Lines -> Fluid Pressure.

After setting both "Results view" and "Deformation" -> Click to see the Animation.

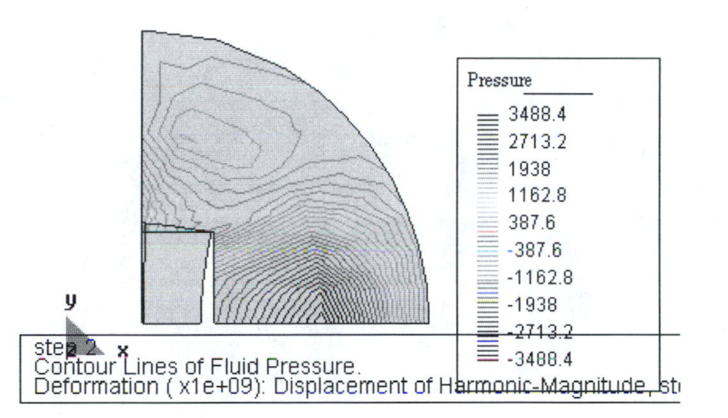

52,000 Hz - Resonance

Back to Top

Back to Home

6.2 UNDERWATER CYMBAL (1)

GEOMETRY/DRAWING MATERIAL ASSIGNMENT

PROBLEM—CYMBAL 12 MM Φ × 2.4 MM T (METAL THICKNESS 0.2 MM, PZT THICKNESS 0.5 MM × TWO LAYER, CAVITY HEIGHT 0.5 MM); 2D AXISYMMETRIC WITH A FULL PORTION

PROBLEM - Cymbal 12 mm φ × 2.4 mm t (Metal thickness 0.2 mm, PZT thickness 0.5 mm × two layer, Cavity height 0.5 mm); 2D Axisymmetric with a full portion

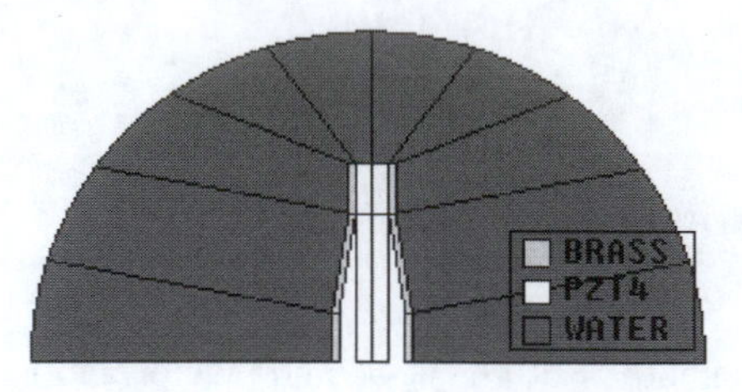

- New underwater transducer
- Learn how to enhance the displacement/effective *d* coefficient
- Learn about underwater sound propagation—monopole and dipole

GEOMETRY/DRAWING

This is the same cymbal geometry as in section **3.4, "Cymbal."**

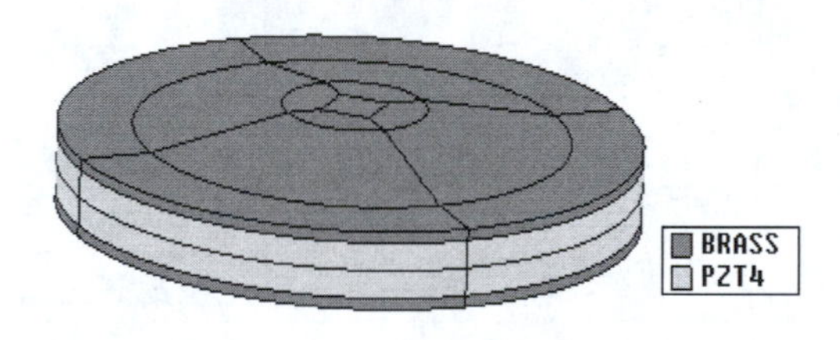

CREATE PZT SURFACE

Geometry -> Create -> Object -> Rectangle -> Enter First and Second Corner Points -> (0, 0); (0.5, 6).

DIVIDE SURFACE

Geometry -> Edit -> Divide -> Surface -> Near point -> Select Surface to Divide -> Choose NURBS sense -> USense -> Near point (0, 4.5).

CREATE METAL ENDCAP

Geometry -> Create -> Line -> Enter Points -> (1.2, 0); (1.2, 1.5); (0.7, 4.5); (0.7, 6).

COPY LINES

Utilities -> Copy -> Lines -> Translation -> First point (0.2, 0, 0) to Second point (0, 0, 0); Do Extrude -> Select Lines -> Finish.

Create Surfaces

Geometry -> Create -> NURBS Surface -> Automatic 4-sided lines.

Copy Lines

Utilities -> Copy -> Surfaces -> Mirror -> First point (0, 0, 0); Second point (0, 1, 0); Third point (0, 0, 1); Do Not Extrude -> Select Surfaces -> Finish.

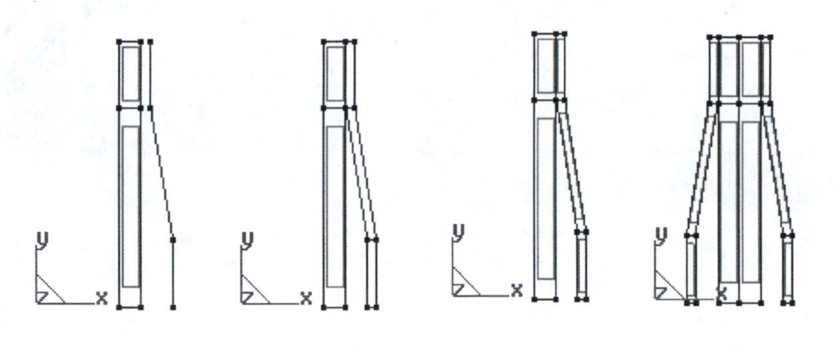

Create Water Area

Geometry -> Create -> Arc -> First, Second, and Third points -> 10, 0; 0, 10; -10, 0 Then, Geometry -> Create -> Line.

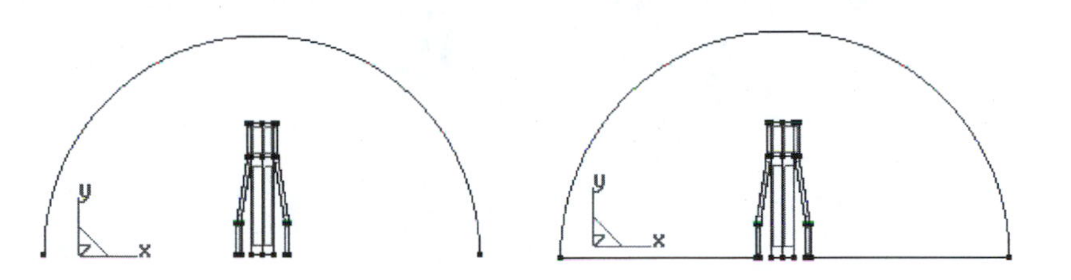

Divide the Water Surface

To simplify the meshing in the water area (small number of mesh), we will use the water surface area division technique.

Geometry -> Edit -> Divide -> Num Divisions -> Select line (water boundary) -> Num Divisions 10.
Then, Geometry -> Create -> Line.
Geometry -> Create -> NUEBS surface -> Automatic -> 4-sided lines.

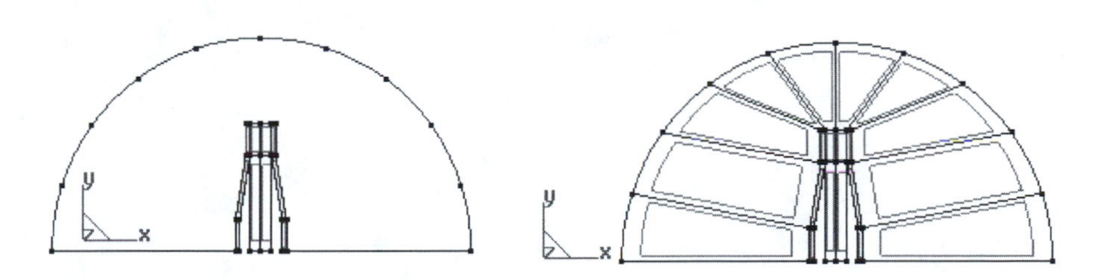

MATERIAL ASSIGNMENT

Piezoelectric Materials

Data -> Materials -> Piezoelectric -> PZT4 -> Assign -> Pick Surface.

Elastic Materials

Data -> Materials -> Elastic -> Brass-> Assign -> Pick Surface.

Fluid Materials

Data -> Materials -> Fluid -> Water-> Assign -> Pick Surface.
Click Draw -> All materials -> Check the materials are assigned as PZT4, Brass, and Water.

Fluid
WATER
General | Losses | Magnetic
BULK MODULUS 2.22e9
DENSITY 1000
Assign | Draw | Unassign | Import/Export
Close

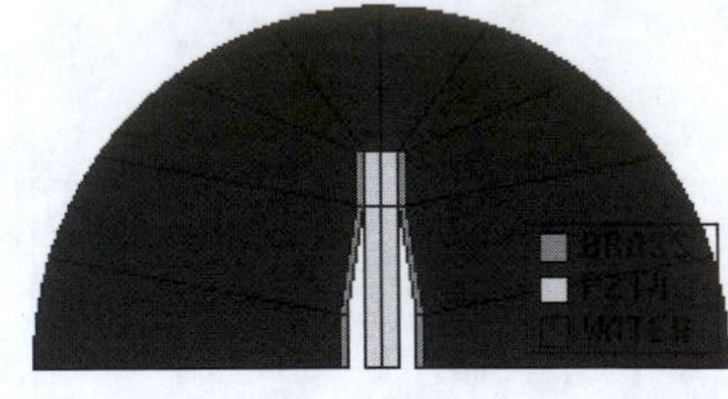

- Note that because we will discuss both symmetric and asymmetric (Homework 9) modes, the whole part is required for the simulation.

Back to Top

Back to Home

6.2 UNDERWATER CYMBAL (2)

BOUNDARY CONDITIONS MESHING

PROBLEM - Cymbal 12 mm ϕ × 2.4 mm t (Metal thickness 0.2 mm, PZT thickness 0.5 mm × two layer, Cavity height 0.5 mm); 2D Axisymmetric with a full portion

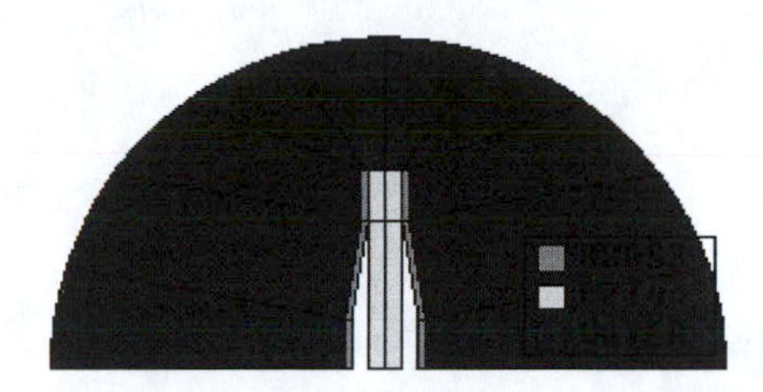

- New underwater transducer
- Learn how to enhance the displacement/effective *d* coefficient
- Learn about underwater sound propagation—monopole and dipole

BOUNDARY CONDITIONS

Polarization/Local Axes

High voltage should be applied on the center surface of the two layers of the PZT, and Ground on the metal side surface. Choose the polarization direction along the electric field direction.

Data -> Conditions -> Surfaces -> Geometry -> Polarization (Cartesian) -> Define Polarization (P1) (1, 0) direction.
Choose 3PointsXZ -> Center (0, 0, 0) -> Define X axis by putting (0, 1, 0) -> Then define Z axis by putting (1, 0, 0) -> Escape.

Data -> Conditions -> Surfaces -> Geometry -> Polarization (Cartesian) -> Define Polarization (P2) (−1, 0) direction.
Choose 3PointsXZ -> Center (0, 0, 0) -> Define X axis by putting (0, 1, 0) -> Then define Z axis by putting (−1, 0, 0) -> Escape.

Assign Polarization (Local-Axes P1 and P2) -> Choose the PZT surface -> Finish.

Potential

Data -> Conditions -> Lines -> Potential 0.0 (or Ground in ATILA 5.2.4 or higher version) -> Assign the center line of the PZT -> Forced Potential 1.0 -> Assign the right- and left-side lines of the PZT.

Surface Constraints

The bottom axisymmetric line of the whole system (including water) - Y component is clamped. The axisymmetric axis is taken along X axis (this is the rule in ATILA).

Data -> Conditions -> Lines -> Line Constraints -> Y & Z components -> Clamped -> Assign the Axisymmetric X axis.

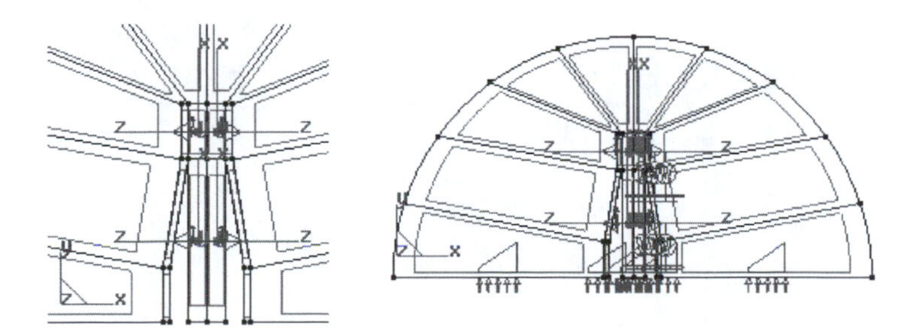

Radiation Boundary

Data -> Conditions -> Lines -> Line Constraints -> X components -> Clamped -> Assign the left-side PZT and water lines. Radiation boundary is applied not to reflect the water-propagating sound at this boundary (like an anechoic wall). In this sense, the radius of this water boundary is not essential.

To check the assignment -> Draw -> All conditions -> Include local axes.

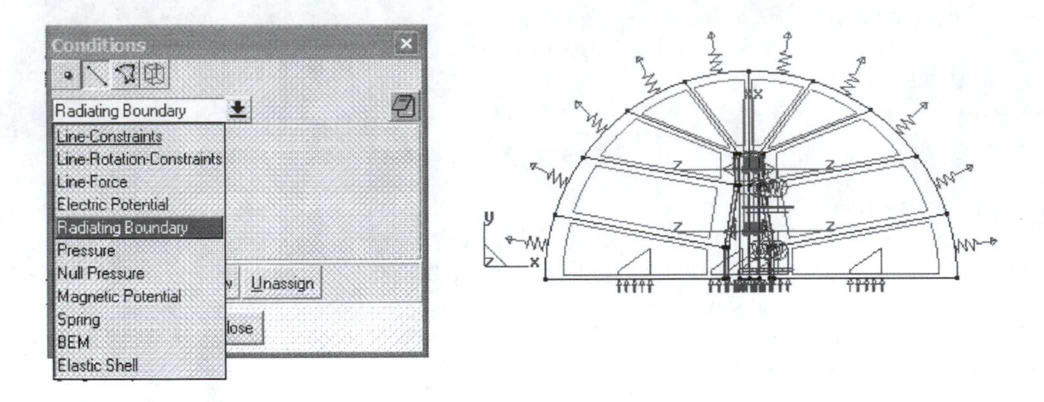

MESHING

Mapped Mesh

Meshing -> Structured -> Surfaces -> Select all Surfaces -> Then Cancel the further process.
Meshing -> Structured -> Lines -> Enter number to divide line -> Choose "1, 2, 4" for the lines -> Generate.

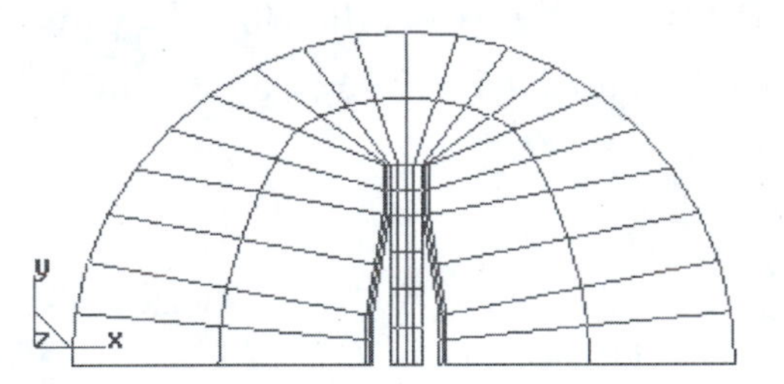

Back to Top

Back to Home

6.2 UNDERWATER CYMBAL (3)

PROBLEM TYPE VIEW RESULTS

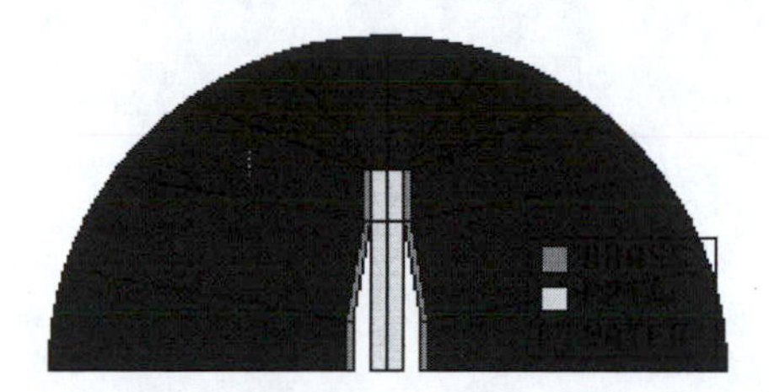

PROBLEM - Cymbal 12 mm ϕ × 2.4 mm t (Metal thickness 0.2 mm, PZT thickness 0.5 mm × two layer, Cavity height 0.5 mm); 2D Axisymmetric with a full portion

- An underwater transducer
- Learn about underwater sound propagation—monopole and dipole

PROBLEM TYPE

Data-> Problem data -> 2D, Axisymmetric, Harmonic, Click both Compute Stress and Include Losses.

FREQUENCY

Data -> Harmonic -> Interval 1 (20,000–30,000 Hz), Interval 2 (23–25 kHz).

CALCULATION

Calculate -> Calculate window -> Start.

POST PROCESS/VIEW RESULTS

Click to Open the Post Calculation (Flavia) Domain.

CLICK ADMITTANCE/IMPEDANCE RESULT

To find the exact resonance frequency, Click "Show Values."

Admittance Peak Point: 24,000 Hz -> This peak corresponds to the First Resonance.

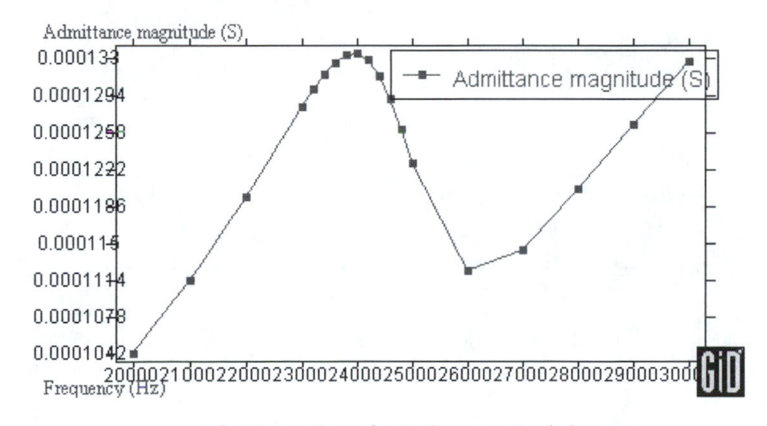

Admittance Curve for Underwater Cymbal

Compared to the no-load Cymbal Impedance curve (see the right figure), water load drops both the resonace frequency and Qm

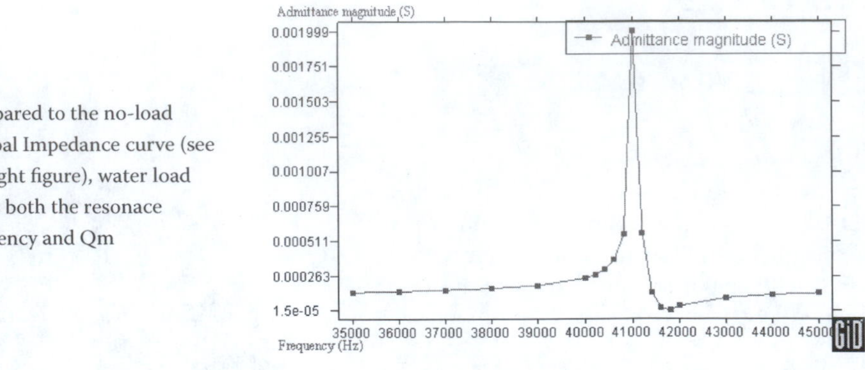

Admittance Curve for Air Cymbal

View Animation—Vibration Mode at a Particular Frequency

View results -> Default Analysis/Step -> Harmonic-Magnitude -> Set a Frequency (Peak Point: 24,000 Hz for Resonance Mode).

Double Click Loading frequency: 24,000 Hz Harmonic Case -> Click OK (single-click and wait for a while).

Set Deformation -> Displacement

View results -> Deformation -> Displacement.

Parameter other than Displacement

Set Results View -> View results -> Contour Lines -> Fluid Pressure.

After setting both "Results view" and "Deformation" -> Click to see the Animation.

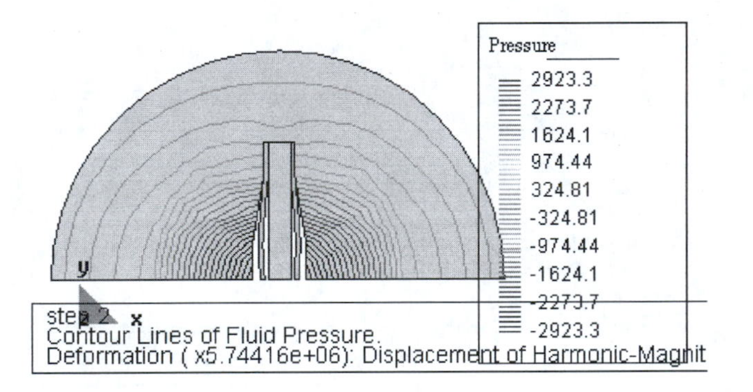

24,000 Hz - Resonance

Transmitting Voltage Ratio (TVR) [as a function of frequency]

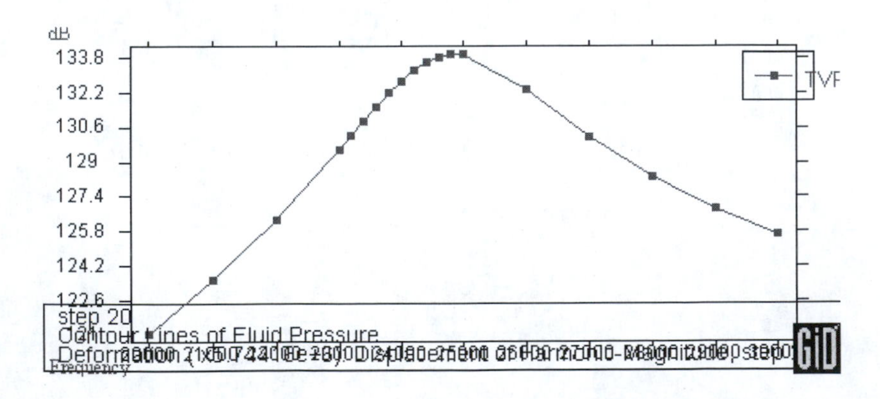

Beam Pattern [as a function of angle]

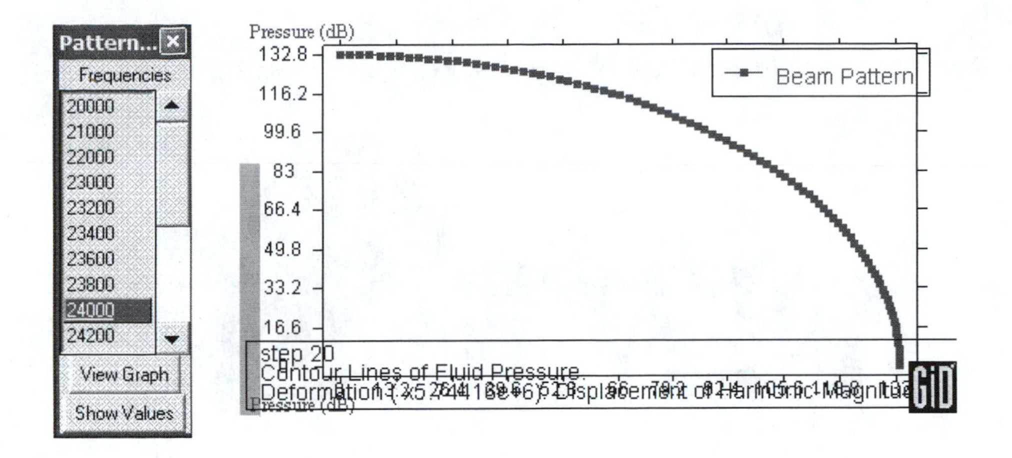

Back to Top

Back to Home

6.3 TONPILZ SONAR (1)

GEOMETRY/DRAWING MATERIAL ASSIGNMENT

PROBLEM—TONPILZ TRANSDUCER UNDER WATER, AXISYMMETRY, PZT8, STEEL, AND ALUMINUM

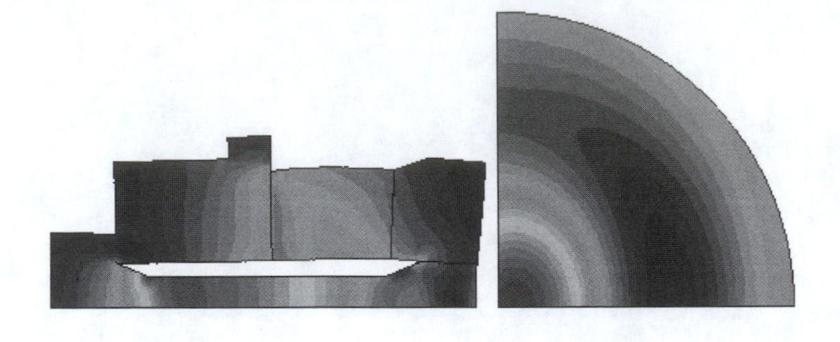

This part describes the harmonic (frequency domain) analysis of a Tonpilz-type transducer. First, the transducer is considered without loading (in-vacuum); then, it is considered to be radiating in water. The transducer is assumed to be axisymmetrical; thus, the model is two-dimensional and only a cross section of the transducer is represented.

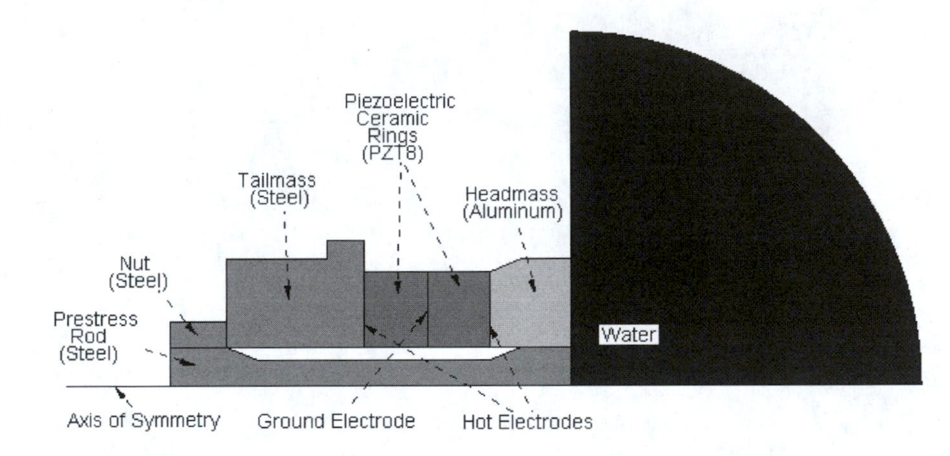

GEOMETRY/DRAWING

The dimensions used for the geometry are (in millimeters) shown in the next table, along with the material type.

Ceramic rings	Inner radius	6	PZT8
	Outer radius	18	
Prestress rod	Length	64	Steel
	Small radius	4	
	Large radius	6	
Head mass	Inner radius	6	Aluminum
	Small outer radius	18	
	Large outer radius	20	
Tail mass	Inner radius	6	Steel
	Small outer radius	20	
	Large outer radius	23	
Nut	Inner radius	6	Steel
	Outer radius	10	
Water	Radius	56	Water

Because this is an axisymmetrical problem, many quantities are affected by a 2π factor of symmetry, such as the impedance, for instance. Always verify that the factor of symmetry is properly taken into account. Generally speaking, all results in the ATILA–GiD interface are provided without correction for the factor of symmetry.

The coordinates of the points you will need to create are shown in the following table, for reference. It is better in acoustic radiation problems to position the origin of the acoustic source (center of the piston face) at the origin of the global coordinate system. Although we do not consider the acoustic radiation problem in the first part of this tutorial, it will save time later to follow the same rules. For this reason, the solid part of the transducer is located in the negative X region, and the fluid domain is in the positive X region.

Note that because ATILA uses the X axis as the axis of axisymmetry, the points are all in the positive Y region of the plane. The coordinates of the points you need to create are arranged in the following table.

			(-39,23)	(-33,23)				
	(-55,20)		(-39,20)				(-8,20)	(0,20)
				(-33,18)	(-23,18)	(-13,18)		
(-64,10)	(-55,10)							
(-64,6)	(-55,6)			(-33,6)	(-23,6)	(-13,6)	(-8,6)	(0,6)
		(-50,4)				(-13,4)		
(-64,0)								(0,0)

Remember that there is more than one way to enter points in GiD in Cartesian and polar coordinates. Please refer to the online help in GiD for further details. Use GiD functions to create points, lines, and surfaces until you obtain the same result as that shown in the next figure.

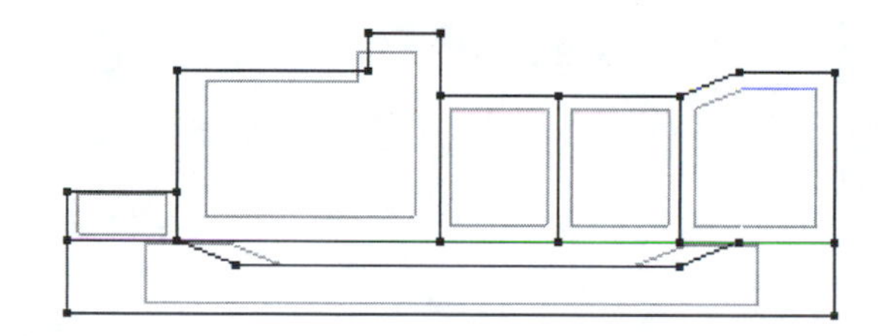

MATERIAL ASSIGNMENT

Piezoelectric Materials

Assign material **PZT8** to the two ceramic surfaces. Note that the definition of the material **PZT8** includes losses. You can verify the loss angles by clicking on the **Losses** tab. The values **Delta_m, Delta_p,** and **Delta_d** correspond to the loss angles for the mechanical, piezoelectric, and dielectric tensors, respectively.

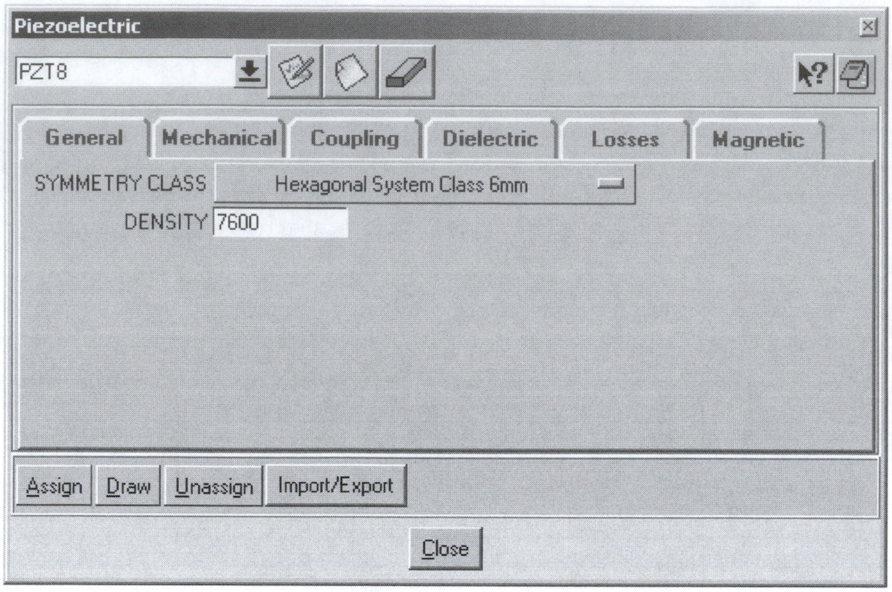

Elastic Materials

Assign material **STEEL1** to the **Tailmass**, **Prestress Rod**, and **Nut** surfaces, and then material **ALUMINUM1** to the **Headmass** surface.

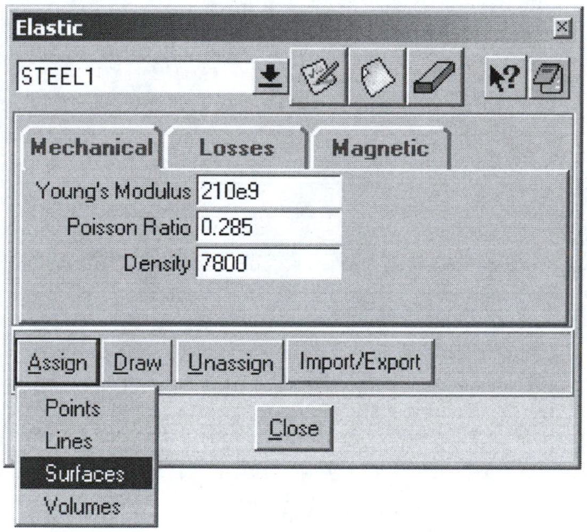

Assign material **STEEL1** to the **Tailmass**, **Prestress Rod**, and **Nut** surfaces, then material **ALUMINUM1** to the **Headmass** surface.

When this is done, the material assignment should be similar to the following image (you can obtain this by selecting, on any material window, the option **Draw | All** materials).

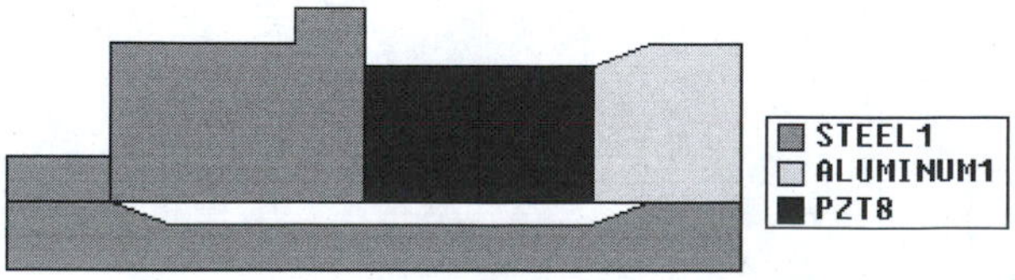

SURROUNDING MEDIUM

Add the required points, lines, and surfaces to create the water domain. The water domain has a radius of 56 mm. Because the geometry of the transducer was created such that the center of the piston would be at the origin of the global coordinate system, the water domain is also centered at the origin. If needed, look in the GiD documentation about how to create an arc.

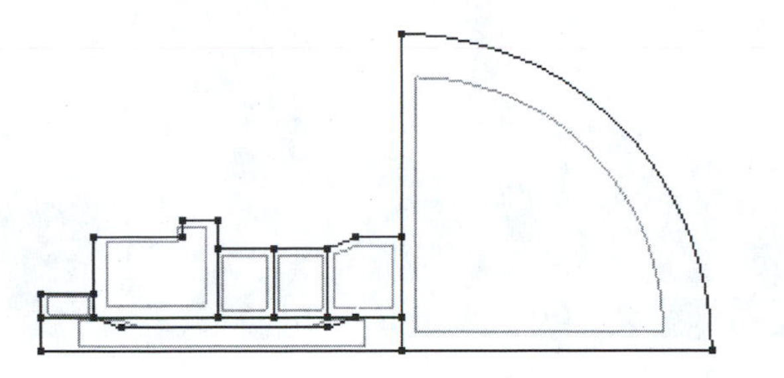

Open the **Fluid** materials window and assign **Water** to the water domain. Verify the material assignment with **Draw | All materials.**

Back to Top

Back to Home

6.3 TONPILZ SONAR (2)

BOUNDARY CONDITIONS

PROBLEM—Tonpilz Transducer under Water, Axisymmetry, PZT8, Steel, and Aluminum

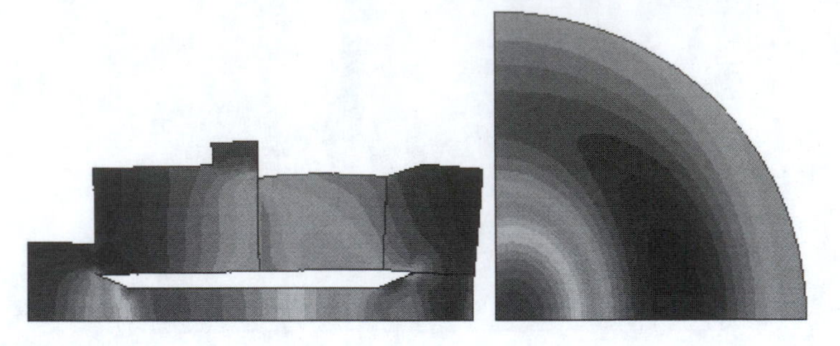

BOUNDARY CONDITIONS

The boundary conditions needed for the problem are the following: electrical conditions on the piezoelectric ceramics, definition of the polarization direction of the ceramics, and the mechanical symmetry conditions.

Potential

Data -> Conditions -> Lines -> Potential 0.0 (or Ground in ATILA 5.2.4 or higher version) -> Assign the side surfaces of the PZT -> Forced Potential 1.0 -> Assign the center surface.

To apply the electrical boundary conditions on the ceramics, select the **Lines** tab, and then **Electric Potential** in the drop-down list. First, apply the **Forced** condition on the hot electrodes. The **Amplitude** and **Phase** of the harmonic signal are 1 V and 0°, respectively.

Conditions
Electric Potential
Potential Forced
Amplitude 1.0
Phase 0.0
Assign Entities Draw Unassign
Close

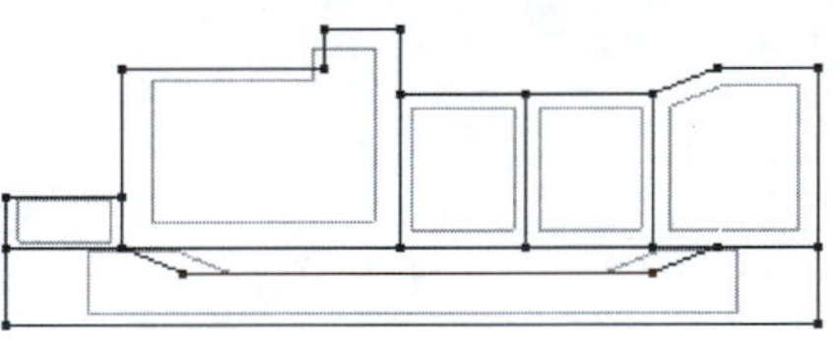

By using a 1 V excitation, the results we will obtain (displacement, stress, and pressure) will be per volt. Note that the results computed by ATILA are proportional to the magnitude of the input signal because a linear formulation is used. This means that an input of 100 V generates displacements, for instance, 100 times larger than with 1 V.

Next, assign a **Ground** (reference) potential on the cold electrode.

Conditions
Electric Potential
Potential Ground
Assign Entities Draw Unassign

Line Constraints

The axisymmetrical nature of the model requires that the displacement of nodes on the axis of axisymmetry (the X axis) be constrained to that axis (nodes on the axis cannot move away from the axis). To assign this condition, select **Line-Constraints** in the drop-down list. Select the **Clamped** option for the Y component of the constraint. Apply this condition to the lines of the solid parts that are on the X axis.

Conditions
Line-Constraints
X-component None
X-Amplitude 0.0
X-Phase 0.0
Y-component Clamped
None
✔ Clamped
Forced
Y-Amplitude 0.0
Y-Phase 0.0
Z-component None
Z-Amplitude 0.0
Z-Phase 0.0
Assign Entities Draw Unassign
Close

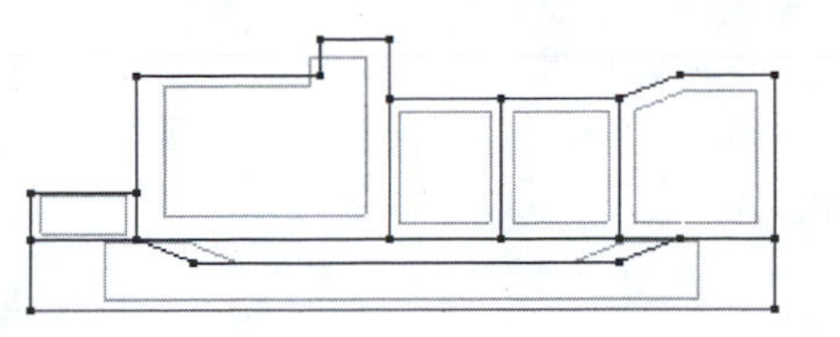

Polarization

The piezoelectric ceramics are poled along their thickness direction such that the transduction occurs in the d_{33} mode. Because the poling of piezoelectric elements such as the rings used in Tonpilz transducers can be assumed to be uniform, we will use a so-called **Cartesian** polarization for each one of the transducing elements. To define **Cartesian** polarizations, **Local** Axes are used. The direction of the polarization is given by the Z axis of the local axes.

First, select the **Surface** tab of the conditions window; then, **Polarization** from the drop-down list. Select **Local-Axes | Define** to create new local axes. Enter *P1* for the name of the local axis.

Select the **3 Points XZ** definition mode, for instance. Here, it is convenient to use geometrical points to define the center and directions of the local axes. Remember that it is also possible to enter point coordinates in the command line.

Conditions
Polarization
GEOMETRY POLARIZATION CARTESIAN
Local-Axes No
✔ No Orientation
Define ...
Assign Entities Draw Unassign
Close

Enter value window
Enter name of new local axes
P1
OK Cancel

Dialog window
Choose definition mode
3 Points XZ Xand Angle Cancel

To use the mouse cursor to select the points, click once on the lower left corner of the PZT region. GiD should prompt you to use the existing point. Click **Join**. Repeat for the top left and lower right points to define the local axes.

Continue in a similar way for the second polarization, *P2*.

Enter value window
Enter name of new local axes
P2
OK Cancel

Once this is done, make sure **Geometry Polarization** is set to **Cartesian.** This will impose a uniform polarization in the selected surfaces. Also, ensure that **Local-Axes** is set to P1. Then assign the condition to one of the piezoelectric surfaces.

Select **P2** and repeat for the second piezoelectric surface.

Conditions
Polarization
GEOMETRY POLARIZATION CARTESIAN
Local-Axes P1
Assign Entities Draw Unassign
Close

Conditions
Polarization
GEOMETRY POLARIZATION CARTESIAN
Local-Axes P2
Assign Entities Draw Unassign
Close

Finally, you may verify that all conditions have been applied by selecting **Draw | All Conditions | Exclude Local** Axes from the conditions window.

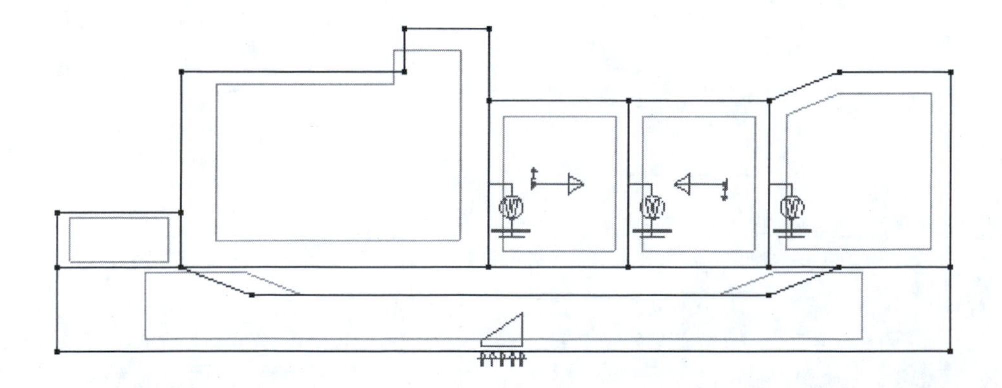

Radiation Boundary

The only additional boundary condition that is needed is the definition of the radiating boundary. From the **Boundary Conditions** window, select the Lines tab, then the **Radiating Boundary** condition. Applying this condition to the arc will cause the pressure flux to be absorbed (and not reflected), thus allowing the acoustic energy to radiate in an infinite medium. Further, the elements of that boundary will perform the computation of the far-field pressure. Rules to position the radiating boundary are complex and vary depending on the problem.

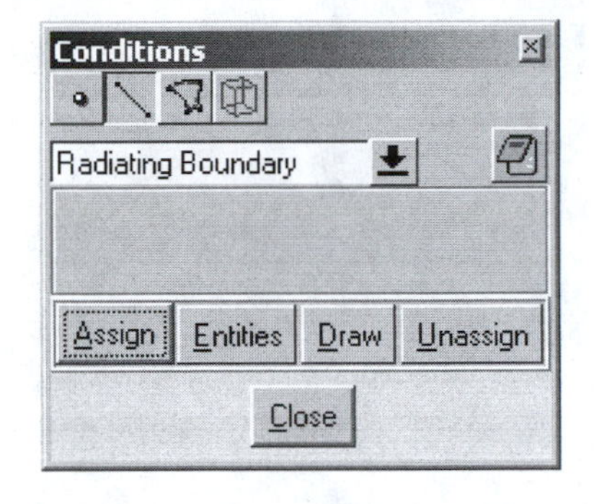

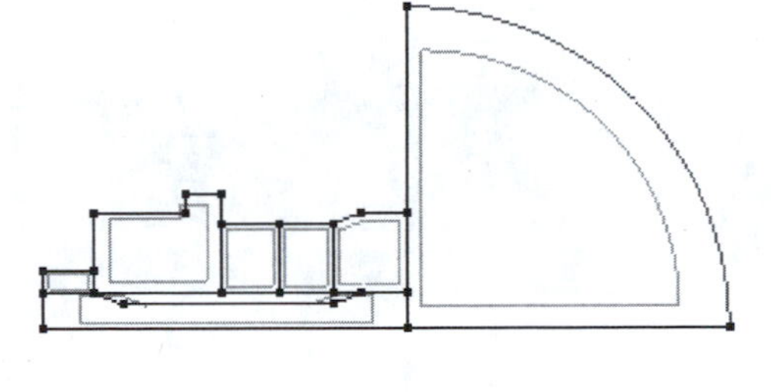

Finally, you may verify that all conditions have been applied by selecting **Draw | All Conditions | Exclude Local Axes** from the conditions window. The **Zigzag Arrow** indicates the location of the radiating boundary.

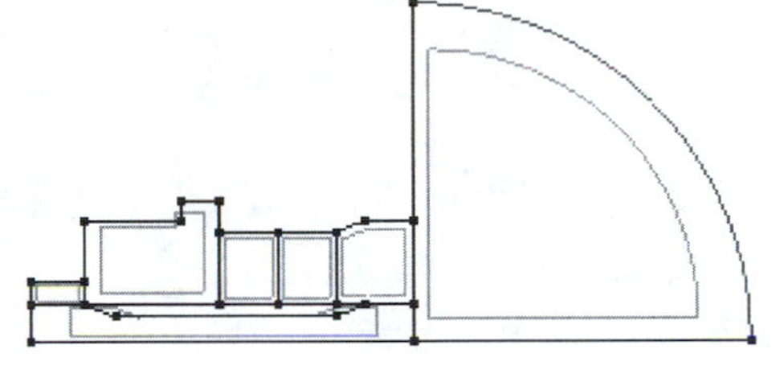

Back to Top

Back to Home

6.3 TONPILZ SONAR (3)

MESHING

PROBLEM—TONPILZ TRANSDUCER UNDER WATER, AXISYMMETRY, PZT8, STEEL, AND ALUMINUM

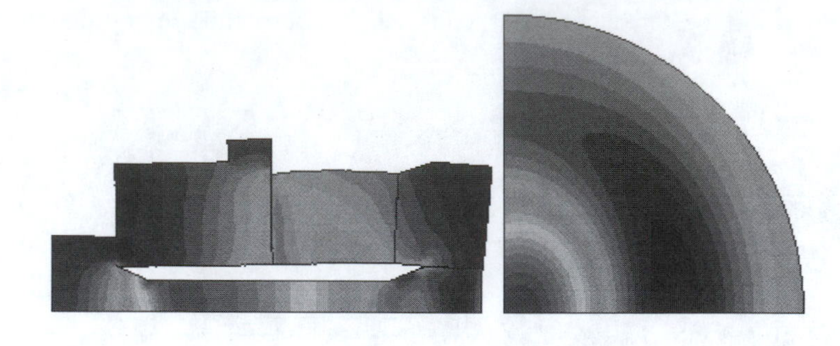

MESHING

ATILA requires that the mesh be composed exclusively of second-order elements. The corresponding setting can be found under the Meshing menu. Select **Quadratic Elements**, and then **Quadratic.** Note that this option is automatically set when you open the ATILA–GiD interface.

We will create a mix of structured rectangles mesh and unstructured triangle mesh. Note that such mixing of element types in GiD is currently only available in 2D. Also, note that only four-sided surfaces can be structured.
First, we define the structured mesh. Select **Meshing | Structured | Surfaces.**

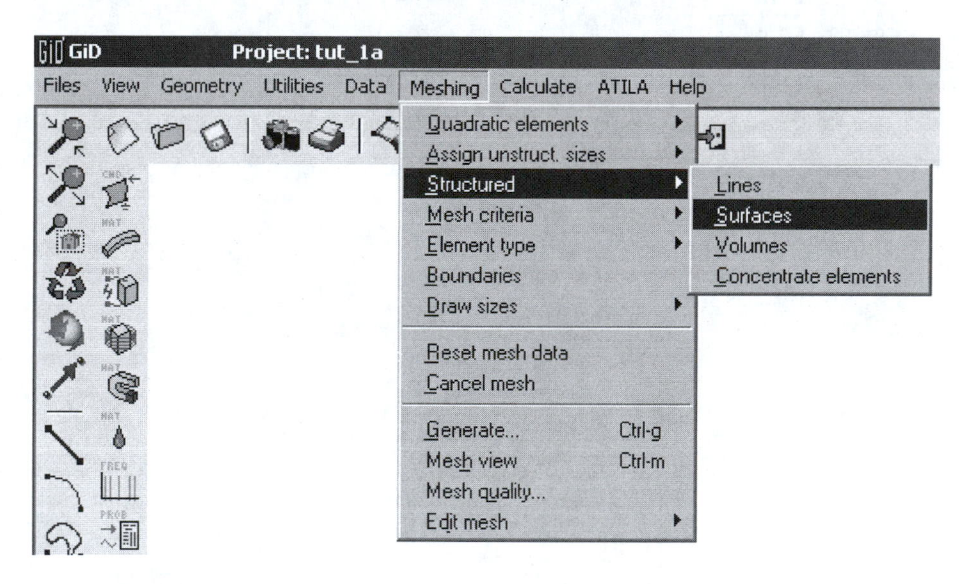

Apply this meshing condition to the piezoelectric surfaces.

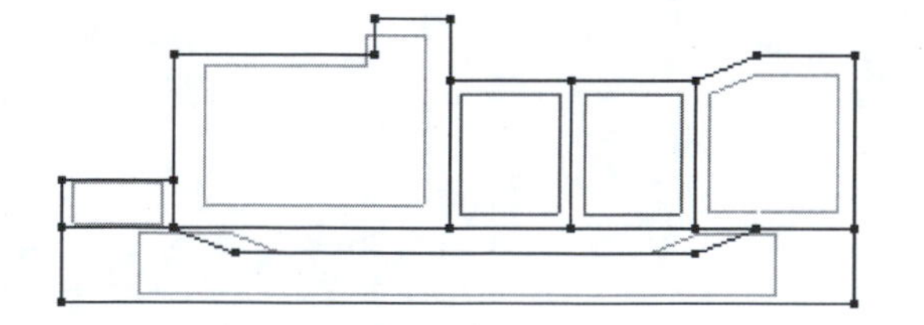

GiD then requests further information about the number of subdivisions that should be applied to lines (this will structure the mesh). Select six mesh subdivisions and apply to all the lines.

Because the remainder of the geometry is meshed automatically, proceed by selecting **Meshing | Generate.** Apply a default size value of 1.9 for the automatic mesh (in the same units as the rest of the model).

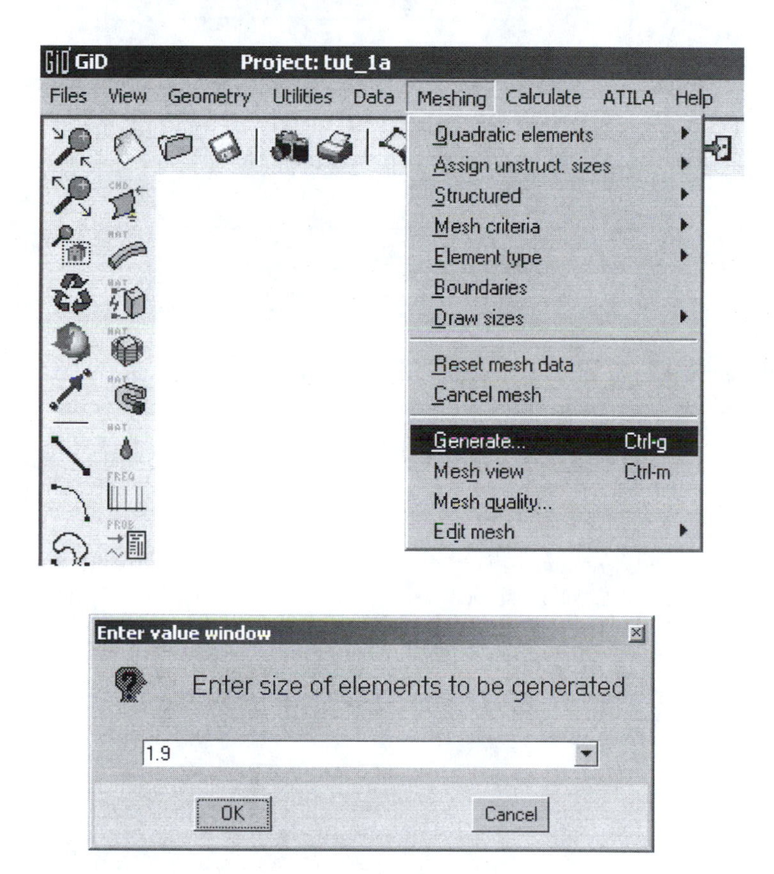

When GiD has terminated the mesh generation, a window is displayed with information about the number of elements that were generated. The mesh is shown in the next figure.

Note that the mesh depends on your settings. To access the settings, select Preferences from the Utilities menu, then open the Meshing tab. The settings used to produce the mesh of this tutorial are shown in the following text.

Preferences
General | Graphical | Meshing | Import | Fonts
Surface mesher:
RFast RSurf 2Dumg
Mesh until end
Automatic correct sizes
Unstructured size transitions
slower
0.2
Smoothing: HighGeom
No mesh frozen layers
Allow automatic structured
Mesh always by default:
Accept Reset Close

Note that we desire to produce a fine mesh here, because it will be desired to compute stresses in the transducer. Stresses are obtained as the first derivative of the finite element problem unknowns (displacements). For these derivatives to be evaluated correctly, a dense mesh is required.

MESHING (WATER)

The mesh can be immediately generated. Previous definitions of the structured mesh are still present in the model. The new water domain will be meshed automatically. The mesh below was obtained with a default element size set to 1.9.

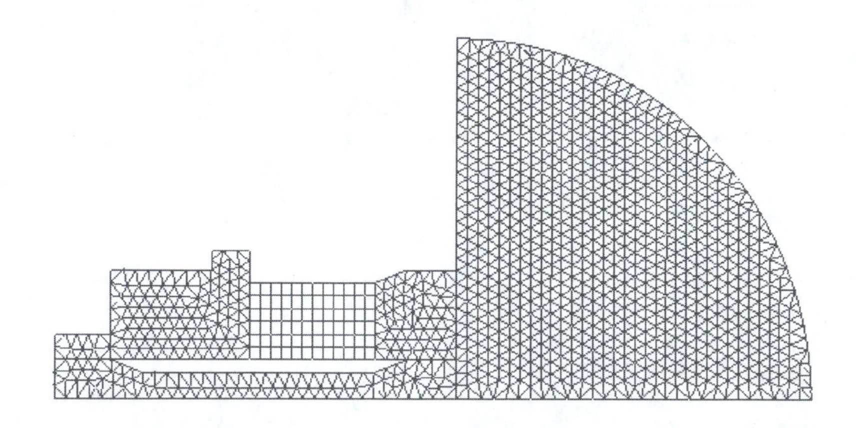

Back to Top

Back to Home

6.3 TONPILZ SONAR (4)

PROBLEM TYPE

PROBLEM—TONPILZ TRANSDUCER UNDER WATER, AXISYMMETRY, PZT8, STEEL, AND ALUMINUM

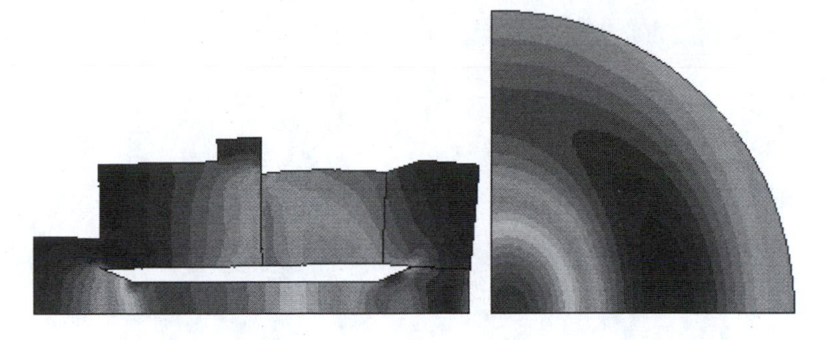

PROBLEM TYPE—MODAL ANALYSIS

PROB Data -> Problem data -> 2D, Axisymmetric.

Make sure **Geometry** is set to **2D**, and that the model **Class** is set to **Axisymmetric.** Next, change the **Analysis** to **Harmonic,** and check both **Compute Stress** and **Include Losses** boxes.

Finally, make sure the **Mesh Units** are set to millimeters.

Do not forget to validate your changes by clicking on **Accept Data.**

Problem Data
Problem Data | Mesh Units
PRINTING 1
GEOMETRY 2D
CLASS AXISYMMETRIC
ANALYSIS HARMONIC
☑ COMPUTE STRESS
☑ INCLUDE LOSSES
Accept data | Close

Problem Data
Problem Data | Mesh Units
MESH UNITS: mm
Accept data | Close

FREQ FREQUENCY INTERVALS

The harmonic analysis is performed for frequency intervals that we must specify. Open the **Frequency Intervals** window, and define a linear frequency sweep between 25 and 55 kHz with increments of 1 kHz.

Interval Data
1
TYPE LINEAR DISTRIBUTION
MIN FREQUENCY 25000
MAX FREQUENCY 55000
NUMBER OF FREQUENCIES 31
Accept data | Close

Next, create a second interval (by copying the first one), and define a frequency interval between 32 and 34 kHz with increments of 100 Hz.

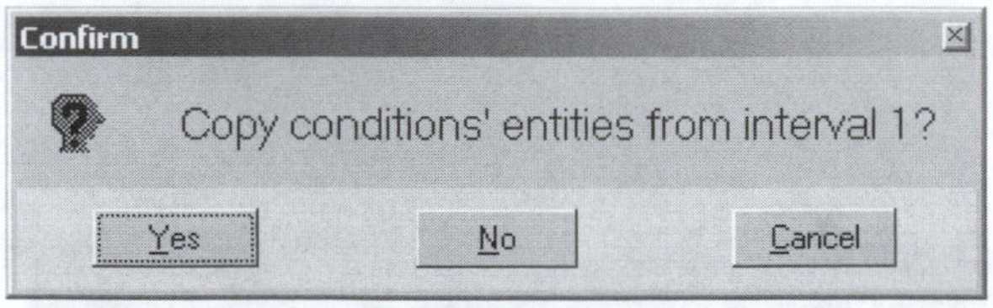

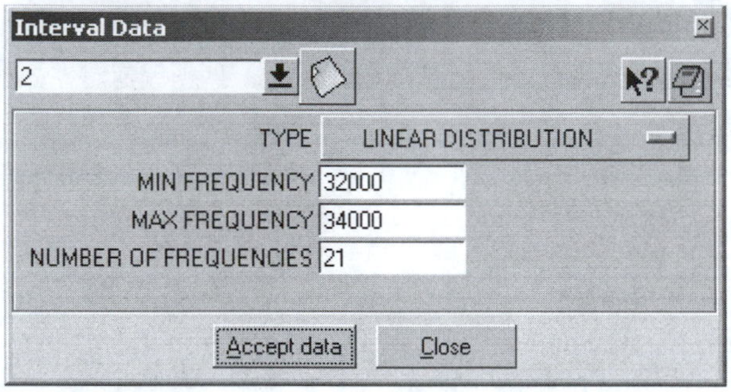

Click on **Accept Data** to validate your changes.

Repeat for a third interval between 36 and 40 kHz with increments of 200 Hz.

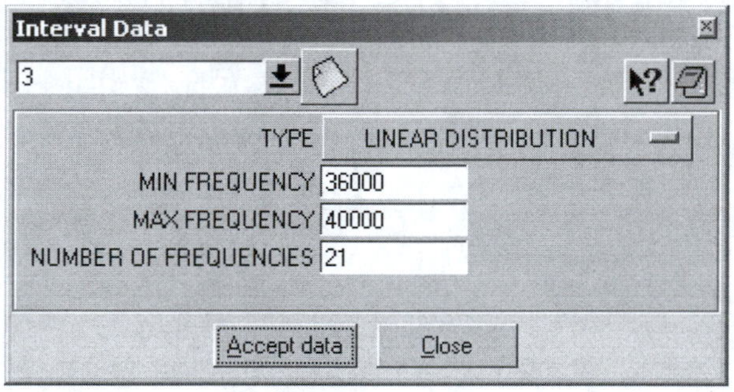

The choice of intervals is dictated by the user's knowledge of the system and desired objectives of the analysis. Here, we use intervals 2 and 3 to obtain a smoother admittance plot around the electrical resonance and antiresonance of the transducer. The frequency intervals are automatically combined and sorted. The list of frequencies that will be used for the harmonic computation is as follows:

25000, 26000, 27000, 28000, 29000, 30000, 31000, 32000, 32100, 32200, 32300, 32400, 32500, 32600, 32700, 32800, 32900, 33000, 33100, 33200, 33300, 33400, 33500, 33600, 33700, 33800, 33900, 34000, 35000, 36000, 36200, 36400, 36600, 36800, 37000, 37200, 37400, 37600, 37800, 38000, 38200, 38400, 38600, 38800, 39000, 39200, 39400, 39600, 39800, 40000, 41000,42000, 43000, 44000, 45000,46000, 47000, 48000,49000, 50000, 51000, 52000, 53000, 54000, 55000

Note that the three lists, composed of 31, 21, and 21 frequencies, add up to one final list of 65 frequencies instead of 73. This is because duplicate frequencies are automatically removed from the final frequency list.

Underwater

The problem data and mesh units have not changed, so there is no need to modify them. However, you may want to modify the frequency intervals 2 and 3 as follows, because the resonance frequency will shift lower due to the water load.

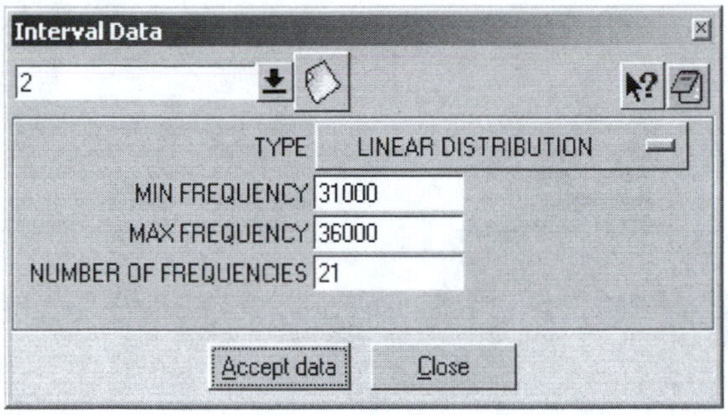

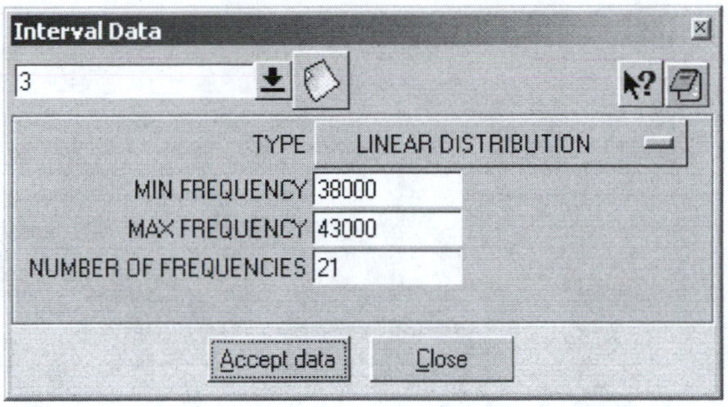

Interval Data
2
TYPE LINEAR DISTRIBUTION
MIN FREQUENCY 31000
MAX FREQUENCY 36000
NUMBER OF FREQUENCIES 21
Accept data Close

Interval Data
3
TYPE LINEAR DISTRIBUTION
MIN FREQUENCY 38000
MAX FREQUENCY 43000
NUMBER OF FREQUENCIES 21
Accept data Close

Back to Top

Back to Home

6.3 TONPILZ SONAR (5)

VIEW RESULTS

PROBLEM—TONPILZ TRANSDUCER UNDER WATER, AXISYMMETRY, PZT8, STEEL, AND ALUMINUM

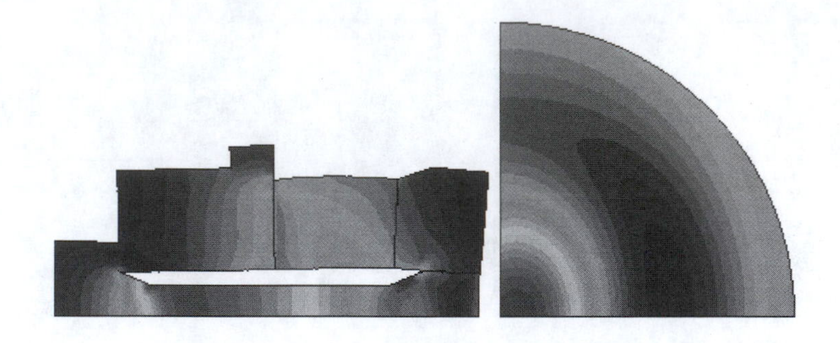

From the **Calculate** menu, select **Calculate.** During the computation, use the **Calculate Window,** as well as **View Process Info** commands to monitor the progression of the computation. The amount of information printed in the **Calculate Window** is controlled by the **Printing** setting **(Problem Data** section). When the computation is finished, GiD displays the following message.

Process info

Process 'spring' started at Sat Sep 27 20:06:33 has finished.

OK Postprocess

POST PROCESS/VIEW RESULTS

Click to Open the Post Calculation (Flavia) Domain.

CLICK ADMITTANCE/IMPEDANCE RESULT

Note that the admittance is given without correction for the model symmetry. The magnitude of the admittance should be multiplied by 2π. This is easily done by using the comma-separated value (CSV) files located in the working directory. Use **Open Shell** from the ATILA menu to access the working directory.

Note that the water load shifts the transducer resonance frequency lower and provides significant damping.

Admittance & Impedance Results

Admittance: Magnitude Phase Conductance Susceptance Circle

Impedance: Magnitude Phase Resistance Reactance Circle

View Graph Show Values

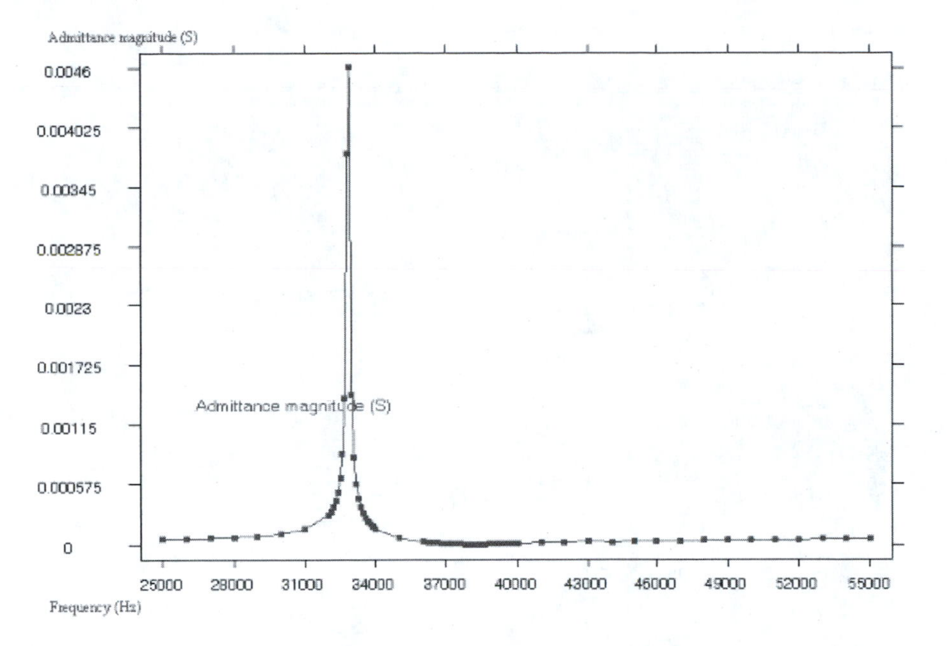

Calculation under Air.

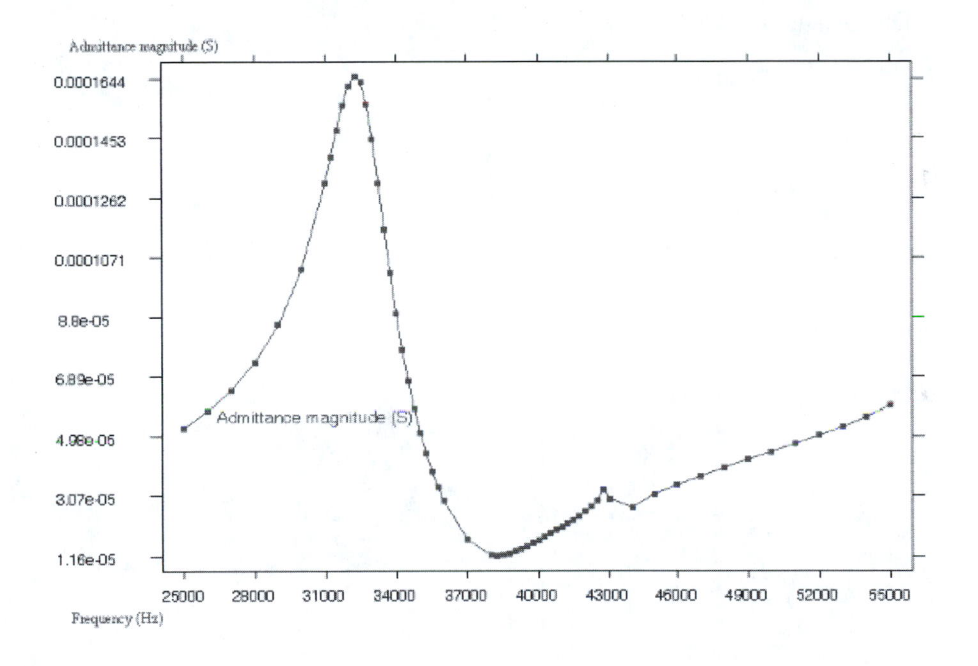

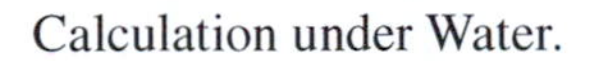
Calculation under Water.

View Animation—Vibration Mode at a Particular Frequency

View results -> Default Analysis/Step -> Harmonic-Magnitude -> Set a Frequency
(Peak Point: 32,900 Hz for the Resonance Mode).

Double Click Loading frequency: 32,900 Hz Harmonic Case -> Click OK (single-click and wait for a while).

A message confirms the frequency and case that you are preparing for animation:

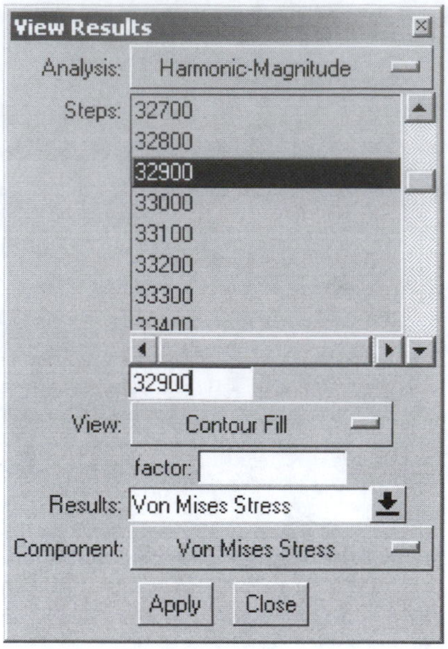

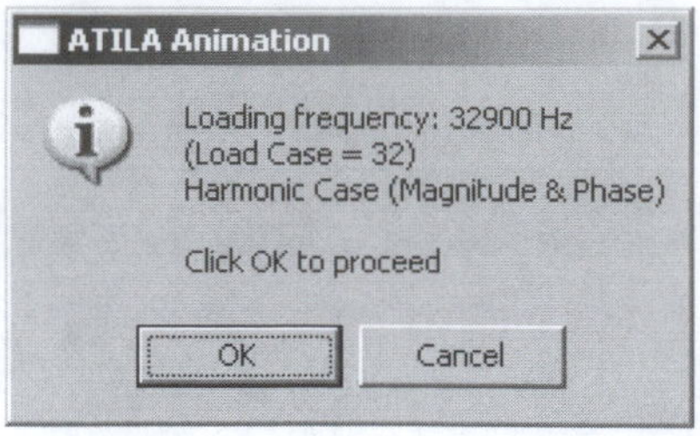

Set Deformation -> Displacement

View results -> Deformation -> Displacement.

Parameter other than Displacement

Set Results View -> View results -> Contour Fill -> Von Mises Stress.

After setting both "Results view" and "Deformation" -> Click  to see the Animation.

Note that we have superposed the magnitude of the Von Mises Stress on the Real Part of the structure deformation. Note that you may want to change the deformation factor (here it is set to $2{\cdot}10^7$).

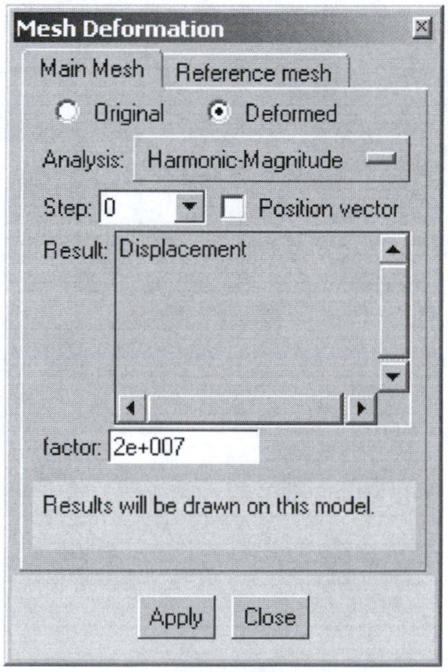

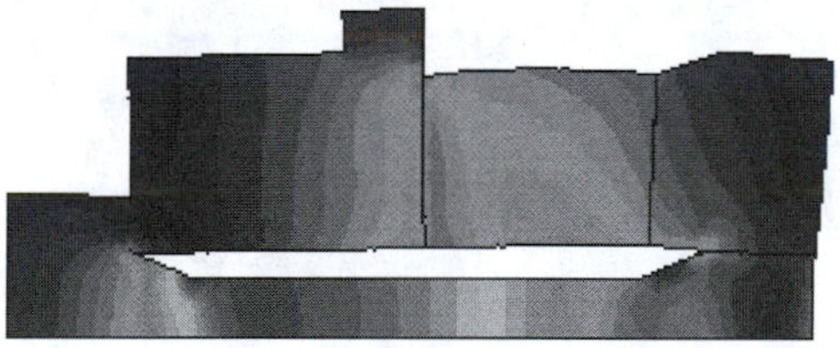

Under Air

Note that the number of frames in the animation is set in the **Preferences (ATILA** menu).

Pressure Observation

You may also want to display the deformation and the pressure Underwater at the frequency of maximum admittance (32.25 kHz).

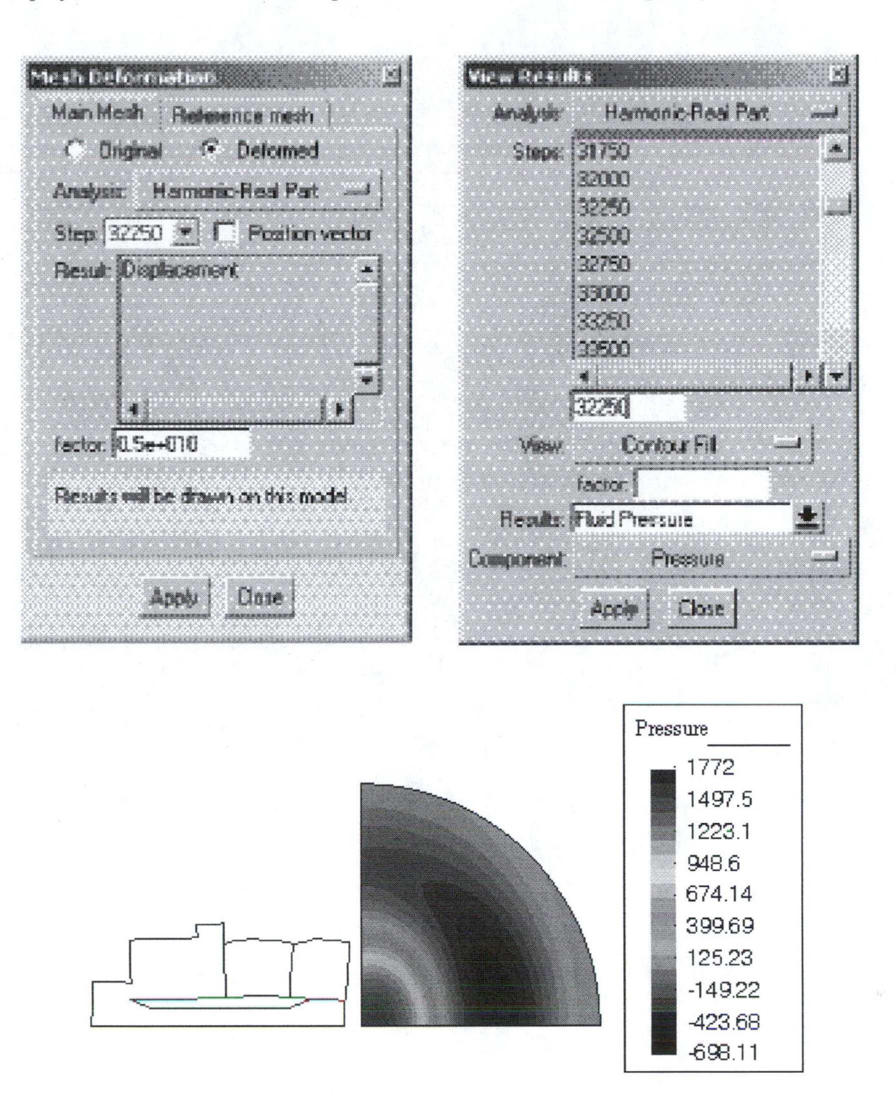

The solid part of the transducers appears disconnected from the water domain. This is normal, because there are no displacements associated with the fluid elements used in this computation. Similarly, the solid part is not colored, because we are plotting pressures, and that there is no pressure degree of freedom in the elastic and piezoelectric domains.

Transmitted Voltage Response (TVR)

Lastly, we can use the acoustic postprocessing features available. For instance, display the **Transmitted Voltage Response (TVR).**

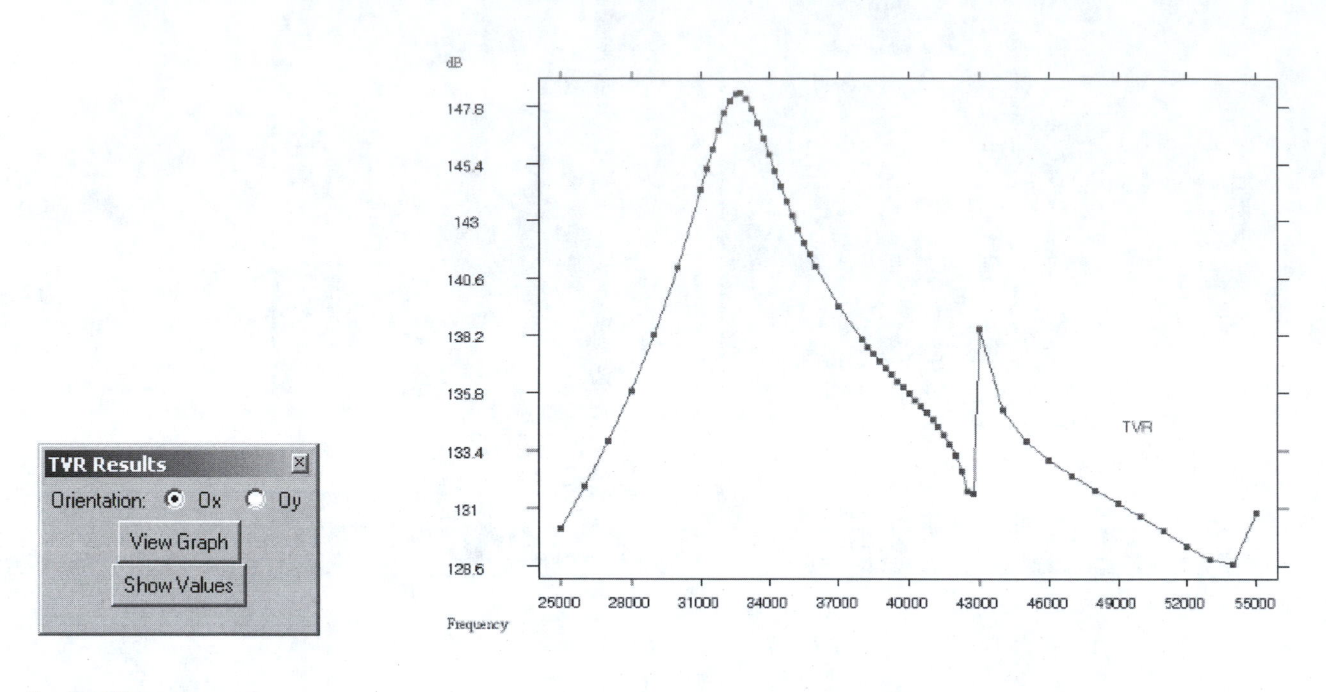

Back to Top

Back to Home

7 Acoustic Lens

- Acoustic impedance

TEXT CONTENT

7. Acoustic Lens

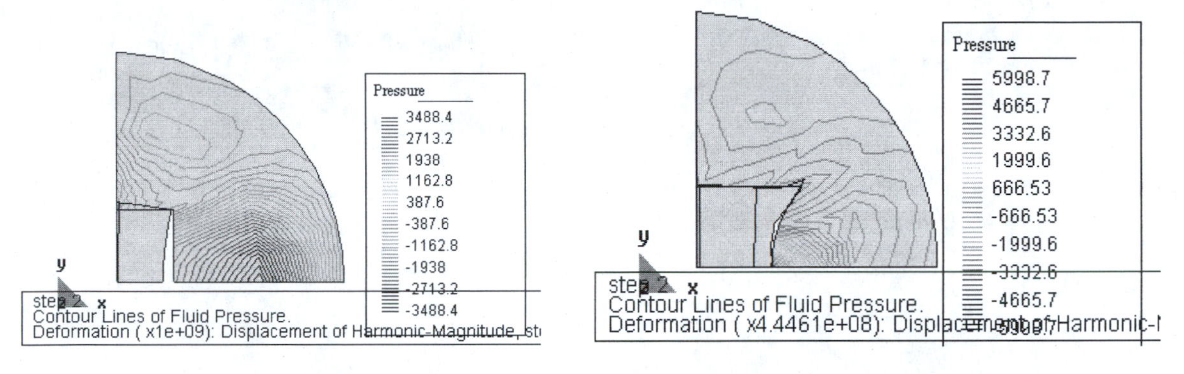

Langevin Transducer (LT)

LT with a plastic acoustic lens. Notice the narrower acoustic beam.

Sound Pressure Calculation in a Tonpilz Transducer

⇨	7. Acoustic Lens
How To Use this GiD file	**Example GiD File** Langevin-Lens.gid

Back to Top

7. ACOUSTIC LENS (1)

GEOMETRY/DRAWING MATERIAL ASSIGNMENT BOUNDARY CONDITIONS MESHING

PROBLEM - Langevin Transducer 50 mm φ × 40 mm t (Two PZT 8 disks with 50 mm φ × 1 mm t are sandwiched by two Steel 1 blocks with 50 mm φ × 19 mm thickness) with an Rubber Acoustic Lens; 2D Axisymmetric with 1/2 portion

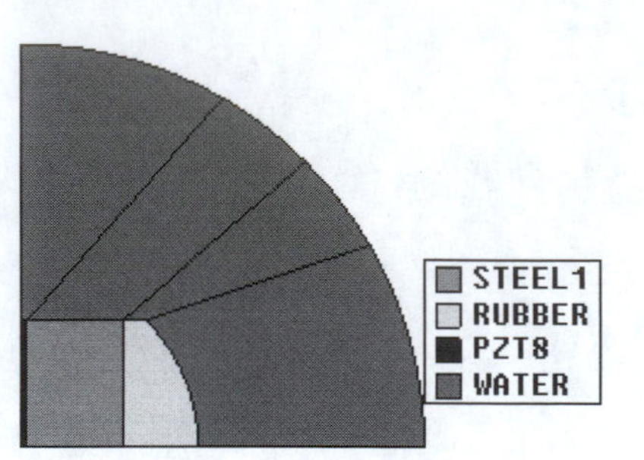

- Learn how to use an acoustic lens
- Check the Beam Patter/Directivity

GEOMETRY/DRAWING

Create Surface

Geometry -> Create -> Object -> Rectangle -> Enter First and Second Corner Points -> 0, 0 ; 20, 25.

Divide Surface

Geometry -> Edit -> Divide -> Surface -> Near point -> Select Surface to Divide -> Choose NURBS sense -> VSense -> Near point 1, 0.

Create Lens

Geometry -> Create -> Object -> Circle -> Enter Center Point-> 0, 0 -> Radius 35.

Divide Surface

Geometry -> Edit -> Divide -> Surface -> Near point -> Select Surface to Divide -> Choose NURBS sense -> USense -> Near point 0, 25 -> Geometry -> Delete -> Surfaces.
Geometry -> Create -> Intersection -> Delete unnecessary lines and points.
Geometry -> Create -> NURB Surfaces -> Automatic -> 4-sided lines.

Create Water Area

Geometry -> Create -> Arc -> First, Second, and Third points -> 80, 0; 0, 80; –80, 0.
Geometry -> Edit -> Divide -> Num Divisions -> Select line -> Num Divisions 2.
Delete -> Line and point.
Then, Geometry -> Create -> Line.

MATERIAL ASSIGNMENT

Piezoelectric Materials

Data -> Materials -> Piezoelectric -> PZT8 -> Assign -> Pick Surface.

Elastic Materials

Data -> Materials -> Elastic -> Steel 1-> Assign -> Pick Surface.
Data -> Materials -> Elastic -> Rubber -> Assign -> Pick Surface.

Fluid Materials

Data -> Materials -> Fluid -> Water-> Assign -> Pick Surface.
Click Draw -> All materials -> Check that the materials are assigned as PZT8, Steel 1, and Water.

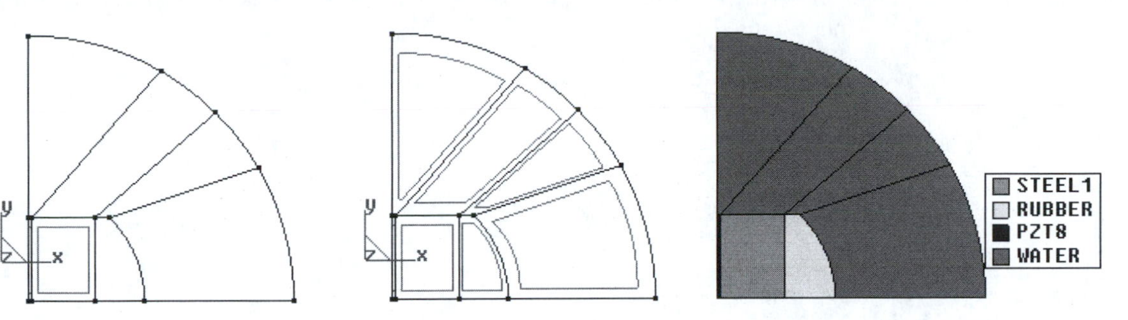

BOUNDARY CONDITIONS

Polarization/Local Axes

High voltage should be applied on the center surface of the two layers of the PZT, and Ground on the metal-side surface. Choose the polarization direction along the electric field direction.

Data -> Conditions -> Surfaces -> Geometry -> Polarization (Cartesian) -> Define Polarization (P1) (1, 0) direction.

Choose 3PointsXZ -> Center (0, 0, 0) -> Define X axis by putting (0,1, 0) -> Then define Z axis by putting (1, 0, 0) -> Escape.

Assign Polarization (Local-Axes P1) -> Choose the PZT surface -> Finish.

Potential

Data -> Conditions -> Lines -> Potential 0.0 (or Ground in ATILA 5.2.4 or higher version) -> Assign the right-side line of the PZT -> Forced Potential 1.0 -> Assign the left-side line of the PZT (which is the center surface of the two PZT plates).

Surface Constraints (half of the whole Cymbal)

We use only half of the whole Langevin transducer, taking symmetry into account; the left-side line of the PZT and water should be clamped (X component is clamped).

The bottom axisymmetric line of the whole system (including water) - Y component is clamped. The axisymmetric axis is taken along X axis (This is the rule in ATILA).

Data -> Conditions -> Lines -> Line Constraints -> X components -> Clamped -> Assign the left-side PZT and Water lines.
Similarly, on the Axisymmetric X axis -> Line Constraints -> Y and Z components -> Clamped -> Assign.

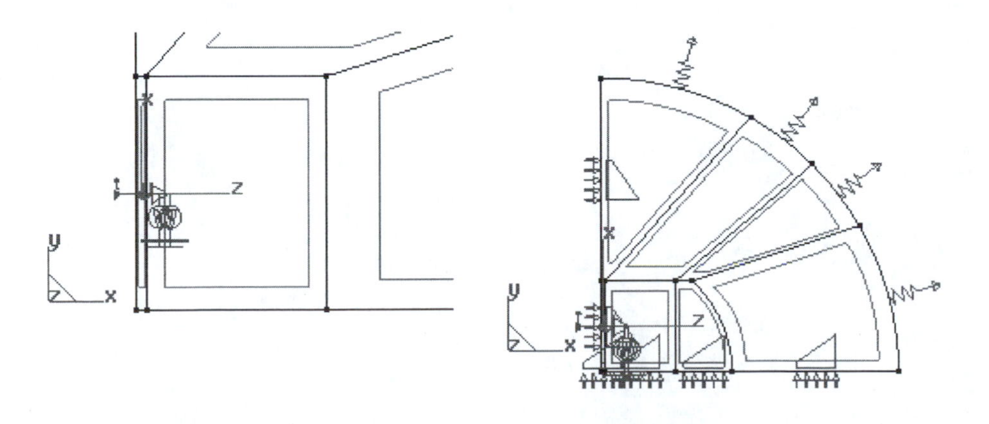

Radiation Boundary

Data -> Conditions -> Lines -> Line Constraints -> X components -> Clamped -> Assign the left-side PZT and Water lines.

Radiation boundary is applied not to reflect the water-propagating sound at this boundary (like an anechoic wall). In this sense, the radius of this water boundary is not essential.

To check the assignment -> Draw -> All conditions -> Include local axes.

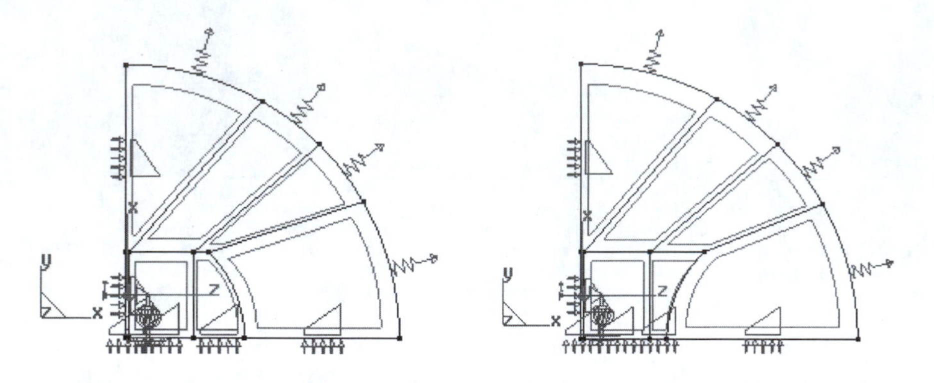

MESHING

Mapped Mesh

Meshing -> Structured -> Surfaces -> Select all Surfaces -> Then Cancel the further process.
Meshing -> Structured -> Lines -> Enter number to divide line -> Choose "1, 2, 4" for the lines -> Generate.

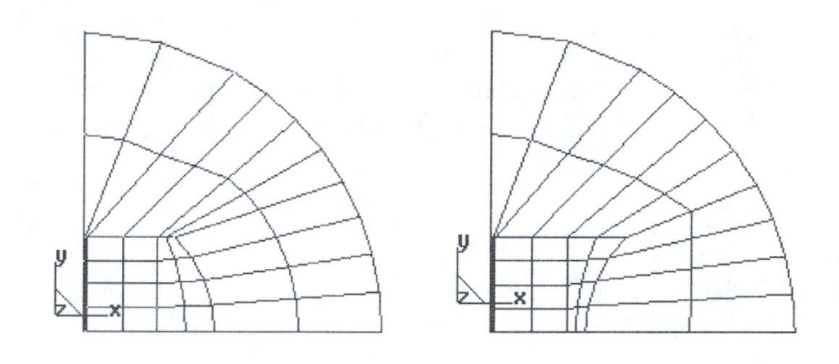

Back to Top

Back to Home

7. ACOUSTIC LENS (2)

PROBLEM TYPE VIEW RESULTS

PROBLEM - Langevin Transducer 50 mm φ × 40 mm t (Two PZT 8 disks with 50 mm φ × 1 mm t are sandwiched by two Steel 1 blocks with 50 mm φ × 19 mm thickness) with a Rubber Acoustic Lens; 2D Axisymmetric with 1/2 portion

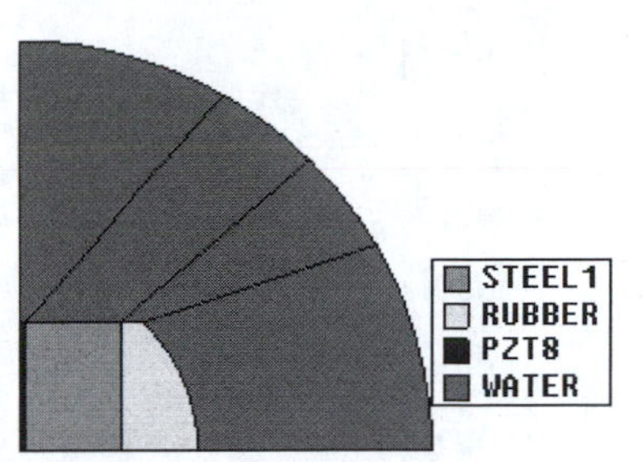

- Learn how to use an acoustic lens
- Check the Beam Patter/Directivity

PROBLEM TYPE

] Data-> Problem data -> 2D, Axisymmetric, Harmonic, Click both Compute Stress and Include Losses.

FREQUENCY

Data -> Harmonic -> Interval 1 (20,000–100,000 Hz), Interval 2 (45,000–55,000 Hz), Interval 3 (80,000–90,000 Hz).

CALCULATION

Calculate -> Calculate window -> Start.

POST PROCESS/VIEW RESULTS

Click to Open the Post Calculation (Flavia) Domain.

Click Admittance/Impedance Result

Admittance Peak Point: 50 kHz -> This peak corresponds to the First split Resonance.

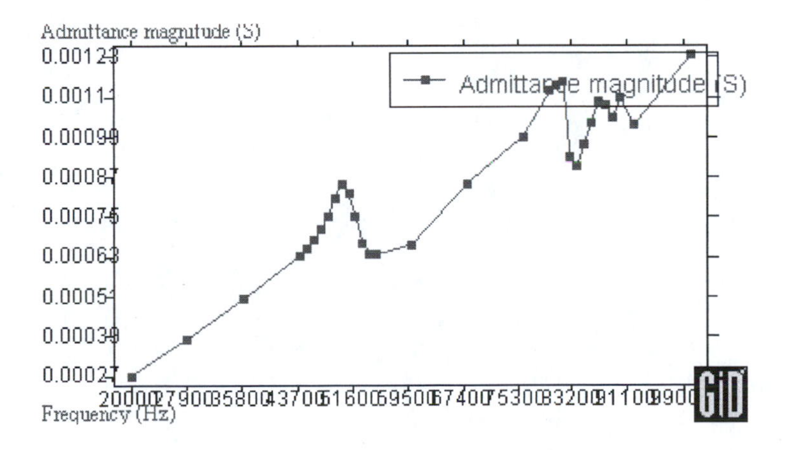

Admittance Curve with a lens (50,000 Hz): Damped compared to the pure Langevin Transducer.

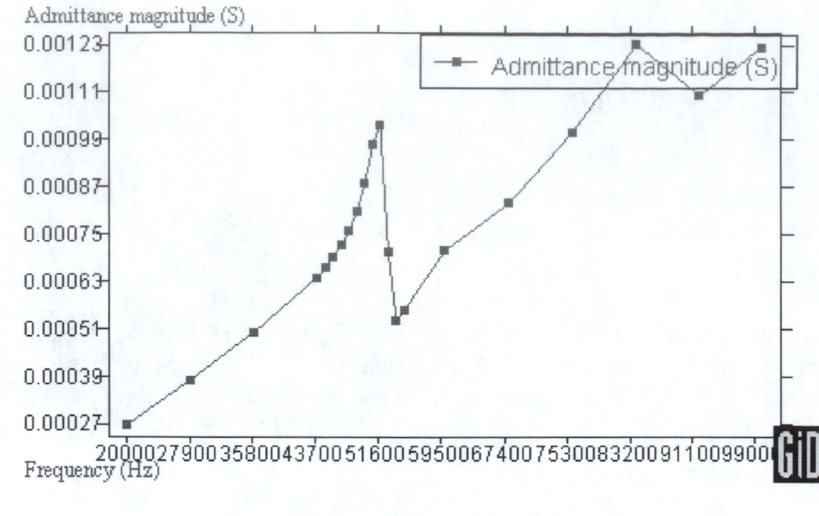

Admittance Curve without a lens (51,000 Hz)

View Animation—Vibration Mode at a Particular Frequency

View results -> Default Analysis/Step -> Harmonic-Magnitude -> Set a Frequency (Peak Point: 50,000 Hz for the Resonance Mode).

Double Click Loading frequency: 50,000 Hz Harmonic Case -> Click OK (single-click and wait for a while).

Set Deformation -> Displacement

View results -> Deformation -> Displacement.

Parameter other than Displacement

Set Results View -> View results -> Contour Fill -> Von Mises Stress.

After setting both “Results view” and “Deformation” -> Click to see the Animation.

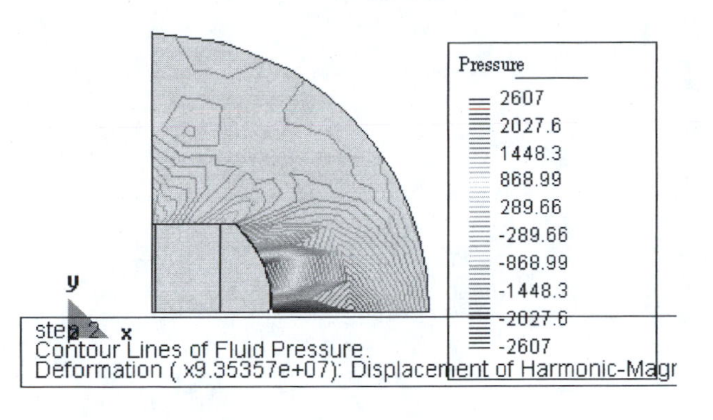

Wider Beam

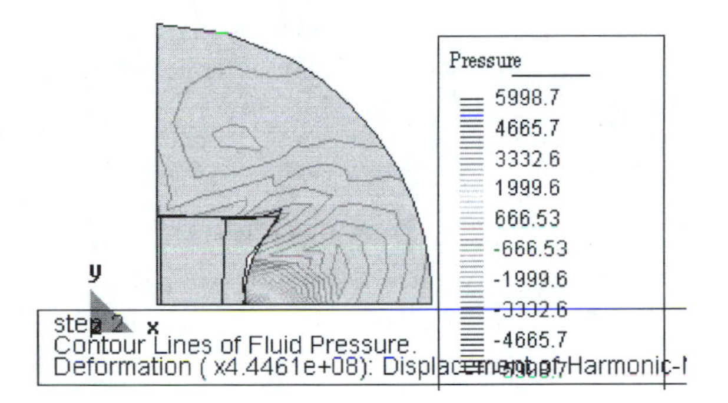

52,000 Hz Narrower Beam

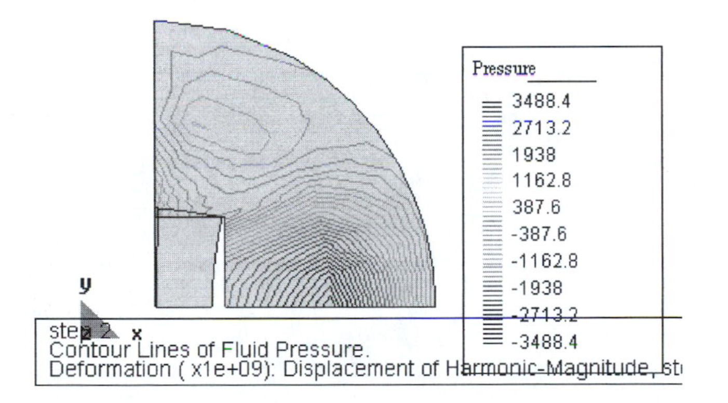

51,000 Hz - First Lower Resonance

NB: A concave lens narrows the acoustic beam; while a convex lens diverges the beam. The situation is opposite in comparison with optical lenses.

Back to Top

Back to Home

APPENDIX: Homework

1.1 RECTANGULAR PLATE (HOMEWORK 1)

SECOND-ORDER HARMONIC VIBRATION MODE

PROBLEM—3D RECTANGULAR PLATE, k_{31} MODE, PZT4 = 40 MM × 6 MM × 1 MM

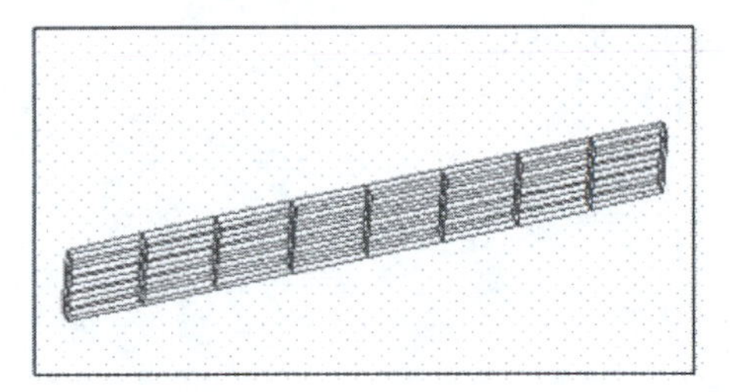

- Note that the second resonance frequency is thrice the first resonance frequency.

1. Loading the frequency range around 120 kHz, calculate the admittance spectrum for the second higher-order harmonic vibration.
2. Loading the maximum frequency, obtain the second-order harmonic vibration mode.
3. Compare the maximum strain and stress levels (under 1 V applied) among the first and second modes.

You are requested to use a different mesh, and more calculation frequency points (21 points minimum). The admittance spectrum, vibration-mode animation, and the maximum strain/stress table (comparing the first and second modes) should be submitted in a PowerPoint presentation file.

HINT: VIEW RESULTS—HIGHER-ORDER HARMONIC

Loading frequency: 122,000 Hz -> Harmonic Case -> OK.
Set Deformation -> Displacement.
Set Results -> View -> View results -> Contour Fill -> Stress - > XX-stress.

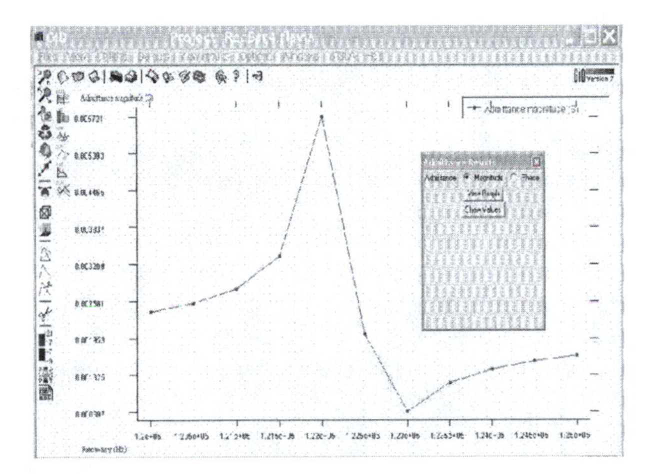

1.2 CIRCULAR DISK (HOMEWORK 2)

RING VIBRATOR

PROBLEM—3D CIRCULAR PLATE, PZT4 = 50 MM OUTER Φ (INNER HOLE 20 MM Φ) × 1 MM *T* RADIALLY POLARIZED

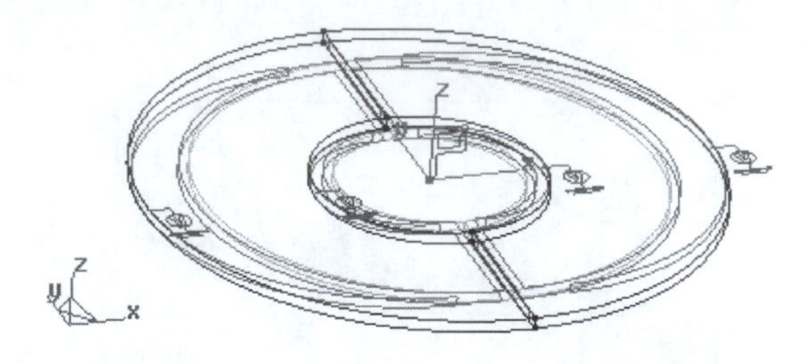

- Learn how to assign the polarization in the radial direction:

1. Draw a PZT4 ring with outer and inner hole diameters of 50 mm and 20 mm, respectively.
2. Polarize the PZT ring in the radial direction, and apply a voltage (1 V) on the inner surface and ground on the outer surface.
3. Calculate the vibration mode for 1 kHz (off-resonance) and around 30–35 kHz (resonance).
4. The admittance spectrum and vibration-mode animation should be submitted in a PowerPoint presentation file. Also, discuss the physical reason why it is largely the inner part that vibrates rather than the outer part.

HINT: CYLINDRICAL POLARIZATION

This type of condition applies to materials that are polarized radially with respect to an axis. Piezoelectric tubes are a very common example. Here, the direction *z* of the local coordinate system defines the axis of the cylinder, as well as the direction M2 of the material. The direction *M*3 of the material is radial with respect to the cylinder. Finally, M2 is in tangential to the cylinder.

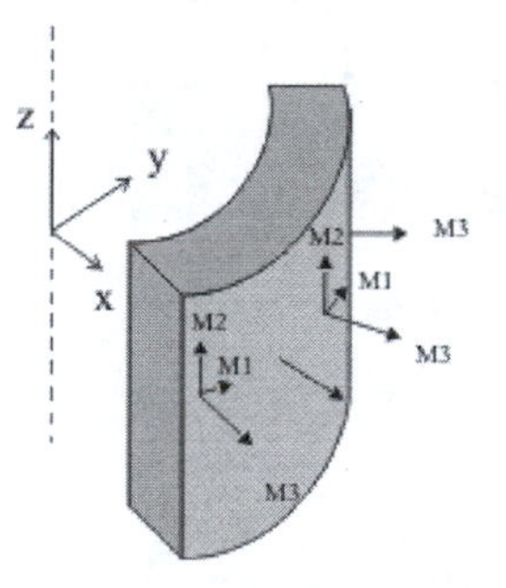

NB: The local Z axis is not the polarization direction in this case.

GEOMETRY/DRAWING

CREATE SURFACE

Geometry -> Create -> Object -> Circle -> Enter a center for the circle -> 0,0 -> Enter a normal for the circle -> Positive Z -> OK -> Enter a radius for the circle -> 25 -> Enter.

Geometry -> Create -> Object -> Circle -> Enter a center for the circle -> 0,0 -> Enter a normal for the circle -> Positive Z -> OK -> Enter a radius for the circle -> 10 -> Enter.

Geometry -> Delete -> Surface.

Now, continue the process by yourself.

POST PROCESS/VIEW RESULTS

Click to Open the Post Calculation (Flavia) Domain.

Click Admittance/Impedance Result

Admittance Peak Point: 32,000 Hz -> This peak corresponds to the First Resonance.

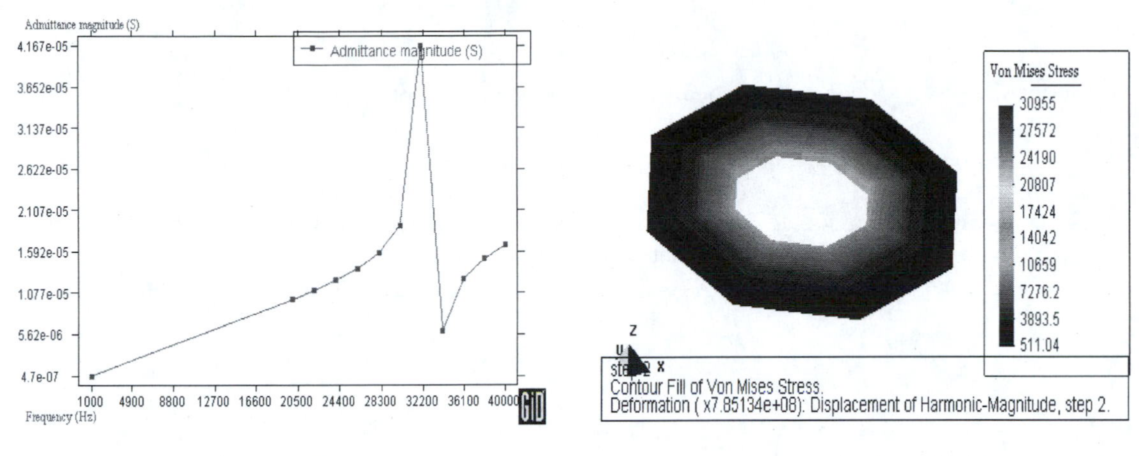

Admittance Curve

Resonance Mode at 32 kHz

Back to Top

Back to Home

3.2 UNIMORPH (HOMEWORK 3)

Clamping/Supporting Configuration

PROBLEM—3D Rectangular Plate, PZT4 and Brass = 40 mm × 6 mm × 1 mm, Two-End Support

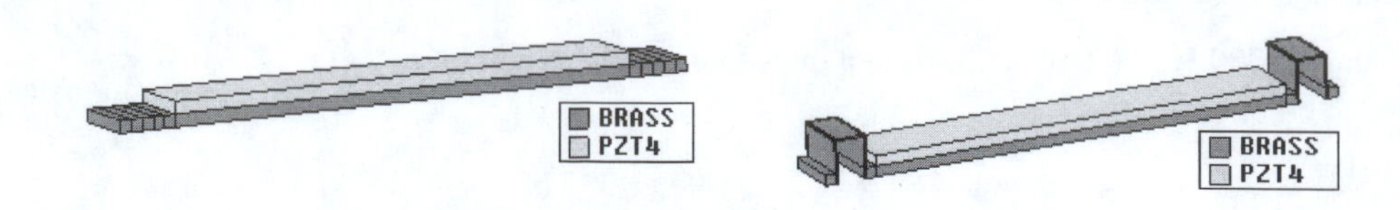

- We will consider a two-end supporting mechanism of the unimorph without losing the displacement:

1. Adding 5-mm-long brass plate (left figure on the top), calculate the center displacement for the both-end support (clamped) unimorph.
2. Adding clipped 5-mm-long brass plate (right figure on the top), calculate the center displacement for the both-end support (clamped) unimorph. Use a thin brass plate (such as 0.2 mm) for the clip parts.
3. Compare the static center displacement (100 Hz) under 1 V applied, and discuss the merit of the supporting mechanism.

HINT:

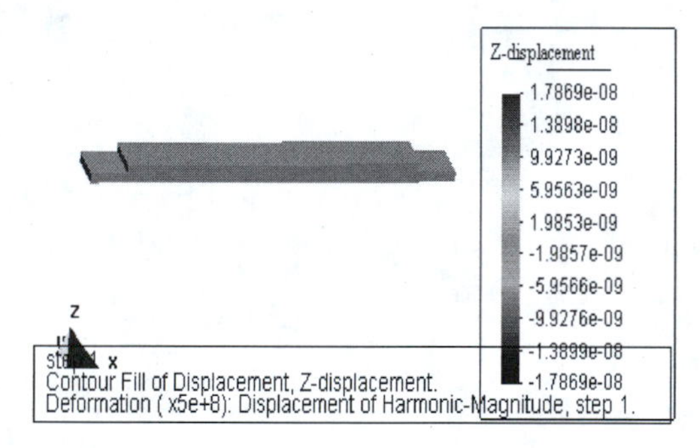

Both-end Clamp Unimorph Under 1 V Applied
(+0.017 μm)

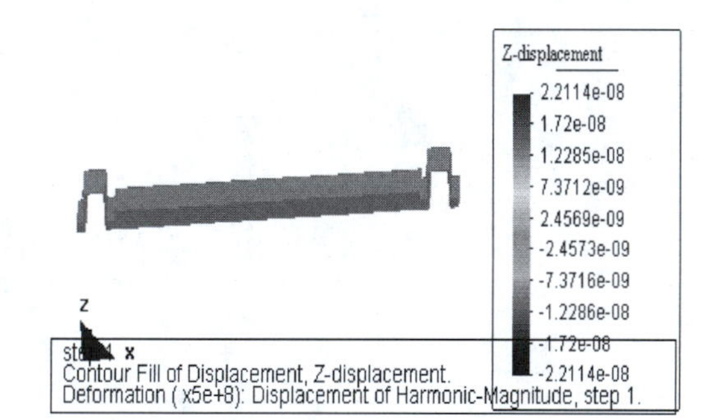

Both-end Clamp Unimorph Under 1 V Applied
(+0.022 μm)

Back to Top

Back to Home

3.2 UNIMORPH (HOMEWORK 4)

Unimorph Optimized Design

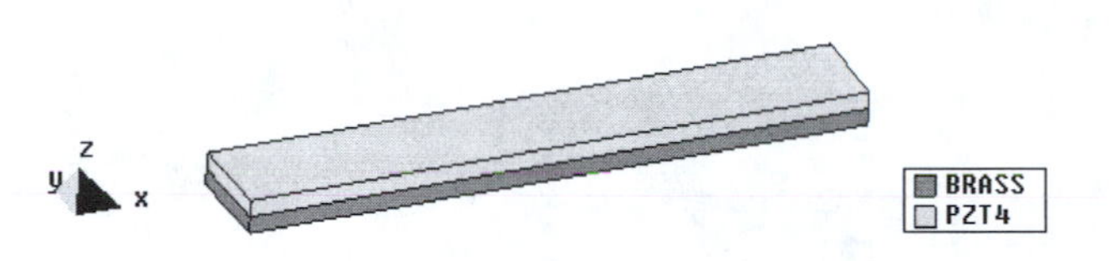

We will decide the brass plate thickness to obtain the maximum tip displacement under the same voltage applied:

1. Retaining the PZT4 thickness as 1 mm, change the brass plate thickness from 0.2 to 1.2 mm (as much as possible).
2. Calculate the tip displacement for a cantilever support configuration, under 1 V applied on the PZT plate.
3. Plot the tip displacement (1 V) as a function of the brass plate thickness, and obtain the optimized thickness for the brass plate.

HINT:

When a piezoceramic plate is bonded to a metallic shim, a unimorph bending device can be fabricated. The tip deflection d of the unimorph supported in a cantilever style is given by:

$$d = (d_{31}E)\,L^2 Y_c\,t_c \,/\, [Y_m\{t_0^2 - (t_0 - t_m)^2\} + Y_c\{(t_0 + t_c)^2 - t_0^2\}].$$

Here, E is the electric field applied to the piezoelectric ceramic; d_{31}, the piezoelectric constant; L, the length of this unimorph; Y_c or Y_m, Young's modulus for the ceramic or the metal; and t_c or t_m is the thickness of each material. In addition, t_0 refers to the distance between the strain-free neutral plane and the bonding surface, which is represented as follows:

$$t_0 = [t_c\,t_m^2(3\,t_c + 4\,t_m)\,Y_m + t_c^4 Y_c] \,/\, [6\,t_c\,t_m(t_c + t_m)\,Y_m].$$

If $Y_c = Y_m$, the maximum displacement is obtained when $t_m = t_c/2$.

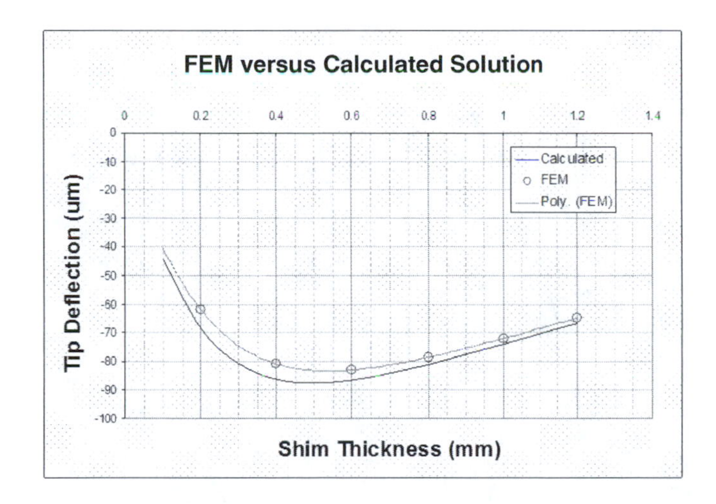

Back to Top

Back to Home

3.3 MULTILAYER (HOMEWORK 5)

2D MULTILAYER

PROBLEM—2D 8-LAYER, PZT4, WIDTH 5 MM × LAYER THICKNESS 1 MM, INTERDIGITAL ELECTRODE

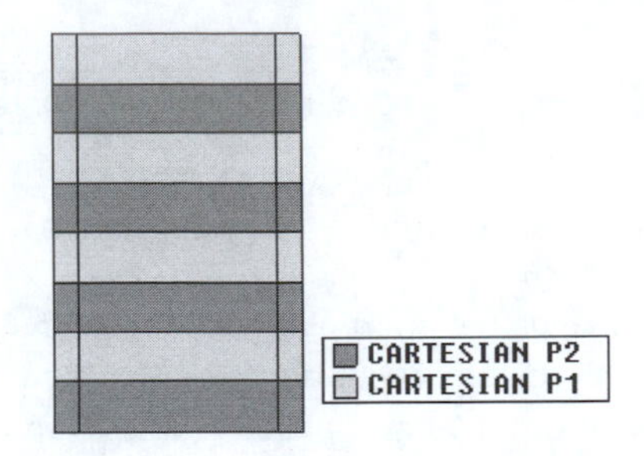

- Composite of two different polarization layers, with an interdigital electrode configuration.
- Learn the internal stress distribution:

1. Calculate the potential and stress distributions in a 2D piezoelectric PZT4 multilayer actuator with the following dimensions under a static and a dynamic drive: width 5 mm × 8 layers of 1 mm thick each layer with an interdigital electrode configuration (0.5 mm embedded from the surface).
2. Show the displacement motion of this actuator (show the nonuniform surface motion of the actuator) under static (100 Hz) drive.
3. Calculate the admittance spectrum as around 190 kHz.
4. Discuss the deformation/stress distribution difference between the off-resonance (100 Hz) and the resonance mode (190 kHz).

HINT: GEOMETRY/DRAWING

CREATE SURFACE

Geometry -> Create -> Object -> Rectangle -> Enter first corner point -> 0, 0 -> Enter second corner point -> 5, 1 -> Enter.

DIVIDE SURFACE

Geometry -> Edit -> Divide -> Surfaces -> Near point -> Select the Surface -> Choose NURBS sense -> Input > Near Points -> 0.5, 0 and 4.5, 0.

CREATE MULTILAYER COPY MODE

Utilities -> Copy -> Surfaces -> Translate by 1 mm (No Extrude) -> Select.

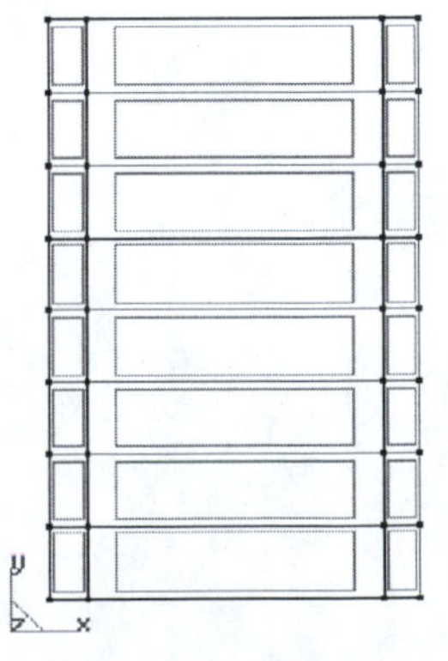

MATERIAL ASSIGNMENT

Piezoelectric Materials

Data -> Materials -> Piezoelectric -> PZT4 -> Assign -> Surfaces.

BOUNDARY CONDITIONS

Polarization/Local Axes

Data -> Conditions -> Surfaces -> Geometry -> Polarization (Cartesian) -> Define Polarization (P1) (0,0,1) direction.
Choose 3PointsXZ -> Define X axis by putting (1, 0, 0) -> Then define Z axis by putting (0, 0, 1) -> Escape.
Assign Polarization (Local-Axes P1) -> Choose every two-layer surface -> Finish.

Data -> Conditions -> Surfaces -> Geometry -> Polarization (Cartesian) -> Define Polarization (P2) (0, 0, −1) direction.
Choose 3PointsXZ -> Define X axis by putting (1, 0, 0) -> Then define Z axis by putting (0, 0, −1) -> Escape.
Assign Polarization (Local-Axes P2) -> Choose every two-layer surface -> Finish.

Potential

Data -> Conditions -> lines -> Potential 0.0 (or Ground in ATILA 5.2.4 or higher version) -> Assign every two-layer line-> Forced Potential 1.0 -> Assign every two-layer line.

MESHING

Mapped Mesh

Meshing -> Structured -> Surfaces -> Select all surfaces -> Then Cancel the further process.
Meshing -> Structured -> Lines -> Enter number to divide line -> Choose necessary numbers -> Generate.

PROBLEM TYPE

Data -> Problem data -> 2D and Plane Strain, Harmonic, Click both Compute Stress and Include Losses.

FREQUENCY

Data -> Interval data Interval 1 (100 Hz), Interval 2 (180,000–200,000 Hz).

CALCULATION

Calculate -> Calculate window -> Start.

POST PROCESS/VIEW RESULTS

Click to Open the Post Calculation (Flavia) Domain.
Note the new Toolbar appearance.

Click Admittance/Impedance Result

Admittance Peak Point: 190,000 Hz -> This peak corresponds to the First Resonance.

View Animation—Vibration Mode at a Particular Frequency

View results -> Default Analysis/Step -> Harmonic-Magnitude -> Set a Frequency (100 Hz for Off-Resonance Mode: Peak Point: 190,000 Hz for Resonance Mode).

Double Click

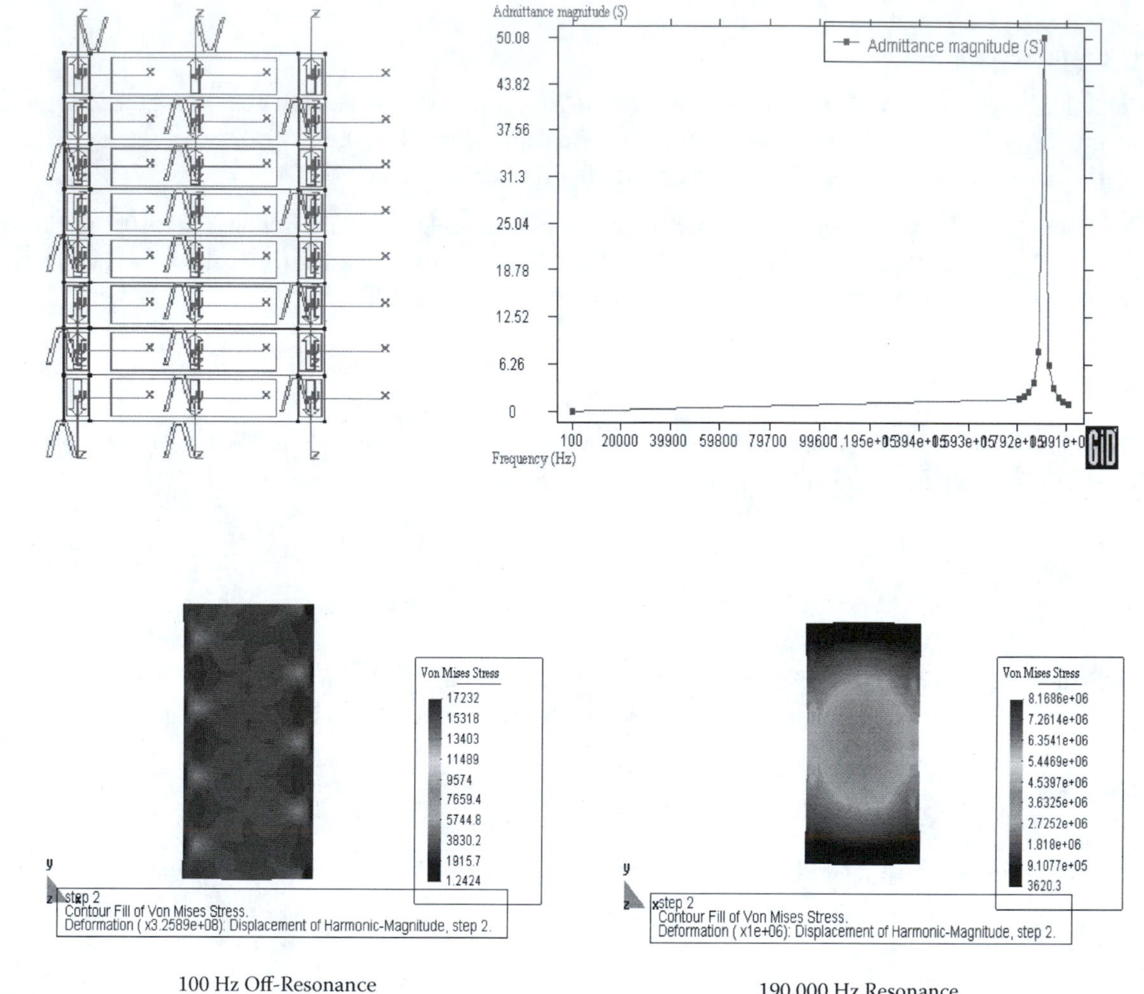

100 Hz Off-Resonance

190,000 Hz Resonance

Back to Top

Back to Home

4.1 ROSEN-TYPE TRANSFORMER (HOMEWORK 6)

Step-Up Ratio

PROBLEM—Rosen Transformer 25 mm × 8 mm × 2.2 mm *T*

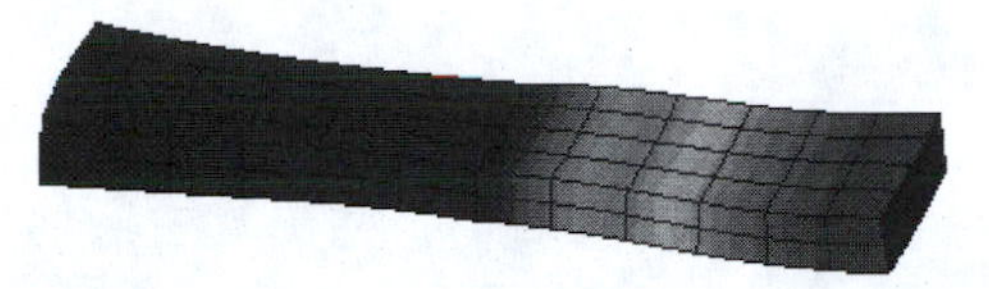

- Depending on the external load, the step-up voltage ratio is different.
- Learn the external load application:

1. Obtain the admittance curves around the first resonance frequency (71 kHz) under the external resistance load for 1 GΩ (almost open) and 1 MΩ. Show the resonance frequency shift and the peak broadening, and then explain the reasons briefly from the physical viewpoint.
2. Calculate the output voltage under the external load (instead of Floating Condition) from 1 kΩ to 1 GΩ under the 1 V input voltage and the frequency (71 kHz) fixed. Draw the step-up voltage ratio as a function of load in a log–log graph.
3. Calculate the output voltage under the external load from 1 kΩ to 1 GΩ under the 1 V input voltage fixed at the peak frequency. Draw the step-up voltage ratio as a function of load in a log–log graph.
4. Calculate the output power under the external load, and show that the maximum is obtained at the load that corresponds to the impedance of the transformer output (Impedance Matching).

HINT: BOUNDARY CONDITIONS

POTENTIAL

Data -> Conditions -> Surfaces -> Loaded Potential: Resistance from 1000 to 1,000,000,000 -> Assign the right-side surface.

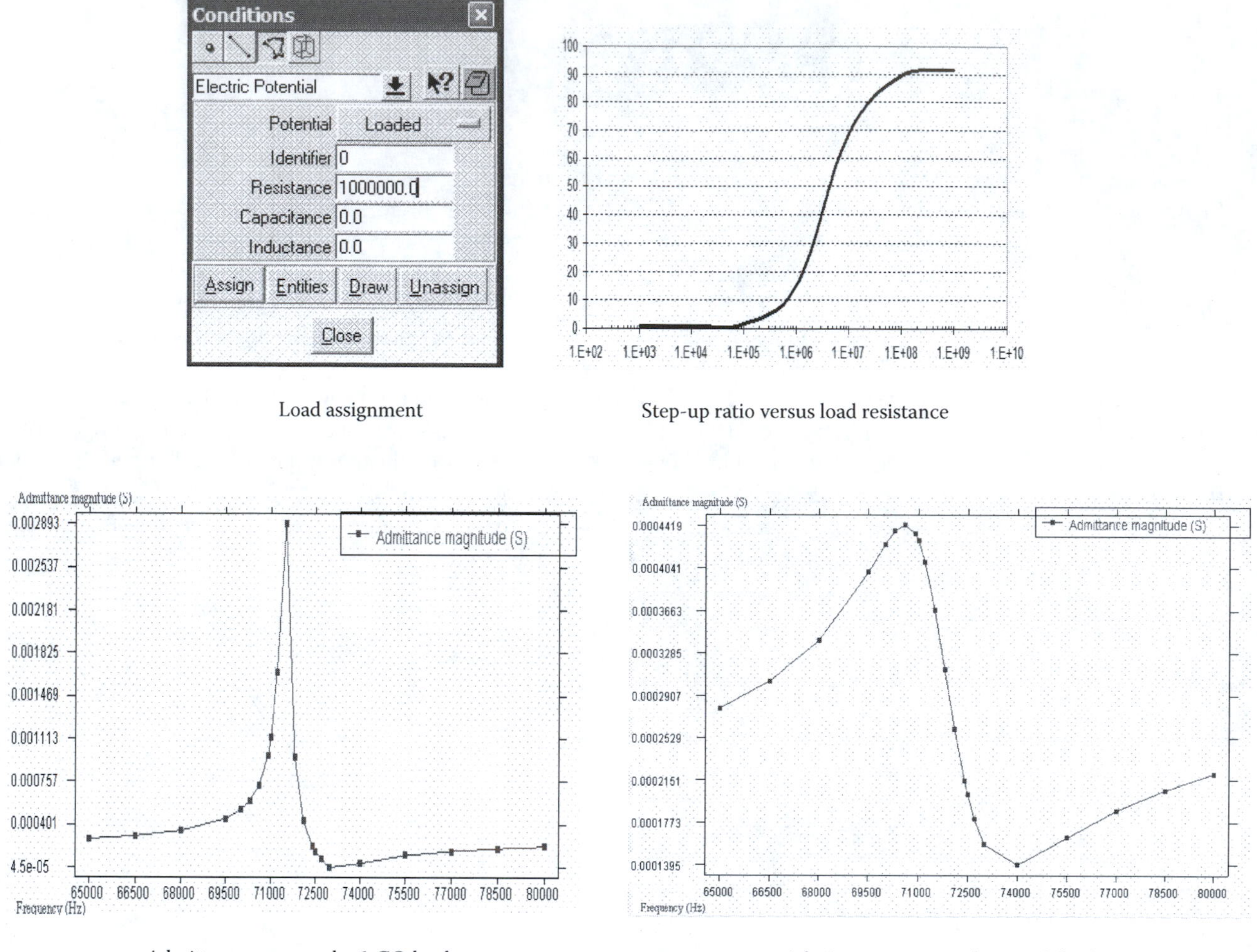

Load assignment

Step-up ratio versus load resistance

Admittance curve under 1 GΩ load

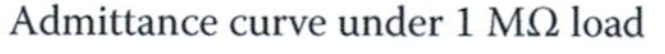

Admittance curve under 1 MΩ load

> NB: When decreasing the load resistance, the maximum voltage is obtained slightly inside of the rectangular bar (not on the right-hand side electrode). To obtain the voltage at a point of the electrode:
>
> 1 Assign a suitable frequency, such as the admittance peak frequency.
> 2 Without clicking "Deformation," click "View Results," "Graphs," "Point Analysis," and "Potential" successively. Then enter the point coordinate (25, 0, 0).
> 3 You will obtain the time dependence of the voltage at this point.

Back to Top

Back to Home

5.1 SHINSEI USM (HOMEWORK 7)

Traveling wave-type motor

PROBLEM - Shinsei Hula-Hoop Type USM 3D Model :

PZT 8 & Steel 1 Ring with
Outer Diameter = 50 mm,
Inner Diameter = 40mm,
Thickness = 1 mm each

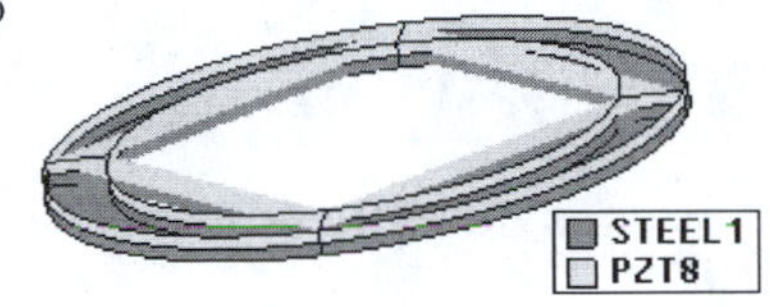

- Learn the most famous traveling wave motor:

1. Draw the Shinsei Hula-Hoop-type USM with PZT8 and Steel 1 with outer diameter 50 mm, inner diameter 40 mm, and thickness 1 mm each.
2. Four-segmented PZT portions should be as follows:

Portion	Polarization	Voltage
1	Up	Sine
2	Up	Cosine
3	Down	Sine
4	Down	Cosine

3. Obtain the admittance curve.
4. Show the resonance-mode animation.

HINT: GEOMETRY/DRAWING/MATERIALS ASSIGNMENT/BOUNDARY CONDITIONS

ADMITTANCE CURVE AND ANIMATION EXAMPLE

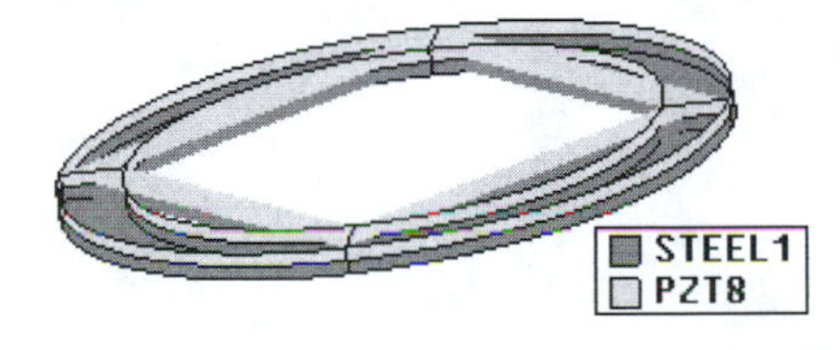

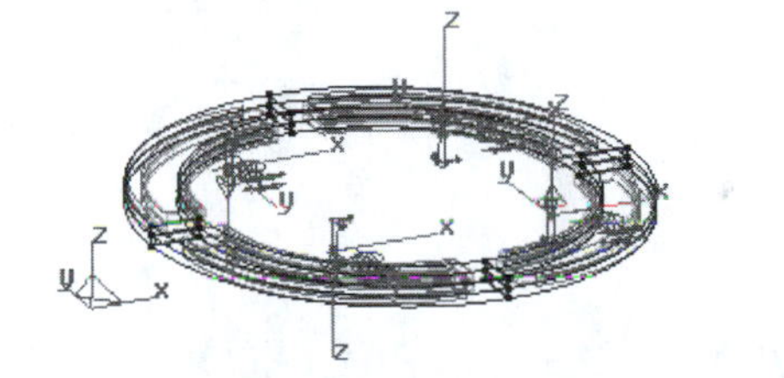

Back to Top

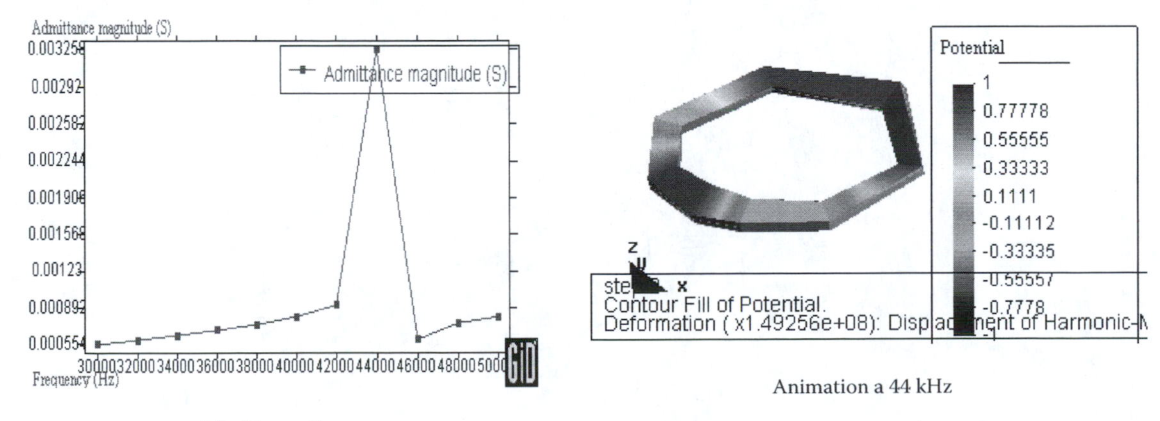

Admittance Curve

Animation a 44 kHz

Back to Home

5.2 Π-SHAPED USM (HOMEWORK 8)

METAL THICKNESS OPTIMIZATION

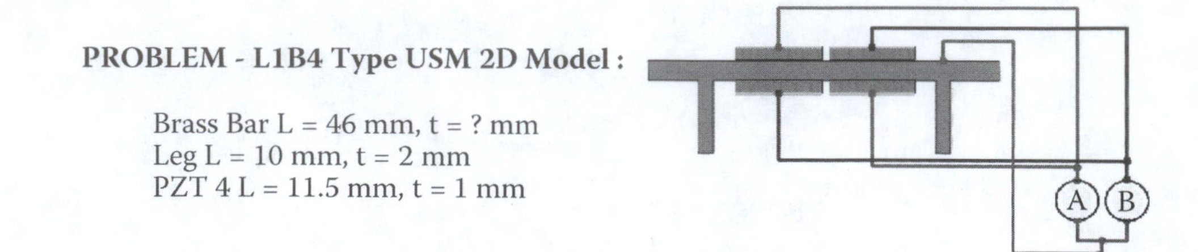

- Learn how to optimize the metal bar thickness:

1. Keeping the Leg size constant and the Bar length constant, obtain the admittance curves by changing the Bar thickness from 2 to 6 mm (2, 3, 4, 5, and 6).
2. Obtain typical mode animations to make sure the L1 and B4 Modes exchange with changing Bar thickness.
3. Obtain the optimized Bar thickness.
4. Show the Leg-tip elliptical locus. Show the *x* and *y* displacement magnitude under 1 V applied.

HINT: GEOMETRY/DRAWING

CREATE SURFACE

Create the coordinate map as illustrated in the following text (you may use 46 mm instead of our original 45.9 mm), and draw the motor in 2D map.

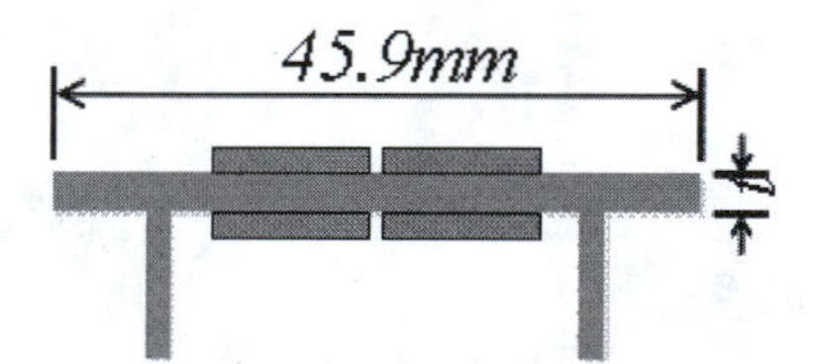

ESTIMATION OF THE RESONANCE FREQUENCIES OF B4 AND L1 MODES

EULER–BERNOULLI BEAM THEORY

$$f_{L_1} = \frac{1}{2l}\sqrt{\frac{E}{\rho}}$$

$$f_{B_n} = \frac{\alpha_n^2 t}{4\pi\sqrt{3}l^2}\sqrt{\frac{E}{\rho}} \qquad \alpha_n \approx \left(\frac{2n+1}{2}\right)\pi \tag{16.38}$$

$$f_{B_n} - f_{L_1}$$

$$l = \frac{\alpha_4^2 t}{2\pi\sqrt{3}} \approx 18.4l$$

Example Data for a Different Design

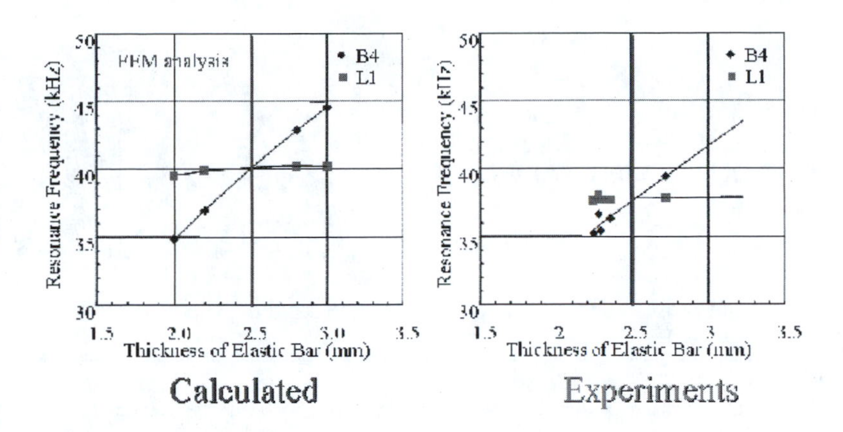

Note that the optimum Thickness difference between the calculation and experiment is less than 3%.

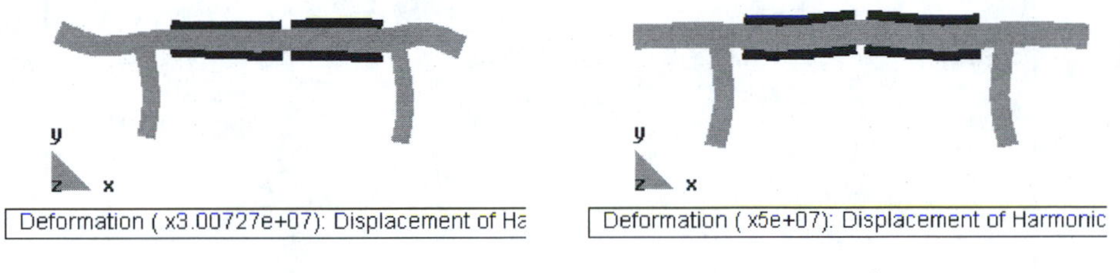

L1 Mode B4 Mode

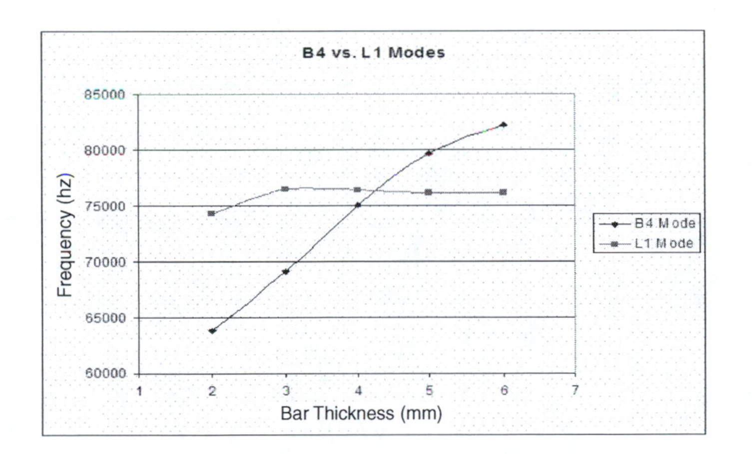

Simulation Example: L1 and B4 mode frequencies coincide at 4.28 mm thickness.

Back to Top

Back to Home

6.2 UNDERWATER CYMBAL (HOMEWORK 9)

Unipolar and Bipolar Sound Sources

PROBLEM - Cymbal 12 mm φ × 2.4 mm t (Metal thickness 0.2 mm, PZT thickness 0.5 mm × two layer, Cavity height 0.5 mm); 2D Axisymmetric with a full portion

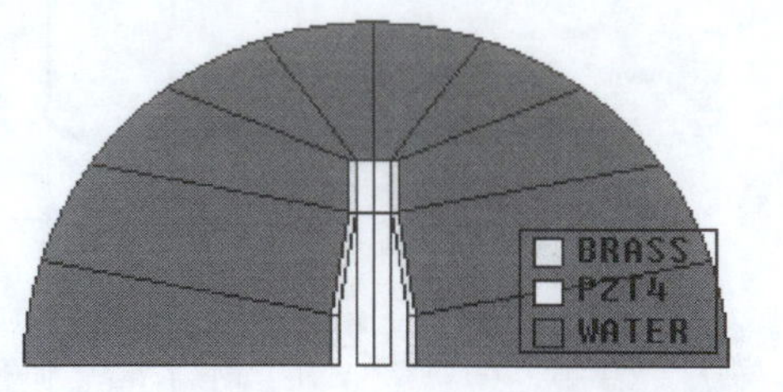

- Learn about underwater sound propagation—Monopole and Dipole, and the superposed "Cardioid" mode.

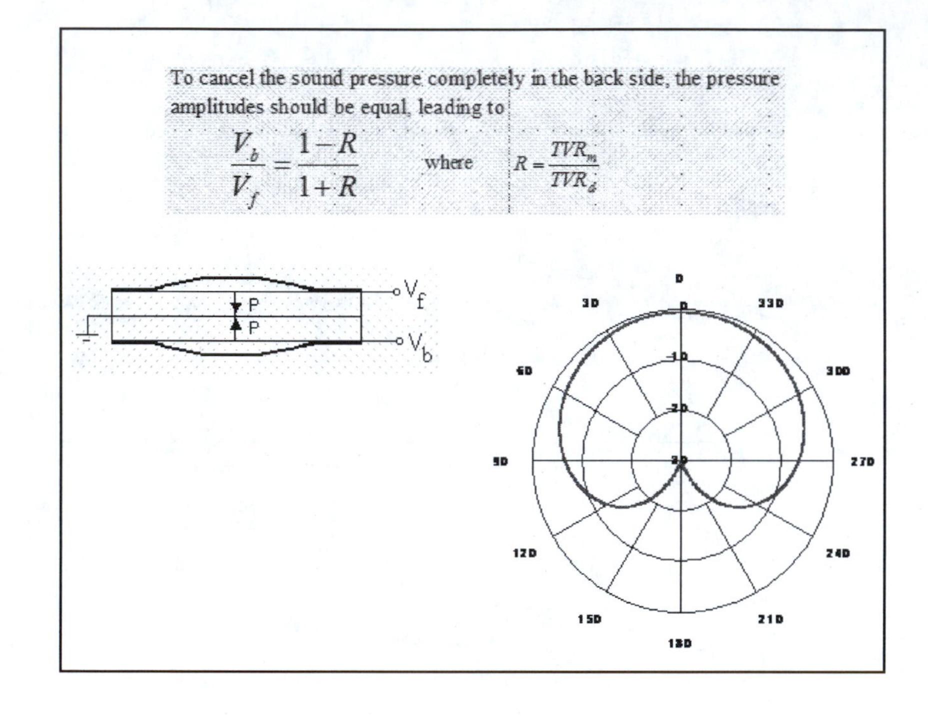

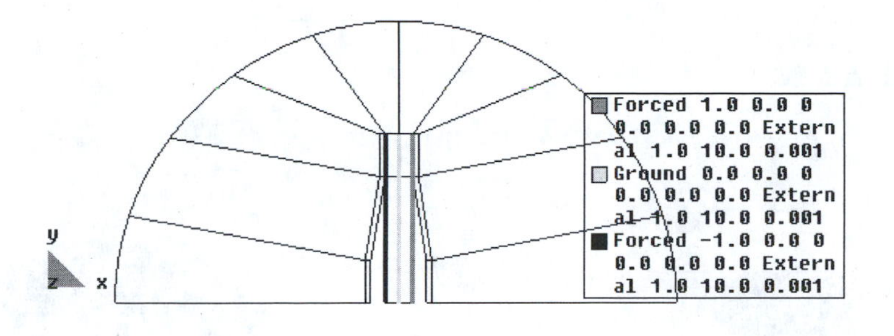

HINT: Boundary Condition

Vf = IV, V_b = IV, phase = 180°

Admittance Curve

The First resonance frequency is 26,400 Hz in this case.

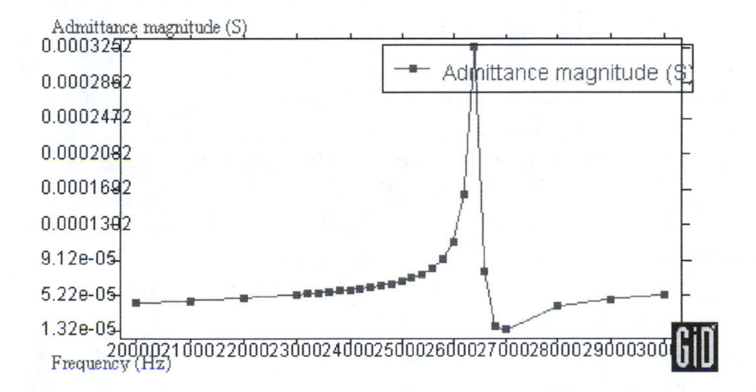

Unipolar and Bipolar Sound Sources

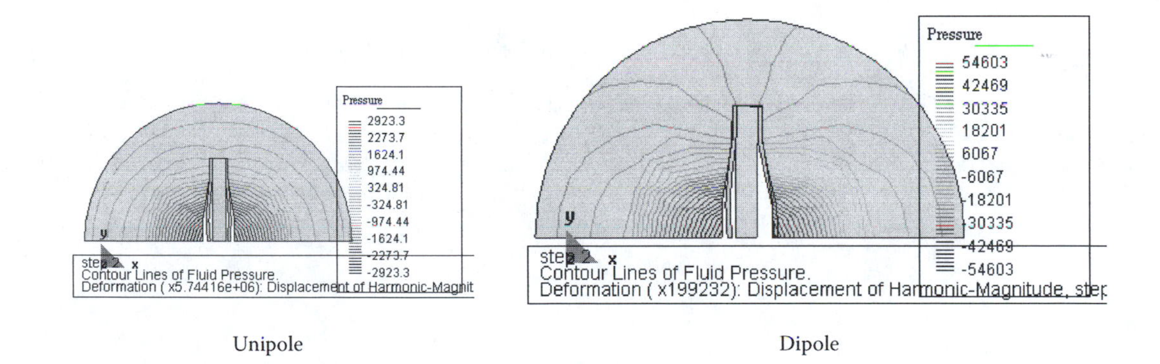

Unipole Dipole

NB: Cardioid may be obtained under Vf = 10V, phase = 0°, V_b = 3V, phase = 55°.

Back to Top

Back to Home

0.4 FINAL EXAM REPORT

Optimization with ATILA FEM

PROBLEM—An optimized piezoelectric/magnetostrictive device

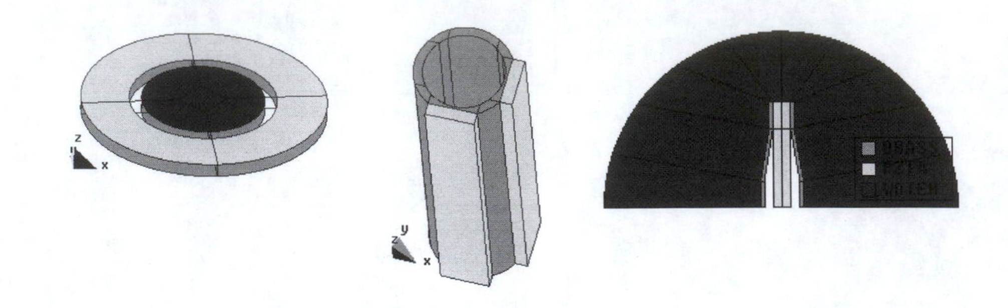

1. Consider a suitable piezoelectric or magnetostrictive device that can be optimized in its design by using ATILA FEM software. Choose a device that has not been treated in this book.
    a. Examples (not limited):
        i. Dot-ring type transformer—By changing the center ring size (keeping the outer diameter of the disk), how can we obtain the maximum electrical power?
        ii. Metal-tube motor—By changing the PZT plate angle (different from the original 90°), how can we increase the wobbling displacement?
        iii. Cymbal Underwater transducer—By changing the metal and/or piezoelectric ceramic thickness (keeping the outer diameter of the cymbal), how can we increase the sound power?
        iv. Magnetostrictive and piezoelectric laminated film—By changing the thickness ratio, how can we increase the magnetoelectric sensor sensitivity?
2. Discuss the obtained optimized values from the physical/analytical viewpoints.
3. Prepare the PowerPoint presentation file for the final presentation exam purpose: a 10 min presentation (7–10 slides) including Background, Changing parameters, Simulation result, and Summary/Conclusion.

REFERENCES

1. P. Laoratanakul, A. V. Carazo, P. Bouchilloux, and K. Uchino: Unipolarized disk-type piezoelectric transformers, *Jpn. J. Appl. Phys.* 41, 1446-1450 (2002).
2. S. Cagatay, B. Koc, and K. Uchino: Piezoelectric ultrasonic micromotors for mechatronic applications, Proc. *9th Int. Conf. New Actuators*, A4.2, pp. 132–134, Bremen, Germany, June 14–16 (2004).
3. J. Ryu, S. Priya, A. V. Carazo, K. Uchino, and H.-E Kim: Effect of the magnetostrictive layer on magnetoelectric properties in Pb(Zr,Ti)O3/Terfenol-D laminate composites, *J. Am. Ceram. Soc.*, 84, 2905–2908 (2001).

Back to Top

Index

Note: Page numbers in bold indicate figures and/or tables.

A

D

E

F

M

Q

R

S

T

U

V